VOLUME NINETY EIGHT

ADVANCES IN FOOD AND NUTRITION RESEARCH

Application of Polyphenols in Foods and Food Models

VOLUME NINETY EIGHT

ADVANCES IN FOOD AND NUTRITION RESEARCH

Application of Polyphenols in Foods and Food Models

Edited by

DANIEL GRANATO

Department of Biological Sciences, Faculty of Science and Engineering, University of Limerick, Limerick, Ireland

ELSEVIER

ACADEMIC PRESS

An imprint of Elsevier

Academic Press is an imprint of Elsevier
50 Hampshire Street, 5th Floor, Cambridge, MA 02139, United States
525 B Street, Suite 1650, San Diego, CA 92101, United States
The Boulevard, Langford Lane, Kidlington, Oxford OX5 1GB, United Kingdom
125 London Wall, London, EC2Y 5AS, United Kingdom

First edition 2021

ISBN: 978-0-12-822506-6
ISSN: 1043-4526

For information on all Academic Press publications
visit our website at https://www.elsevier.com/books-and-journals

Publisher: Zoe Kruze
Acquisitions Editor: Sam Mahfoudh
Developmental Editor: Jhon Michael Peñano
Production Project Manager: Denny Mansingh
Cover Designer: Christian J. Bilbow

Typeset by STRAIVE, India

Contents

Contributors

Luciana Azevedo
Federal University of Alfenas, Nutrition Faculty, Alfenas, MG, Brazil

Elisa Rafaela Bonadio Bellucci
Department of Food Technology and Engineering, UNESP—São Paulo State University, São José do Rio Preto, SP, Brazil

Soottawat Benjakul
International Center of Excellence in Seafood Science and Innovation, Faculty of Agro-Industry, Prince of Songkla University, Hat Yai, Songkhla, Thailand

Lalita Chotphruethipong
International Center of Excellence in Seafood Science and Innovation, Faculty of Agro-Industry, Prince of Songkla University, Hat Yai, Songkhla, Thailand

Andrea Carla da Silva Barretto
Department of Food Technology and Engineering, UNESP—São Paulo State University, São José do Rio Preto, SP, Brazil

Mariana Araújo Vieira do Carmo
Federal University of Alfenas, Nutrition Faculty, Alfenas, MG, Brazil

Rubén Domínguez
Centro Tecnológico de la Carne de Galicia, rúa Galicia n° 4, Parque Tecnológico de Galicia, Ourense, Spain

Marina Fidelis
Food Processing and Quality, Production Systems Unit, Natural Resources Institute Finland (Luke), Helsinki, Finland

Daniel Granato
Department of Biological Sciences, Faculty of Science and Engineering, University of Limerick, Limerick, Ireland

Zisheng Han
Department of Food Science, Rutgers University, New Brunswick, NJ, United States

Eila Järvenpää
Natural Resources Institute Finland (Luke), Jokioinen, Finland

Minna Kahala
Natural Resources Institute Finland (Luke), Jokioinen, Finland

José M. Lorenzo
Centro Tecnológico de la Carne de Galicia, rúa Galicia n° 4, Parque Tecnológico de Galicia; Facultad de Ciencias de Ourense, Área de Tecnología de los Alimentos, Universidad de Vigo, Ourense, Spain

Sari Mäkinen
Natural Resources Institute Finland (Luke), Jokioinen, Finland

Pirjo Mattila
Natural Resources Institute Finland (Luke), Turku, Finland

Ajay Mittal
International Center of Excellence in Seafood Science and Innovation, Faculty of Agro-Industry, Prince of Songkla University, Hat Yai, Songkhla, Thailand

Paulo E.S. Munekata
Centro Tecnológico de la Carne de Galicia, rúa Galicia n° 4, Parque Tecnológico de Galicia, Ourense, Spain

Juanying Ou
Institute of Food Safety & Nutrition, Jinan University, Guangzhou, China

Nora Pap
Natural Resources Institute Finland (Luke), Jokioinen, Finland

Mirian Pateiro
Centro Tecnológico de la Carne de Galicia, rúa Galicia n° 4, Parque Tecnológico de Galicia, Ourense, Spain

Avtar Singh
International Center of Excellence in Seafood Science and Innovation, Faculty of Agro-Industry, Prince of Songkla University, Hat Yai, Songkhla, Thailand

Qian Wu
Hubei University of Technology, Wuhan, China

Liang Zhang
State Key Laboratory of Tea Plant Biology and Utilization, Anhui Agricultural University, Hefei, China

Jie Zhou
Northwest Agriculture & Forestry University, Yangling, China

Preface

This volume of *Advances in Food and Nutrition Research* comprises eight chapters written by an international board of authors and reports the latest developments in relevant and interesting topics about the technological applications of phenolic compounds in food models. Topics of the book chapters are related to the classification, methods of identification, and general beneficial effects of phenolic compounds in human health, application of polyphenols in food preservation, for the development of nondairy foods and beverages, meat-based foods, baked products, and conjugates (phenolic compounds—proteins). In addition, one chapter is dedicated to the cell-based evaluation of the bioactivity (cellular antioxidant activity) and some initial toxicological evaluation (i.e., cytotoxicity) of phenolic compounds.

Chapter 1 reviews published articles and database of dietary polyphenols to draw a profile for the classification, structural identification, quantification methods, and biological activities mainly based on enzymes, cell bioassay, and animal models, as well as the population-based investigation about the alleged health effects of polyphenols. Chapter 2 describes the use of phenolic compounds in different food preservation (i.e., extension of shelf life) applications and the mechanisms of action. As some phenolic compounds have antioxidant and antimicrobial activities, plant-derived polyphenols can be a suitable alternative to synthetic additives used in the food industry. Chapter 3 is an integration between food chemistry, food technologies with a glimpse of biochemistry, and pharmacology to produce phenolic-rich extracts. The application of these extracts in the development of nondairy foods and beverages is described together with practical guidance on the toxicological safety and bioactivity of polyphenol-rich extracts in different foods and beverages. Chapter 4 presents the properties of by-products from the industrial processing of different berry fruits and the environmentally friendly technologies for the pretreatment of the raw materials and includes the description of new green-based extraction procedures to obtain new ingredients rich in phenolic compounds. In addition, a focus is given on the modification and recovery of the polyphenols, as well as the formulation and stabilization of the ingredients. Chapter 5 is focused on two main streams: meat products added with phenolic-rich extracts and active and edible packaging materials. Authors put emphasis on the use of encapsulation techniques to deliver and protect compounds from chemical degradation and the combined use of polyphenols and nonthermal technologies to improve the shelf life of meat products. Chapter 6 focuses on the use of extracts rich in different phenolic compounds to manufacture bakery foods, including breads, cakes, cookies, muffins, rolls, buns, crumpets, pancakes, doughnuts, waffles, and bagels. The role of these extracts in mitigating reactive carbonyl species and the subsequent formation of advanced glycation end products, antioxidant, and antimicrobial activities are also discussed.

Chapter 7 describes the interconnection between oxidative stress, cancer, phenolic compounds, and antiproliferative activity as a way to increase the use of cell-based protocols to screen the bioactivity and toxicological safety of phenolic-rich extracts and foods. In vitro techniques and their biological applications, such as in the evaluation of cell viability and intracellular measure of reactive oxygen assays, are also debated. Finally, Chapter 8 focuses on the methodology for the production and characteristics of protein-phenolic compounds conjugates and their applications in various food systems and nutraceutical field, such as emulsions, edible films, hydrogels, controlled-release nutraceuticals, and nanoparticles.

In summary, this volume presents the combined efforts of 24 professionals developing their research in 6 countries (USA, Brazil, Finland, China, Thailand, Spain) with a variety of background and expertise. The Editor personally thanks the production staff and all the contributors for sharing their experience and for making this book possible.

Daniel Granato
Editor

CHAPTER ONE

Polyphenols in foods: Classification, methods of identification, and nutritional aspects in human health

Liang Zhang[a,*], Zisheng Han[b], and Daniel Granato[c,*]
[a]State Key Laboratory of Tea Plant Biology and Utilization, Anhui Agricultural University, Hefei, China
[b]Department of Food Science, Rutgers University, New Brunswick, NJ, United States
[c]Department of Biological Sciences, Faculty of Science and Engineering, University of Limerick, Limerick, Ireland
*Corresponding authors: e-mail address: zhli2091@sina.com; granatod@gmail.com

Contents

Abstract

Polyphenols widely exists in various foods, including main crops, fruits, beverages and some wines. Famous representatives of polyphenols, such as resveratrol in red wine, (−)-epigallocatechin gallate in green tea, chlorogenic acid in coffee, anthocyanins in colored fruits, procyanidins in grape seed have become hot research topics in food science and nutrition. There have been thousands of papers on the biochemistry, chemistry, nutritional values and population-based investigations of dietary polyphenols. In this chapter, we reviewed the published articles and database of dietary polyphenols to draw a profile for the classification, structural identification, and biological activities mainly based on enzymes, cell bioassay and animal models, as well as the population-based investigation results. The typical compound and its health benefits for each category of polyphenols was also introduced. The identification of dietary polyphenols could be solved by combined spectroscopy methods, of which the liquid chromatography tandem mass spectrometry is highlighted to greatly increase the efficiency

Advances in Food and Nutrition Research, Volume 98
ISSN 1043-4526
https://doi.org/10.1016/bs.afnr.2021.02.004

on structural identification. Although the population-based investigation showed some controversial results for health benefits, the multi-functions of dietary polyphenols on preventing metabolic syndromes, various cancers and neurodegenerative disease have attracted much attention.

Abbreviations

ACC	acetyl-CoA carboxylase
ACE	angiotensin-converting enzyme
AMPK	AMP-activated protein kinase
CcO	cytochrome *c* oxidase
CD	circular dichroic
EC	(−)-epicatechin
ECG	(−)-epicatechin gallate
EGC	(−)-epigallocatechin
EGCG	(−)-epigallocatechin gallate
HDL-C	high density lipoprotein cholesterol
HOMA-IR	homeostasis model assessment of insulin resistance
IFN-γ	interferon gamma
IUPAC	International Union of Pure and Applied Chemistry
LC-MS	liquid chromatography tandem mass spectrometry
LDL-C	low density lipoprotein cholesterol
NMR	nuclear magnetic resonance
PPAR	peroxisome proliferator-activated receptors

1. Introduction

Polyphenols is a type of functional ingredients in foods, beverages and wines. Using "polyphenols" as keyword, we can see the tendency of developing of food polyphenols worldwide as shown in Fig. 1. The phenolic compounds are mainly structural components of cell walls, and most of them are toxins and antifeedants of plant defense, coloring ingredients of flowers and fruits, antioxidants of bark and seeds. They are mainly derived from phenylpropanoid and phenylpropanoid-acetate backbones. During the growth of plants, the secondary metabolism produces a large and diverse class of flavonoids and suberin to protect the plant from environmental stresses, such as UV irradiation and desiccation.

With the development of food science and technology, food polyphenols have gradually walked into the industrial production and health application. The basic theory of food polyphenols involves organic chemistry, nutrition, analytical chemistry and medicine. In this chapter, we

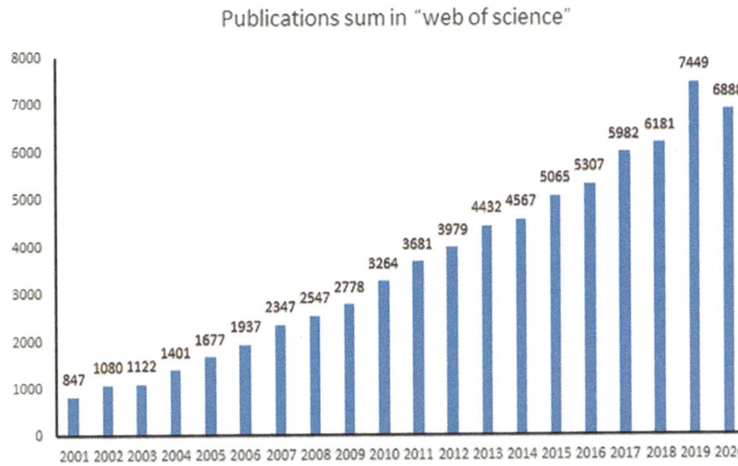

Fig. 1 The publication sum with the topic "polyphenol" in web of science.

mainly summarized and introduced the basis including classification, structural identification and health benefits of dietary polyphenols. The health benefits of dietary polyphenols have attracted much attention, because it is accessible by daily food intake. Thousands of papers have explored the biological activities of dietary polyphenols, like antioxidant, anti-inflammation, promoting apoptosis of tumor cells, estrogen-like action and neuroprotective effects. Based on the activities studies, some population-based studies including randomized, double-blind placebo-controlled clinical trial, cohort investigation, retrospective study and meta-analysis have been carried out to reveal the health benefits of dietary polyphenols.

2. Classification

Polyphenols are the critical secondary metabolites of many tea plants, apart from the primary metabolites such as carbohydrates, fatty acids and proteins. They usually do not provide the energy for human body, but possess various health benefits through many biological activities such as anti-oxidant and anti-inflammation. With the developing of polyphenols, the food polyphenols have garnered the most public interest because of their high presence in various food, including vegetable, fruits, seeds and beverages (Tylewicz, Nowacka, Martín-García, Wiktor, & Gómez Caravaca, 2018). Polyphenols have common structural characteristics, with at least one hydroxyl moiety, sometimes also with carboxyl moiety. Usually, food polyphenols are classified into hydroxybenzoic acids, hydroxycinnamic acids, flavonoids, stilbenes, lignans, flavonols, flavones, isoflavones, flavanones, anthocyanidins, flavanols, hydrolysable tannins and condensed tannins. Different from other natural products, polyphenols' structures are not complicated and usually have at least one hydroxyl moiety in the molecule.

So far, over 50,000 diverse polyphenols have been identified in plants. According to the regulation of IUPAC, one hydroxyl group on a benzene ring or other arene ring is defined as "phenols." It was also suggested that "polyphenols" was defined as plant secondary metabolites derived from the shikimate derived phenylpropanoid and/or the polyketide pathways. Among these polyphenol compounds, http://phenol-explorer.eu/foods reviewed almost all types of polyphenols (Neveu et al., 2010). In total, 501 polyphenols in various foods were reviewed. Basically, these polyphenols mainly contain 279 flavonoids, 108 phenolic acids, 10 stilbenes, 29 lignans, and 80 other polyphenols (Rothwell et al., 2012). If classified based on food types, most polyphenols were detected in fruit and fruit products, beverages, seeds and vegetables. Based on the chemical structures, flavonoids have 15-carbon skeletons, represented as the C6-C3-C6 system. Their skeletons are constructed by two benzene rings and a chain of three carbon atoms of their heterocyclic systems. Flavonoids could be further classified into anthocyanins, chalcones, dihydrochalcones, dihydroflavonols, flavanols, flavanones, flavones, flavonols and isoflavonoids.

Usually, the flavonoids have different color because of their structural characteristics. For example, the anthocyanins often show the color of blue, red or violet, while flavones and flavonols present pale yellow. Other flavonoids including flavanols, isoflavones and flavanones are colorless. The spectrum information of these flavonoids are similar. The representative flavonoids aglycones are quercetin, kaempferol, myricetin and their glycosides. These compounds have been widely detected in onion, blueberry, clack currant, apple, black grape, tomato, and tea infusions.

Phenolic acids is also widely presented in berries, fruits and beverages. The phenolic acids could be divided into soluble and insoluble (bound form) types. The first type is low-molecular weight phenolic acids with water-soluble properties during processing and digestion. The phenolic acids can be condensed into insoluble high-molecular weight molecules with glucose, caffeic and quinic acids. The most common phenolic acid is hydroxycinnamic acids, including *p*-coumaric, caffeic, ferulic, and sinapic acids. The other usual phenolic acids is hydroxybenzoic acid, which could be

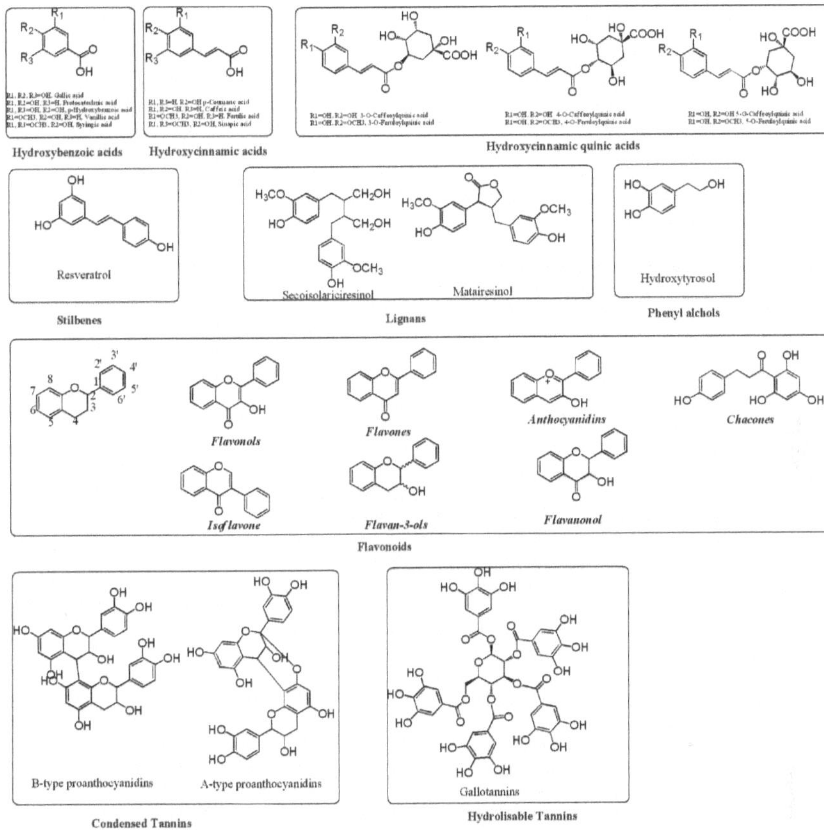

Fig. 2 The classification of dietary polyphenols based on chemical structures.

further named as *p*=hydroxybenzoic, protocatechuic, vanillic and syringic acids. Mattila et al. analyzed the content of phenolic acids in various plant-sourced foods. Chokeberry has the highest content of phenolic acids among all berries, of which dark plum and cherry are rich in caffeic acids (Mattila, Hellström, & Törrönen, 2006) (Fig. 2).

There are mainly three types of non-alcoholic beverages worldwide, tea, coffee and cocoa. In tea and cocoa, flavanols are very important polyphenols such as (+)-catechin, (−)-epicatechin, procyanidins dimer B2 and procyanidin trimer C1. In dark chocolate, the content of (+)-catechin is about 20 mg/100 g fresh weight, which is higher than that of tea. In the beverages of green tea and oolong tea, the typical flavanols are 2R, 3R-configuration flavanols, such as (−)-epigallocatechin gallate. For coffee, the typical polyphenols are hydroxycinnamic acids, including dicaffeoylquinic acid, mono-caffeoylquinic acid and geruloylquinic acid. The 5-caffeoylquinic acid, also named as neochlorogenic acid is also distributed in blueberry, apple, loquat, plum and sunflower seeds. Meanwhile, the typical flavanols epicatechin also abundantly exists in wine, raw broad ben seed and raw broad ben pod.

Flavan-3-ols and their oligomers are the most commonly consumed type of polyphenols, and are widely distributed in different foods. Tea is a typical beverage rich in flavan-3-ols in Asian countries, but in many Western countries, the flavan-3-ols are consumed from fruits. The catechins, especially (−)-epigallocatechin gallate is the abundant polyphenol in un-fermented teas, like green, white and yellow teas, but the oxidation products of flavan-3-ols, theaflavins were main components of black tea, which is the main form of tea consumed in Europe and USA. Cacao is very rich in flavan-3-ols monomers and oligomers, such as procyanidins.

Under the catalog of flavonoids, isoflavonoids are a unique subgroup of plant polyphenols with over 5000 compounds, mainly exists and reported in soybean. The main isoflavonoids in soybean are genistein and daidzein, which can bind with estrogen receptor so that they are referred as phyto-estrogens. Al-Maharik reviewed the novel isoflavonoids from 2012 to 2017 (Al-Maharik, 2019). Isoflavonoids are particularly concentrated in the legumes (Fabaceae) such as kudzu (*Pueraria lobata* (Willd.) Ohwi) or red clover (*Trifolium pratense* L.) or of course soy (*Glycine* max (L.) Merr.) (Franke, Halm, Kakazu, Li, & Custer, 2009).

Lignans is different from lignins which are polymers as the cell wall components of plant. They are a group of diphenolic compounds derived from shikimic acid biosynthetic pathway, structurally constructed by the combination of two phenylpropanoid C6–C3 units at the β and β′ carbon, and can be linked to additional ether, lactone, or carbon bonds. The plant lignans are a type of food polyphenols widely distributed in many foods, including flaxseed, rapeseed, sesame oils, whole-grain cereals, legume, vegetables and fruits (Milder, Arts, Putte, Venema, & Hollman, 2005; Peñalvo, Haajanen, Botting, & Adlercreutz, 2005). Secoisolariciresinol, lariciresinol, matairesinol, pinoresinol, medioresinol and syringaresinol are common lignans.

Stilbenes, is another class of phenolic compounds in various plants and foods. Like other polyphenols, it mainly acts as defensive secondary metabolite for protecting plant from pathogens and ultraviolet radiation. The basic skeleton is C6-C2-C6, in which two aromatic rings are linked by ethylene bridge (Neveu et al., 2010). Compared to other polyphenols, such as flavan-3-ols and chlorogenic acid, the content of stilbenes acquired from foods is limited. Some herbal medicines from *Polygonum* genus, like *Polygonum cuspidatum*, contains high content of stilbenes, but have not been used as dietary sources (Nonaka, Miwa, & Nishioka, 1982). The red wine is usually recognized as source for stilbenes, which are resveratrol, piceid, piceatannol, astringin and pterostilbene and their dimers (Vitrac et al., 2005; Vitrac, Monti, Vercauteren, Deffieux, & Mérillon, 2002).

3. Identification

In the history of polyphenols, the chromatographic separation contributed to the isolation and purification of many diverse polyphenols from the complicated food matrix or other secondary metabolites. In the past 100 years, the separation of polyphenols went through paper and silica gel partition chromatography, thin layer chromatography, molecular-exclusion chromatography, affinity chromatography, ion exchange chromatography and high speed countercurrent chromatography. The natural polyphenols have different color because of their structural properties. For example, the flavan-3-ols including catechin, galloylated catechins are colorless, but the oligomers and oxidation products of catechins are yellow, such as procyanidins and theaflavins, flavanones, flavonols and flavones are pale yellow or yellow (Balentine, Wiseman, & Bouwens, 1997; Senanayake, 2013).

The identification techniques for structures of polyphenols are not complicated and could be identified by UV-visible spectrum, infrared spectroscopy, mass spectrum, optical rotatory dispersion, circular dichroism and nuclear magnetic resonance spectrum (Power, Chapman, Chandra, & Cozzolino, 2019). With the developing of mass spectrum, more and more polyphenols could be directly identified by liquid-chromatography tandem mass spectrometry (Kontogianni, 2014). Furthermore, some liquid chromatography combining nuclear magnetic resonance (NMR) technique was also used in polyphenols structural identification.

Because of the basic skeleton of polyphenols, the novel structures of food polyphenols were less than other secondary metabolites of plants or marine organism. However, more and more novel quinic acid derivatives were isolated and identified. Hofmann group identified a series of *N*-phenylpropenoyl amino acids as the key astringent compounds of cocoa beans (Stark & Hofmann, 2005). Some caffeoylquinic acid is converted into bitter-tasting chlorogenic acid lactones by liquid chromatography tandem mass spectrometry (LC-MS) and NMR experiments (Frank, Blumberg,

Krümpel, & Hofmann, 2008). Some flavan-3-ol oligomers are not stable and it is difficult to obtain purified compounds, but with high resolution mass spectrometry, ethyl-bridged flavan-3-ol dimers and trimers could be identified by specific fragment ions (Fayeulle et al., 2018). It was suggested that condensation of flavan-3-ols and acetaldehyde produced by microbial fermentation led to the formation of ethyl-bridged flavan-3-ols dimers and oligomers. Furthermore, flavanl-3-ol glycosides, including (+)-catechin-7-O-β-glucoside, (+)-catechin 3-O-glucoside, (+)-catechin 5-O-, 7-O-, 3′-O-, and 4′-O-glucosides, (+)-catechin 7-O-β- and 4′-O-β-glucosides, (+)-catechin 5-O-β-glucoside and (−)-catechin 7-O-β-glucosides have been identified in many foods, like buckwheat, cowpea, barley, lentils, and some herbal medicines (Castillo-Muoz, Gómez-Alonso, García-Romero, Gómez, & Hermosín-Gutiérrez, 2009; Cui et al., 2012; Delcambre & Saucier, 2012). The specific MS/MS fragmentations of flavan-3-ol dimer hexosides were explored and confirmed in grade and red wines by mass spectrometry (Marie, et al., 2018).

The tea polyphenols is one of the hottest research topics. The identification of tea catechin was firstly isolated and identified as (−)-epicatechin. Its separation techniques mainly relied on the chemical precipitation method and crystallization process. After that, other tea catechins were separated and purified by silica gel. Davis reviewed the ^{1}H and ^{13}C NMR assignments of main green tea polyphenols (Davis, Cai, Davies, & Lewis, 1996). The typical chemical shift data for all catechins are similar, but the C-2, C-3, C-4α and C-4β of (2*R*, 3*R*)-epicatechins and (2*R*, 3*S*)-catechins showed minor differences, which was caused by the different spatial shielding effect.

The NMR was also applied in the characterization, quality control, and metabolomics of other food polyphenols. Anthocyanins contain a flavylium cation on the C-ring, and attract much attention for being natural colorant and health-promoting benefits. The NMR is not only used in the identification of unknown compounds, but also in the quantitative analysis, metabolomics, geographical origin traceability, and authentication of food (Ye, Wang, & Xu, 2015). Anastasiadi et al. studied the NMR-based metabolomics to discriminate the wineries, production zones and vintages (Anastasiadi, Zira, Magiatis, Haroutounian, & Mikros, 2009). The quercetin glycosides, syringic, *trans*-caffeic acid, *trans*-catraric acid were the critical factors associated with different wineries. Apple have many cultivars with different features and flavor. Metabolome analysis and NMR spectroscopy also revealed the cultivar-specific differences.

Using NMR, the phenolic compounds of wine could be detected to identify the production region of German white wines. Furthermore, the main natural colorants of wine, anthocyanins could be identified by 1D and 2D NMR spectroscopy. Ferrari et al. reported the multivariate statistical analysis, Fourier transform near-infrared spectroscopy and NMR spectroscopies could discriminate the wines containing anthocyanins from black rice or grapevine (Ferrari, Foca, Vignali, Tassi, & Ulrici, 2011). NMR was also applied in the research of wine-making, cultivars, geographical origin and vintage (Anastasiadi et al., 2009). However, the sensitivity of NMR technique limited its application in the trace or minor metabolites' qualification and quantitative analysis. Only few compounds could be identified by referencing the chemical shift of standards. Other compounds were hard to be detected or characterized by NMR.

Therefore, Jaitz et al. used LC-MS/MS to explore the chemical differences of red wine according to geographic origin, grape cultivars and vintage (Jaitz et al., 2010). The detection limit of LC-MS could achieve the level of μg/mL. Ma et al. also applied the ultrafast polyphenol metabolomics of red wine to successfully find the different metabolites of six red wines (Ma, Tanaka, Vaniya, Kind, & Fiehn, 2016). LC-MS has a main advantage of allowing much shorter time consumed, and more detectable metabolites. In this study, 165 polyphenols were detected by ultrafast LC-MS metabolomics. Because of the speed, efficiency, selectivity and sensitivity of ultra-high performance liquid chromatography coupled to mass spectrometry, the qualitative and quantitative analysis of food polyphenols have been significantly improved.

Kuhnert used the LC-MS^n to study the mass profiling of chlorogenic acids and hydroxycinnamoyl shikimate esters (Jaiswal, Sovdat, Vivan, & Kuhnert, 2010). In total, 51 phenolic acids were identified by parent ion, MS^2 fragments and MS^3 fragments. The original structure of quinic acid has three hydroxyls in C-3, 4, 5 and one pair hydroxyl and carboxyl in C-1 position. Usually, the C-3,4,5 hydroxyls could be esterified by caffeoyl, feruloyl, coumaroyl and sinapoyl. To identify these quinic acid derivatives, the parent ion could be observed at $[M\text{-}H]^-$ of [quinic acid + n × caffeic acid or other moiety-n × 18–1]. For example, monocaffeoyl, dicaffeoyl, tricaffeoyl of quinic acid presented the negative parent ions m/z at 353 $[192+180-18-1]$, $[192+2\times180-2\times18-1]$ and $[192+3\times180-3\times18-1]$ (Clifford, Stoupi, & Kuhnert, 2007). To further explore the structure, the parent ion m/z at 677 of tri-caffeoyl quinic acid could continuously produce MS^2 ion at 515 by losing 162, MS^3 ion at 353 by losing 162 and MS^4 ion at 191. Other hydroxycinnamic acids could also be identified through the similar way.

The glycosylated flavonoids and phenolic compounds were identified by a same analytical scheme based on aqueous methanol extraction, reverse phase liquid chromatography, diode array and mass spectrometric detection (Lin & Harnly, 2007). According to the structural differences, the λ_{max} UV/vis absorbance of hydroxycinnamoyl quinic acids were 240, 298, 326 nm, while the glycosylated flavonols including quercetin, isorhamnetin, kaempferol were 256, 266, 352 nm.

Lin did a more comprehensive study on the identification of flavanols, proanthocyanidins, isoflavones, flavanones, dihydrochalcones, stilbenes, benzoic acids by LC-MSn (Lin & Harnly, 2012). All these polyphenols have a maximum absorbance wavelength at 260–288 nm in ultraviolet spectrum analysis. To identify the structures of flavonoids, usually, the LC-MS/MS has been widely applied in structural analysis. The regular flavanols include 2*R*,3*R*- or 2*R*,3*S* configuration catechins and their gallate (Zhou et al., 2018). During the mass fragmentation of these compounds, the galloyl moiety could be firstly lost to produce a neutral loss of 152. Furthermore, the flavan-3-ols dimers, trimers even oligomers has a typical fragments ions at 425[−152], 407 [−170] (Zhang, Li, Ma, & Tu, 2011; Zhang, Tai, Wang, Meng, & Wan, 2017). Therefore, the mono and oligomer flavan-3-ols could be distinguished and identified by their different parent ion and fragment ions.

However, during the processing of black tea, fermentation produces more complicated flavan-3-ol oxidation products. For example, the B-ring of flavan-3-ols couple was oxidized into benzotropolone moiety, in the form of theaflavins, isotheaflavins, theaflagallins, theaflavates, theaflavic acids and their derivatives (Drynan, Clifford, Obuchowicz, & Kuhnert, 2010).

In terms of the structure of stilbenes, *trans*- and *cis*-resveratrol were both presented in red wine. The chemical shift of the two isomers are slightly different in C-7,8, in which *trans*-molecule has a higher chemical shift value for protons at C-7,8 (Wu et al., 2004). LC-MS was applied to qualitatively and quantitatively determine the content of resveratrol in grapes juice, cranberry juice and wine (Wang, Catana, Yang, Roderick, & van Breemen, 2002). The red wine contained higher content of resveratrol than other fruit juices. Tisserant et al. used ^{13}C NMR, 2D-NMR and LC-MS to identify the resveratrol dimer, hydroxylated resveratrol dimer, dihydroxylated resveratrol dimer and resveratrol trimer in elicited grapevine hairy root cultures (Tisserant, Hubert, Lequart, Borie, & Courot, 2016) (Table 1).

Table 1 The mass fragmentation ions of various dietary polyphenols under soft ionization techniques.

Classification type	Compounds	Molecular weight	Parention (-MS)	Productions
Hydroxybenzoic acid	Benzoic acid	122	121	77
	Gallic acid	170	169	125, 79
Hydrolisable Tannins	HHDP-glucose	482	481	301, 275
	Mono-galloyl glucose	332	331	169, 151
	HHDP-galloyl glucose	634	633	375, 301
	Bis-HHDP glucose	785	786	301, 481
	Galloyl-bis-HHDP glucose	936	935	633, 783, 301
	Tris-galloyl-HHDP glucose	952	951	907, 783, 605, 301
Hydroxycinnamic acids	Caffeic acid	180	179	135
	Caffeoylquinic acid	354	353	191, 179, 135, 173
	Feruloylquinic acid	368	367	193, 191, 173, 149
	Coumaroylquinic acid	338	337	191, 163, 119, 173
	Trihydroxycinnamoylquinic acid	370	369	195, 191, 151
	Di-caffeoylquinic acid	516	515	353, 191, 173, 161, 135
	Tri-caffeoylquinic acid	678	677	515, 497, 353, 191, 179
Flavan-3-ols	(+)-catechin	290	289	245, 205
	(−)-catechin	290	289	245, 205, 179
	(−)-gallocatechin	306	305	261, 221, 219, 179
	(−)-epigallocatechin	306	305	261, 221, 219, 179
	(−)-epicatechin gallate	442	441	331, 289, 169
	(−)-epigallocatechin gallate	458	457	331, 305, 169

Continued

Table 1 The mass fragmentation ions of various dietary polyphenols under soft ionization techniques.—cont'd

Classification type	Compounds	Molecular weight	Parention (-MS)	Productions
Isoflavone	Genistein	270	269	269, 195, 133
Flavonone	Hesperetin	302	301	163.7
	Naringenin	272	271	151, 119
Flavones	Luteolin	286	285	217, 151, 133
Flavonols (glycosides)	Quercetin-3-galactoside	464	463	301
	Quercetin-rhamnosylgalactoside	610	609	301
	Quercetin-rutinoside	610	609	301
	Quercetin-rhamnose-hexose-rhamnose	756	755	609, 301
	Kaempferol-3-glucoside	448	447	285
	Kaempferol-3-rutinoside	594	593	285
	Kaempferol-rhamnose-hexose-rhamnose	740	739	593, 431, 285
Theaflavins	Theaflavin-3-monogallate	716	715	697, 577, 563, 545, 527, 483, 407, 389
	Theaflavin-3′-monogallate	716	715	577, 563, 545, 527, 483, 389
	Theaflavin-3,3′-digallate	868	867	697, 577, 563, 545, 527, 483, 407, 389
Condensed tannins	Procyanidins B2	578	577	451, 425, 407, 289, 287
	Procyanidin trimer	866	865	847, 695, 739, 577, 425, 289
	A-type procyanidin dimer	576	575	539, 449, 423, 407, 289
Anthocyanidins	Pelargonidin-3-glucoside	433	433(+MS)	271.6
	Peonidin-3-O-glucoside	463	463(+MS)	301
	Cyanidin-3-O-glucoside	449.4	449.2(+MS)	287

The circular dichroic (CD) is a necessary and powerful tool for the configuration confirmation of many flavonoids, like flavan-3-ols. The flavan-3-ols exhibited two absorption bands in the 240 and 280 nm range of UV spectra. The 1La and 1Lb transition was correlated to the absolute configuration of C2, C3 of the C-ring of flavan-3-ols (Rensburg, Steynberg, Burger, Heerden, & Ferreira, 1999). The natural catechins of tea leaves are mainly divided into two types, (2*R*, 3*R*)-flavan-3-ols including (−)-EGCG, (−)-EC, (−)-EGC and (−)-ECG, (2*S*, 3*R*)-flavan-3-ols including (−)-GCG and (−)-GC, (2*R*, 3*S*)-flavan-3ols including (+)-catechin and (−)-CG. For the flavan-3-ol derivatives, the configuration of new carbon could be clarified by subtracting one CD spectrum from the flavan-3-ols without substituent group (Wang et al., 2014).

4. Health benefits

Food polyphenols are mainly accessed by the types, frequency, and amount of food consumed. The average total polyphenol intake could be calculated by questionnaires and chemical analysis. In Japan, elderly population consumed about 1.5 g/day total polyphenols (Chie et al., 2015). There have been many reports on the health benefits of food polyphenols, such as chronic degenerative diseases, metabolic syndrome, cardiovascular diseases, cancer prevention, anti-aging and anti-diabetes. The well-known effects of food polyphenols are antioxidant, anti-inflammatory, anti-hypertensive, anti-diabetic. Furthermore, polyphenols also involved in carcinogenesis and then showed chemo preventive effects.

Dihydrochalcone is a typical polyphenol in Malus species, in total about 265 diverse dihydrochalcones have been identified in many plants and foods (Gutierrez, Zhong, & Brown, 2018). In the apple fruits, although hydroxycinnamic acids and flavnonols accounted for large proportion of total polyphenols, dihydrochalcones are also representative compounds in apple. Phloretin 2′-xyloglucose and phloridzin are two main dihydrochalcones in apple (Puel et al., 2005). Phlorizin inhibits sugar transport, improve insulin sensitivity to prevent diabetes 2. It normalizes the insulin sensitivity without changing the insulin levels, and strongly inhibits the sodium-coupled glucose transporter 1 (SGLT1) and increases the renal glucose loss (Schulze et al., 2014).

Lariciresinol and Secoisolariciresinol are the most reported lignans with many biological activities, such as anti-cancer, antifungal, anti-inflammatory, and anti-oxidant activities. Both oral administration and subcutaneous injection of (−)-secoisolariciresinol suppressed the gain of body weight, increased serum adiponectin level and decreased gene expression of fatty acid synthase and sterol regulatory element-binding protein-1c, promoted gene expression of acyl-CoA oxidase, carnitine palmitoyl transferase-1, and peroxisome proliferator-activated receptor α (Tominaga et al., 2012). Secoisolariciresinol diglucoside also improved the vascular endothelial function, endothelial NO synthase and haeme oxygenase-1-mediated myocardial angiogenesis, reduced low density lipoprotein cholesterol (LDL-C), maalnodialdehyde and increased high density lipoprotein cholesterol (HDL-C), so that to prevent the progression of atherosclerosis (Penumathsa et al., 2008; Prasad, 2005, 2007).

Except for the tea catechins, the methyl-catechins also showed many health benefits including anti-allergy, anti-obesity and prevention of cardiovascular disease risks. Methylated tea catechins inhibit mast cell degranulation by suppressing FcεRI expression in the cell surface, inhibit extracellular signal-regulated kinase 1/2 phosphorylation, and bind to the surface of KU812 cell to alleviate allergic disorders (Man, Xin, Qingrong, & Huang, 2018). Methyl catechin demonstrates stronger inhibitory effect on ACE compared with other major tea catechins, such as EC, ECG and EGC. Methyl catechins also influence the transcriptional factors and expression of related protein in adipocytes in preventing obesity.

In coffee and tea, chlorogenic acids are main phenolic acids compounds with many biological activities including antioxidant activity, antibacterial, hepatoprotective, cardioprotective, anti-inflammatory, antipyretic, neuroprotective, anti-obesity, antiviral, anti-microbial, anti-hypertension, free radicals scavenger and a central nervous system stimulator (Naveed et al., 2018). They include 3/4/5-caffeoylquinic acid, 3/4/5-feruloylquinic and 3,4/3,5/4,5-dicaffeoylquinic acid, 4/5-O-*p*-coumaroylquinic acid. The absorbed chlorogenic acids has a higher oral-bioavailability than other flavan-3-ols. The metabolites of chlorogenic acid are ferulic, isoferulic and caffeic acid in the systemic circulation after consumption of chlorogenic acid. Chlorogenic acid exhibited anti-inflammatory and anti-tumor activity by inhibiting matrix metalloproteinase-9. They also reduced fasting plasma glucose level and increased insulin sensitivity by inhibiting α-amylase and α-glucosidase (Williamson, 2013). It was suggested that chlorogenic acid inhibit glucose-6-phosphate translocase. To sum up, chlorogenic acid has three main biological activities, which are enhancing antioxidant properties, regulating metabolic syndrome and glucose metabolism. Chlorogenic acid upregulated the expression of peroxisome proliferator-activated receptors (PPAR)α, free fatty acid β-oxidation, adiponectin, AMP-activated protein kinase (AMPK) phosphorylation and downregulated Liver X Receptor α, PPARγ, acetyl-CoA carboxylase (ACC) and 3-hydroxy-3-methylglutaryl coenzyme A reductase (Naveed et al., 2018). Furthermore, chlorogenic acid showed preventive effects of rat hepatotoxicity induced by methotrexate, an anti-cancer drug used in clinical applications. The mechanism was related to its antioxidant activities (Ali et al., 2017). Cheng et al. also reported the protective effects on aluminum chloride induced hepatotoxicity and hepatotoxicity at dose of 30 mg/kg (Cheng et al., 2017). It also showed that 40 mg/kg BW of chlorogenic acid protected against Isoproterenol-induced myocardial infarction by restore the activities of heart mitochondrial enzymes isocitrate dehydrogenase, α-ketoglutarate dehydrogenase, succinate dehydrogenase and malate dehydrogenase. The activities of lysosomal enzymes (β-glucosidase, β-glucuronidase, α-galactosidase, β-galactosidase, cathepsin-B and cathepsin-D) were increased significantly in the heart tissue (Akila, Asaikumar, & Vennila, 2017).

Wine is widely reported for its resveratrol with respect to anti-oxidant capacity. However, the real functions of resveratrol in red wine is still in debate. In addition, other phenolic acids including gallic acid, caffeic acid, *p*-coumaric acid, flavonoids were also distrusted in red wine (Anli, Vural, & Kizilet, 2008). There have been hundreds of papers about the biological activities, mainly regarding anti-cancer, anti-obesity, preventing non-alcoholic fatty liver disease, cardiovascular diseases, diabetes, metabolic syndrome, blood pressure. It is concluded that resveratrol may have positive effects on cellular proliferation, migration and apoptosis, act as inflammatory markers, lower LDL-cholesterol and restore insulin sensitivity. The most probable effects of resveratrol may focus on cardiovascular by population intervention study.

In red wine, resveratrol is also a famous compound in red wine with many potential health benefits, including antiaging and antidiabetogenic activities (Pastor et al., 2017). In 2003, Howitz et al. reported resveratrol can promote cell survival by stimulating sirtuin 1 (Howitz et al., 2003). Further study indicated resveratrol increased cAMP levels by activation of AMPK. Epac1 is essential for resveratrol to activate AMPK in HeLa cells. Resveratrol activates CamKKb-AMPK via the Epac1-PLC-Ryr pathway as a nonselective phosphodiesterase inhibitor (Park et al., 2012). To support the health benefits of red wine, resveratrol, at concentrations attainable moderate wine intake, activates platelet eNOS and in this way blunts the proinflammatory pathway linked to 38MAPK, thus inhibiting ROS production and ultimately platelet function (Gresele et al., 2008).

There are two isoforms of resveratrol, *cis* and *trans*, but the *trans* form of resveratrol presented more potent antiproliferative activities than its *cis* form (Anisimova et al., 2011). Like flavan-3-ol compounds, the oral bioavailability is low but they can be rapidly absorbed within 0.5 h. To regulate diabetes, resveratrol mainly target pancreases, adipose tissue, muscles and liver. The mechanisms are correlated with regulating glucose transporter-4 mediated glucose transport and IR phosphorylation, SIRT-1 and AMPK activation, mitochondrial β-oxidation, mitochondrial biogenesis, glycogen synthase and downregulating macrophage infiltration, adiponectin, lipogenic enzymes, oxidative stress and cytokine mediated inflammation, AMPK mediated FAS, ACC, gluconeogenesis. Like many polyphenols, resveratrol inhibit the formation of bovine serum albumin and advanced glycation end products (Shen, Xu, & Sheng, 2017). Resveratrol declined the cerebrospinal fluid Aβ 40 and plasma Aβ 40 levels in a randomized placebo controlled double blind phase 2 trial (Turner et al., 2015).

In terms of the anthocyanidins in plants, over 600 anthocyanins have been identified, with the most mentioned anthocyanidins, pelargonidin, cyaniding, peonidin, delphinidin, petunidin and malvidin (He & Giusti, 2010). The epidemiological evidences between anthocyanin-rich food and cancers have been convinced. Anthocyanins can inhibit the initiation, promotion and progression of many cancers, like breast, prostate, liver, colorectal and intestinal cancers (Wang et al., 2019). Anthocyanins are also colored water-soluble pigments widely used in food additives. Cyanidin-3-glucoside is the main anthocyanin in many plants. The color of anthocyanin is transformed from red to blue with the increase of pH value. Anthocyanidins and anthocyanins possess antioxidative and antimicrobial activities, improve visual and neurological health, and protect against various non-communicable diseases.

Hydroxycinnamic acids are more common than hydroxybenzoic acids, as being the predominate metabolites in many foods. In the *Rosaceae* fruits, hydroxycinnamic acids widely exist in almond, cherries, peach, pear, berries, apples, and plums (Määttä-Riihinen, Kamal-Eldin, & Törrönen, 2004). In apples, the main compounds are 3, 4, and 5-caffeoylquinic acid and 3-coumarylquinic acid. Flavonols and their glycosides are also detected in these Rosaceae fruits, with the main aglycones kaempferol, quercetin, and isorhamnetin, the last one of which is mainly detected in pears. However, most of these phenolic acids exist in fruits with the conjugated forms, like condensed and hydrolysable tannins. Therefore, actually, most of these phenolic acids are not released and absorbed in intestinal tract during digestion, mainly recognized as prebiotics.

Although flavonoids are scarcely absent in rice, biologist introduced phenylalanine ammonia lyase and chalcone synthase genes into rice and then highly increased the content of naringenin, kaempferol, genistein and other flavonoids in rice. This strategy is an effective tool to increase the accessibility of polyphenols by main staple food grain (Ogo, Ozawa, Ishimaru, Murayama, & Takaiwa, 2013).

Flavan-3-ols is one of the most important polyphenols in many foods and beverages. In tea, cocoa, grapes and berries, flavan-3-ols is the main compound. During the absorption and metabolism of flavan-3-ols, the oral-bioavailability is very limited because most of absorbed flavan-3-ols were transformed into glucuronidated and sulfonated conjugates; therefore, most of flavan-3-ols were less directly detected in the plasma or liver. However, during the digestion and microbial metabolism of flavan-3-ols, most of flavan-3-ols could be further metabolized or catabolized into low-molecular weight phenolic acids (Zhang et al., 2016). Many articles reported that colonic bacteria play a critical role in the absorption and catabolism of flavan-3-ols. The presence of flavan-3-ols could also affect the colon health, such as anti-inflammatory, also its microbiota.

The antioxidant effects of polyphenols have been widely confirmed by cell and animal models *in vivo* and *in vitro*. The common reactive oxygen species include superoxide anion ($O_2^{\bullet}$), hydrogen peroxide (H_2O_2), nitrogen species (NO), hydroxyl (OH−), and alkoxyl radicals which are hazardous factors to cells, which can active the natural until antioxidant enzyme systems to catabolized reactive oxygen species into its less toxic substances. The polyphenols present antioxidant effects by attenuating the free radicals-induced damage by scavenging these radicals, and binding with the anti-oxidant enzymes to activate their activities. Furthermore, lipids especially polyunsaturated fatty acids (PUFA) produce 4-hydroxy-3-nonenal (HNE), which was involved in the onsets of degenerative diseases like diabetes.

The most reported mechanism should be regulating nuclear transcription factors, lipids biosynthesis, lipids oxidation, modulating inflammatory mediators including cytokines tumor necrosis factor α, interleukin

(IL)-1β. Furthermore, most of polyphenols could inhibit the bioactivities of glycosidase, amylase and lipase, leads to the decrease of the co-administered food main nutrients and the increase of blood glucose level. This strategy has been recognized as an effective way for many Asian diabetes 2 patients.

Food polyphenols also showed protective effects on delaying neuronal cell loss in many progress neurodegenerative diseases, such as Parkinson's disease and Alzheimer's disease (Singh, Mandal, & Khan, 2016). Up to now, the factors that trigger these neurodegenerative diseases still remain unknown. Oxidative stress and the abnormal antioxidant defense system may contribute to the formation of toxic proteins in the neuros. Polyphenols synthesized by plant protect plants against reactive oxygen species, and also have same benefits on human body. Most polyphenols can enhance the activities of antioxidant enzymes such as superoxide dismutase, glutathione peroxidase and catalase, thus catalyze the reactive oxygen species into less toxic molecules. Similarly, polyphenols can protect the neuronal cells from undergoing neurodegeneration.

As a typical flavan-3-ol compound, (−)-epigallocatechin gallate presents as an agonist of the main cellular energy sensor, adenosine monophosphate-activated protein kinase. EGCG can increase cell viability, scavenge reactive oxygen species and inhibit the expression of endoplasmic reticulum stress markers and apoptotic markers. Abib et al. reported EGCG also form chemical complex with Cd^{2+} and then inhibit Cd^{2+}-induced mitotoxicity (Abib et al., 2011). The oral-bioavailability of EGCG was still in debate, but Lin et al. reported that EGCG can partly penetrate the blood–brain barriers at a low rate (Lin, Wang, Tseng, Sung, & Tsai, 2007). Thus, it suggested the possibility of EGCG on preventing neurodegenerative diseases by directly affect the neurocyte. Castellano-González et al. reported that EGCG can induce energy turnover of adenine triphosphate purine by targeting cytochrome *c* oxidase (CcO) activity in cultured human neurons and astrocytes (Castellano-González, Pichaud, Ballard, Bessede, & Guillemin, 2016). Beyond the common antioxidant enzymes, EGCG firstly showed effects on CcO activity. It was suggested that EGCG can interact with its physiological regulators or directly act as electron donor. On the other side, Wang et al. also reported that EGCG can significantly increase the number of 5-bromo-2′-deoxyuridine (BrdU)-labeled cells *in vivo* and *in vitro*. The mechanism of EGCG in enhancing adult hippocampal neurogenesis was related to triggering robust upregulation of sonic hedgehog (Shh) receptor mRNA and protein expression and downstream Shh transcriptional target Gli1 (Wang et al., 2012).

The published studies and prospective cohort studies reported that flavan-3-ols exerts positive effects on cardiovascular results for diabetes, hypertension and hyperlipidemia (Osakabe, 2013). The epidemiological researches are important to assess the health benefits of flavan-3-ols and other polyphenols. Although there have been many correlation analysis between flavan-3-ols and diseases, the interpretation and analysis on the published results, biomarkers are necessary. The known health benefits mainly relied on the lipids, glucose, inflammatory factors, etc. analysis, but lack more confirmed disease risk reduction or prevention data, though the epidemiological investigations were usually unstable and affected by lifestyles or food intake. For example, to calculate the flavan-3ol intake by food composition databases is very limited, but clinical intervention experiment is short time and hard to provide more risk-related data of diseases.

Epidemiologic and experimental studies suggested the preventive effects of isoflavones on osteoporosis, cardiovascular diseases, and breast, prostate, colorectum cancers. As estrogen-like molecule, genistein has been recently shown to bind to the estrogen receptor beta (ERβ) with an affinity as that of 17β-estradiol (Kuiper et al., 1997).

Many studies have reported dark chocolate can inhibit the metabolic syndromes risk factors, like hypertension, hyperlipidemia and glucose intolerance. Recent epidemiological evidence suggested that flavan-3-ols could prevent the cardiovascular diseases, the relationship analysis revealed that intake of catechins are inversely correlated to the death of cardiovascular diseases. It also indicated dark chocolate reduced the risk of stroke and other cardio metabolic diseases. A series of meta-analysis on cocoa and its products on cardiovascular diseases indicated that cocoa can improve the homeostasis model assessment of insulin resistance (HOMA-IR), low density lipoprotein (LDL), systolic Blood Pressure, low density lipoprotein (HDL), diastolic blood pressure and total cholesterol.

5. Population based investigation of polyphenols

Most of population investigation studies of food polyphenols belong to epidemiological studies, but the dietary polyphenols frequencies, intake content, polyphenols types, interaction with other co-consumed compounds may affect the epidemiological results. In some studies, the strict double-blind, cross-over, placebo-controlled studies may provide more convincing results by designed protocol. Therefore, for the same type of food polyphenols, the results of population investigation may give different statics, especially for the significance between treated or non-treated groups.

For example, resveratrol is one typical dietary polyphenol by red wine which has been widely consumed in Europe. A large population investigation over 40,685 subjects aged from 35 to 64 years reported that resveratrol and piceid intake were 100 and 933 μg/day, respectively, mainly by wines and grape and its juices (Zamora-Ros et al., 2008). Because resveratrol is less directly used in population studies, many studies focused on the Mediterranean diet. The recent meta-analysis and cohort investigation indicated that Mediterranean diet could decrease the overall cancer risk by 4.7% for men and 2.4% for women (Couto et al., 2011). Other large longitudinal investigation for healthy individuals reported that Mediterranean diet was not correlated to the protection of cognitive decline (Cherbuin & Anstey, 2012). The correlation of obesity and Mediterranean are in debate, because part of cohort and intervention studies could decrease weight loss, but other did not show evidence of this association (Buckland, Bach, & Serra-Majem, 2010). Mediterranean-type diet had 6%–16% lower risk for the cardiovascular disease than other people with poor adherence (Wise, 2016).

As one of the most consumed beverages in Western countries, coffee has been evaluated regarding its health benefits on population. Because coffee is usually consumed in combination with sugar, and some subjects have the habit of smoking, so the evaluation of coffee is also controversial to some extent. An investigation in 1985 reported coffee consumption did not affect the plasma cholesterol (Mathias, Garland, Barrett-Connor, & Wingard, 1985). A similar result was confirmed by a 12-week double-blind trial (van Dusseldorp, Katan, & Demacker, 1990). A dose-response meta-analysis reported that every 2 cups/day increment in coffee intake can decrease 12% of incidence of type 2 diabetes, but it may be related to the caffeine intake (Jiang, Zhang, & Jiang, 2014). Qi et al. reviewed the coffee, tea and caffeine consumption and Parkinson's disease. They also reported that 3 cups/day coffee consumption can decrease Parkinson's diseases (Qi & Li, 2014).

A prospective study also revealed that intake of coffee was negatively associated with breast cancer (Oh et al., 2015).

Grape seed extract is rich in flavan-3-ol monomers, and proanthocyanidins oligomers. Martins et al. reported a randomized, double-blind placebo-controlled clinical trial on grape juice. Consumption of grape juice significantly reduced lipid peroxidation and deoxyribonucleic acid damage, which are indicator of oxidation stress by intermittent physical exercise (Martins et al., 2020). Grape seed supplement also increased the antioxidant capacity to reduce urine redox potential for healthy volunteers (Grases et al., 2015; Zhu & Du, 2020). Recently, we reviewed the health benefits of various teas, and found tea consumption can effectively decrease the risk of colorectal, gastric, and prostate cancer (Zhang et al., 2019).

In some investigations, it showed green coffee extract was beneficial to the body weight, body mass index (BMI) and waist circumference, but chlorogenic acid, one of the main phenolic of coffee was not correlated with these eanthropometric measures (Asbaghi et al., 2020). For subjects with BMI $> 25\,kg/m^2$ significant reduction in body weight and BMI was observed (Gorji et al., 2019). A randomized, double-blind, placebo-controlled trial reported that chlorogenic acids increased Congnitrax domain scores for motor speed, psychomotor speed and executive function, as well as the levels of apolipoproteins A1 and transthyretin (Saitou et al., 2018). A meta-analysis revealed that chlorogenic acids statistically decreased the systolic and diastolic blood pressure, although the mean differences were -4.31 and -3.68, which were not significant (Onakpoya, Spencer, Thompson, & Heneghan, 2015).

Quercetin is the typical flavonoid existed in many fruits and beverages. There have been many dietary supplements or health-care foods containing quercetin. A systematic review and meta-analysis studied the effects of quercetin on the plasma lipid profiles, blood pressure and glucose levels. It reported that quercetin significantly lowered systolic and diastolic blood pressure by -3.09 and $-2.86\,mmHg$, respectively, but its effects on the levels of total triglycerides, high-density lipoprotein cholesterol (HDL-C) is limited (Huang, Liao, Dong, & Pu, 2020). A double-blind randomized clinical trial also suggested that quercetin had statistically significant improvement on high sensitivity C-reactive protein, iron, transferrin saturation and transferrin, but the total iron binding capacity was not affected (Sajadi Hezaveh, Azarkeivan, Janani, Hosseini, & Shidfar, 2019).

When quercetin is in combination with α-linolenic acid, they showed significant decreasing on total cholesterol, apolipoprotein B, low-density lipoprotein cholesterol (Burak et al., 2019).

As a mainly consumed food, soybean containing isoflavones is also studied regarding its health benefits to human. In the Pubmed database, there are in total 580 research publications about the clinical researches on isoflavones. It was reported that 8 weeks of isoflavones at the dose of 100 mg of soy isoflavones would reduce serum lipoprotein by 10%, and increase HDL-C by 11.5% (Yari, Tabibi, Najafi, Hedayati, & Movahedian, 2020). A randomized, double-blind, placebo-controlled trial used 108 mg/day of isoflavones genistein for postmenopausal women. It showed genistein supplement significantly reduced fasting blood glucose, glycated hemoglobin, serum triglyceride, increased serum high-density lipoprotein cholesterol and other antioxidant indexes (Braxas, Rafraf, Karimi Hasanabad, & Asghari Jafarabadi, 2019).

Anthocyanin mainly exists in colorful fruits, flowers and leaves. The pure anthocyanin single compound is hard to obtain, so most of anthocyanin is co-existed with other polyphenols like flavan-3-ols, procyanidins, hydroxycinnamic acids. The purple potatoes extract containing acylated anthocyanin alleviates postprandial glycemia and insulinemia, and affects postprandial inflammation (Jokioja et al., 2020). Açaí (*Euterpe oleracea* Mart.) berries predominantly contain anthocyanins, and consumption of 325 mL of berry juice for 12 weeks can decrease plasma level of interferon gamma and urinary level of 8-isoprostane, but other lipids and glucose related biomarkers were not changed (Kim et al., 2018). Blueberry is also a famous fruit. Curtis et al. designed and carried out a comprehensive double-blind, randomized controlled trail based on adults with metabolic syndromes. After 1/2 and 1 cup consumption of blueberry for 6 months, the main lipids and glucose metabolism indexes including serum glucose, insulin, blood pressure, NO, overall plasma were not affected, but endothelial function, high-density lipoprotein cholesterol, apolipoprotein A-I was improved (Curtis et al., 2019) (Table 2).

Based on the population researches, a clearer health benefits of various dietary polyphenols will impress us. Most of dietary polyphenols had effects on the lipids, glucose metabolism, and some biomarkers of tumorigenesis, but the risk reduction data of polyphenols is hard to evaluate under clinical or cohort investigations.

Table 2 The health benefits of various dietary polyphenols based on population investigation.

Compounds	Subjects	Dosage and duration	Main biomarkers	
Hydroxytyrosol, Punicalagin	Healthy adults	9.9 mg and 195 mg; 20 weeks	oxLDL (−28.74 ng/mL); Systolic pressure (−15.75 ± 9.9 mmHg); diastolic pressure (−6.36 ± 8.7 mmHg)	Quirós-Fernández, López-Plaza, Bermejo, Palma-Milla, and Gómez-Candela (2019)
Flavonoids coco drink	Subjects with Fatigue	350 mg, 120 mg GAE/day	No effects	Coe et al. (2017)
Gallic acid	Diabetic patients	15 mg/day, 7 days	Oxidized purines −31%; pyrimidines −2%; oxidized-LDL −24%; C-reactive protein −39%	Ferk et al. (2018)
Procyanidins (French maritime pine bark extract)	Postmenopausal osteopenic women	250 mg/day, 12 weeks	Procollagen type 1 amino-terminal propeptide; C-terminal telopeptide of type I collagen	Panahande et al. (2019)
Procyanidins (apple polyphenols)	Glucose tolerance in high-normal and borderline human subjects	600 mg/day, 12 weeks	Plasma glucose level	Shoji et al. (2017)
Procyanidins (French Maritime Pine bark extract)	Stage-1-hypertension subjects	150 mg/day, 5 weeks	Increase HDL-C 14.06%, apolipoprotein A-1 8.12%; decrease ratio of apolipoprotein B-100/A-1, 10.26%; systolic blood pressue −6.36 mmHg, decrease oxLDL-C;	Valls et al. (2016)

Continued

Table 2 The health benefits of various dietary polyphenols based on population investigation.—cont'd

Compounds	Subjects	Dosage and duration	Main biomarkers	
Green tea extract	Patients with malignant oral disorders	800 mg/day, 3 months	Downregulation of p53, Ki67, and cyclin D1 expression	Neetha, Panchaksharappa, Pattabhiramasastry, Shivaprasad, and Venkatesh (2020)
Green tea extract	T2MD patients	400 mg/day, 12 weeks	Decreased cAIx75, other biomarkers of lipids, glucose and blood pressure no changes	Quezada-Fernández et al. (2019)
EGCG (Polyphenon E)	Patients with bladder cancer	800, 1200 mg/day, 2–4 weeks	Reduced proliferating cell nuclear antigen and clusterin	Gee et al. (2017)
Chlorogenic acid, luteolin (Altilix®)	Metabolic syndrome patients	150 mg/day, 6 months	Weight, BMI, waist circumference, HbA1c, HOMA-IR, HOMA-β, total cholesterol, triglycerides, LDL-C, fatty liver index, AST, ALT, Flow-mediated dilation, carotid intima-media thickness	Castellino et al. (2019)
Anthocyanins (blackcurrant extract)	Old adults	300 mg/g, 7 days	Reduces central arterial stiffness and central blood pressure	Okamoto, Hashimoto, Kobayashi, Nakazato, and Willems (2020)
Blackcurrant Anthocyanins (apple polyphenols (AE), blackcurrant anthocyanins (BE))	Healthy volunteers	1200AE mg/day, 600AE + 600BE mg/day	Incremental areas under the curve (iAUC) of plasma glucose	Castro-Acosta et al. (2017)

Resveratrol	T2MD patients	1000 mg/day, 8 weeks	Decreased serum levels of asymmetric de-methyl-arginine (ADMA) and paraoxonase-1 (PON1) activity	Tabatabaie et al. (2020)
Resveratrol	T2MD patients	1000 mg/day, 8 weeks	Decreased fasting blood sugar (-7.97 ± 13.6 mg/dL) and increased high density lipoprotein (3.62 ± 8.75 mg/dL) levels	Abdollahi et al. (2019)
Genistein	Obese patients	50 mg/day, 2 months	Decreased metabolic endotoxemia, increased 5′-adenosine monophosphate-activated protein kinase phosphorylation, increased fatty acid oxidation and glucose tolerance	

6. Perspectives and conclusions

Polyphenols has become one of the hottest topics of food science and nutrition. More and more scientists and food technologists concern the biological activities, health benefits of dietary polyphenols. Some pre-clinical, clinical, population investigations also confirmed the preventative effects of food extracts, polyphenol-rich extracts and single compound on metabolic syndromes including hyperglycemia, hyperlipidemia, hypertension, neurodegenerative diseases and various cancers. Furthermore, the utilization and recycling of polyphenols from agricultural wastes, like fruits peels also expands the research and application of dietary polyphenols. Polyphenols not only affects the nutritional values, but also influences the food color, aroma and taste. Especially, the polyphenol-protein, polyphenol-Maillard reaction, can contribute to the flavor chemistry of various foods containing polyphenols.

References

Abdollahi, S., Salehi-Abargouei, A., Toupchian, O., Sheikhha, M. H., Fallahzadeh, H., Rahmanian, M., et al. (2019). The effect of resveratrol supplementation on cardio-metabolic risk factors in patients with type 2 diabetes: A randomized, double-blind controlled trial. *Phytotherapy Research*, *33*(12), 3153–3162.

Abib, R. T., Peres, K. C., Barbosa, A. M., Peres, T. V., Bernardes, A., Zimmermann, L. M., et al. (2011). Epigallocatechin-3-gallate protects rat brain mitochondria against cadmium-induced damage. *Food and Chemical Toxicology*, *49*(10), 2618–2623.

Akila, P., Asaikumar, L., & Vennila, L. (2017). Chlorogenic acid ameliorates isoproterenol-induced myocardial injury in rats by stabilizing mitochondrial and lysosomal enzymes. *Biomedicine & Pharmacotherapy*, *85*, 582–591.

Ali, N., Rashid, S., Nafees, S., Hasan, S. K., Shahid, A., Majed, F., et al. (2017). Protective effect of Chlorogenic acid against methotrexate induced oxidative stress, inflammation and apoptosis in rat liver: An experimental approach. *Chemico-Biological Interactions*, *272*, 80–91.

Al-Maharik, N. (2019). Isolation of naturally occurring novel isoflavonoids: An update. *Natural Product Reports*, *36*(8), 1156–1195.

Anastasiadi, M., Zira, A., Magiatis, P., Haroutounian, S. A., & Mikros, E. (2009). 1H NMR-based metabonomics for the classification of Greek wines according to variety, region, and vintage. Comparison with HPLC data. *Journal of Agricultural and Food Chemistry*, *57*(23), 11067–11074.

Anisimova, N. Y., Kiselevsky, M. V., Sosnov, A. V., Sadovnikov, S. V., Stankov, I. N., & Gakh, A. A. (2011). Trans-, cis-, and dihydro-resveratrol: A comparative study. *Chemistry Central Journal*, *5*, 88.

Anli, R. E., Vural, N., & Kizilet, E. (2008). An alternative method for the determination of some of the antioxidant phenolics in varietal Turkish red wines. *Journal of the Institute of Brewing*, *114*(3), 239–245.

Asbaghi, O., Sadeghian, M., Rahmani, S., Mardani, M., Khodadost, M., Maleki, V., et al. (2020). The effect of green coffee extract supplementation on anthropometric measures in adults: A comprehensive systematic review and dose-response meta-analysis of randomized clinical trials. *Complementary Therapies in Medicine*, *51*, 102424.

Balentine, D. A., Wiseman, S. A., & Bouwens, L. C. M. (1997). The chemistry of tea flavonoids. *Critical Reviews in Food Science and Nutrition*, *37*(8), 693–704.

Braxas, H., Rafraf, M., Karimi Hasanabad, S., & Asghari Jafarabadi, M. (2019). Effectiveness of genistein supplementation on metabolic factors and antioxidant status in postmenopausal women with type 2 diabetes mellitus. *Canadian Journal of Diabetes*, *43*(7), 490–497.

Buckland, G., Bach, A., & Serra-Majem, L. (2010). Obesity and the Mediterranean diet: A systematic review of observational and intervention studies. *Obesity Reviews*, *9*(6), 582–593.

Burak, C., Wolffram, S., Zur, B., Langguth, P., Fimmers, R., Alteheld, B., et al. (2019). Effect of alpha-linolenic acid in combination with the flavonol quercetin on markers of cardiovascular disease risk in healthy, non-obese adults: A randomized, double-blinded placebo-controlled crossover trial. *Nutrition*, *58*, 47–56.

Castellano-González, G., Pichaud, N., Ballard, J. W. O., Bessede, A., & Guillemin, G. J. (2016). Epigallocatechin-3-gallate induces oxidative phosphorylation by activating cytochrome c oxidase in human cultured neurons and astrocytes. *Oncotarget*, 7(7), 7426–7440.

Castellino, G., Nikolic, D., Magán-Fernández, A., Malfa, G. A., Chianetta, R., Patti, A. M., et al. (2019). Altilix(®) supplement containing chlorogenic acid and luteolin improved hepatic and cardiometabolic parameters in subjects with metabolic syndrome: A 6 month randomized, double-blind, placebo-controlled study. *Nutrients*, *11*(11), 2580.

Castillo-Muoz, N., Gómez-Alonso, S., García-Romero, E., Gómez, M. V., & Hermosín-Gutiérrez, I. (2009). Flavonol 3-O-glycosides series of Vitis vinifera Cv. Petit Verdot red wine grapes. *Journal of Agricultural and Food Chemistry*, *57*(1), 209–219.

Castro-Acosta, M. L., Stone, S. G., Mok, J. E., Mhajan, R. K., Fu, C. I., Lenihan-Geels, G. N., et al. (2017). Apple and blackcurrant polyphenol-rich drinks decrease postprandial glucose, insulin and incretin response to a high-carbohydrate meal in healthy men and women. *Journal of Nutritional Biochemistry*, *49*, 53–62.

Cheng, D., Zhang, X., Xu, L., Li, X., Hou, L., & Wang, C. (2017). Protective and prophylactic effects of chlorogenic acid on aluminum-induced acute hepatotoxicity and hematotoxicity in mice. *Chemico-Biological Interactions*, *273*, 125–132.

Cherbuin, N., & Anstey, K. J. (2012). The Mediterranean diet is not related to cognitive change in a large prospective investigation: The PATH through life study. *The American Journal of Geriatric Psychiatry*, *20*(7), 635–639.

Chie, T., Yoichi, F., Yoshimi, K., Norie, S. S., Emi, S., Yoshinari, T., et al. (2015). Estimated dietary polyphenol intake and major food and beverage sources among elderly Japanese. *Nutrients*, 7(12), 10269–10281.

Clifford, M. N., Stoupi, S., & Kuhnert, N. (2007). Profiling and characterization by LC-MSn of the galloylquinic acids of green tea, tara tannin, and tannic acid. *Journal of Agricultural and Food Chemistry*, *55*(8), 2797–2807.

Coe, S., Axelsson, E., Murphy, V., Santos, M., Collett, J., Clegg, M., et al. (2017). Flavonoid rich dark cocoa may improve fatigue in people with multiple sclerosis, yet has no effect on glycaemic response: An exploratory trial. *Clinical Nutrition ESPEN*, *21*, 20–25.

Couto, E., Boffetta, P., Lagiou, P., Ferrari, P., Buckland, G., Overvad, K., et al. (2011). Mediterranean dietary pattern and cancer risk in the EPIC cohort. *British Journal of Cancer*, *104*(9), 1493–1499.

Cui, E. J., Song, N. Y., Shrestha, S., Chung, I. S., Kim, J. Y., Jeong, T. S., et al. (2012). Flavonoid glycosides from cowpea seeds (Vigna sinensis K.) inhibit LDL oxidation. *Food Science and Biotechnology*, *21*(2), 619–624.

Curtis, P. J., van der Velpen, V., Berends, L., Jennings, A., Feelisch, M., Umpleby, A. M., et al. (2019). Blueberries improve biomarkers of cardiometabolic function in participants with metabolic syndrome-results from a 6-month, double-blind, randomized controlled trial. *American Journal of Clinical Nutrition*, *109*(6), 1535–1545.

Davis, A. L., Cai, Y., Davies, A. P., & Lewis, J. R. (1996). 1H and 13C NMR assignments of some green tea polyphenols. *Magnetic Resonance in Chemistry*, *34*(11), 887–890.

Delcambre, A., & Saucier, C. (2012). Identification of new flavan-3-ol monoglycosides by UHPLC-ESI-Q-TOF in grapes and wine. *Journal of Mass Spectrometry*, *47*(6), 727–736.

Drynan, J. W., Clifford, M. N., Obuchowicz, J., & Kuhnert, N. (2010). The chemistry of low molecular weight black tea polyphenols. *Natural Product Reports*, *27*(3), 417–462.

Fayeulle, N., Vallverdu-Queralt, A., Meudec, E., Hue, C., Boulanger, R., Cheynier, V., et al. (2018). Characterization of new flavan-3-ol derivatives in fermented cocoa beans. *Food Chemistry*, *259*(SEP.1), 207–212.

Ferk, F., Kundi, M., Brath, H., Szekeres, T., Al-Serori, H., Mišík, M., et al. (2018). Gallic acid improves health-associated biochemical parameters and prevents oxidative damage of DNA in type 2 diabetes patients: Results of a placebo-controlled pilot study. *Molecular Nutrition & Food Research*, *62*(4), 1700482.

Ferrari, E., Foca, G., Vignali, M., Tassi, L., & Ulrici, A. (2011). Adulteration of the anthocyanin content of red wines: Perspectives for authentication by Fourier transform-near InfraRed and 1H NMR spectroscopies. *Analytica Chimica Acta*, *701*(2), 139–151.

Frank, O., Blumberg, S., Krümpel, G., & Hofmann, T. (2008). Structure determination of 3-O-Caffeoyl-epi-γ-quinide, an orphan bitter lactone in roasted coffee. *Journal of Agricultural and Food Chemistry*, *56*(20), 9581–9585.

Franke, A. A., Halm, B. M., Kakazu, K., Li, X., & Custer, L. J. (2009). Phytoestrogenic isoflavonoids in epidemiologic and clinical research. *Drug Testing and Analysis*, *1*(1), 14–21.

Gee, J. R., Saltzstein, D. R., Kim, K., Kolesar, J., Huang, W., Havighurst, T. C., et al. (2017). A phase II randomized, double-blind, presurgical trial of polyphenon E in bladder cancer patients to evaluate pharmacodynamics and bladder tissue biomarkers. *Cancer Prevention Research (Philadelphia, Pa.)*, *10*(5), 298–307.

Gorji, Z., Varkaneh, H. K., Talaei, S., Nazary-Vannani, A., Clark, C. C. T., Fatahi, S., et al. (2019). The effect of green-coffee extract supplementation on obesity: A systematic review and dose-response meta-analysis of randomized controlled trials. *Phytomedicine*, *63*, 153018.

Grases, F., Prieto, R. M., Fernández-Cabot, R. A., Costa-Bauzá, A., Sánchez, A. M., & Prodanov, M. (2015). Effect of consuming a grape seed supplement with abundant phenolic compounds on the oxidative status of healthy human volunteers. *Nutrition Journal*, *14*(1), 94.

Gresele, P., Pignatelli, P., Guglielmini, G., Carnevale, R., Mezzasoma, A. M., Ghiselli, A., et al. (2008). Resveratrol, at concentrations attainable with moderate wine consumption, stimulates human platelet nitric oxide production. *The Journal of Nutrition*, *138*(9), 1602–1608.

Gutierrez, B. L., Zhong, G. Y., & Brown, S. K. (2018). Genetic diversity of dihydrochalcone content in Malus germplasm. *Genetic Resources and Crop Evolution*, *65*, 1485–1502.

He, J., & Giusti, M. M. (2010). Anthocyanins: Natural colorants with health-promoting properties. *Annual Review of Food Science and Technology*, *1*, 163–187.

Howitz, K. T., Bitterman, K. J., Cohen, H. Y., Lamming, D. W., Lavu, S., Wood, J. G., et al. (2003). Small molecule activators of sirtuins extend Saccharomyces cerevisiae lifespan. *Nature*, *425*(6954), 191–196.

Huang, H., Liao, D., Dong, Y., & Pu, R. (2020). Effect of quercetin supplementation on plasma lipid profiles, blood pressure, and glucose levels: A systematic review and meta-analysis. *Nutrition Reviews*, *78*(8), 615–626.

Jaiswal, R., Sovdat, T., Vivan, F., & Kuhnert, N. (2010). Profiling and characterization by LC-MSnof the chlorogenic acids and hydroxycinnamoylshikimate esters in Maté (Ilex paraguariensis)†. *Journal of Agricultural and Food Chemistry*, *58*(9), 5471–5484.

Jaitz, L., Siegl, K., Eder, R., Rak, G., Abranko, L., Koellensperger, G., et al. (2010). LC–MS/MS analysis of phenols for classification of red wine according to geographic origin, grape variety and vintage. *Food Chemistry*, *122*(1), 366–372.

Jiang, X., Zhang, D., & Jiang, W. (2014). Coffee and caffeine intake and incidence of type 2 diabetes mellitus: A meta-analysis of prospective studies. *European Journal of Nutrition*, *53*(1), 25–38.

Jokioja, J., Linderborg, K. M., Kortesniemi, M., Nuora, A., Heinonen, J., Sainio, T., et al. (2020). Anthocyanin-rich extract from purple potatoes decreases postprandial glycemic response and affects inflammation markers in healthy men. *Food Chemistry*, *310*, 125797.

Kim, H., Simbo, S. Y., Fang, C., McAlister, L., Roque, A., Banerjee, N., et al. (2018). Açaí (Euterpe oleracea Mart.) beverage consumption improves biomarkers for inflammation but not glucose- or lipid-metabolism in individuals with metabolic syndrome in a randomized, double-blinded, placebo-controlled clinical trial. *Food & Function*, *9*(6), 3097–3103.

Kontogianni, V. G. (2014). Chapter 8—Novel techniques towards the identification of different classes of polyphenols. In R. R. Watson (Ed.), *Polyphenols in plants* (pp. 159–185). San Diego: Academic Press.

Kuiper, G. G. J. M., Carlsson, B., Grandien, K., Enmark, E., Häggblad, J., Nilsson, S., et al. (1997). Comparison of the ligand binding specificity and transcript tissue distribution of estrogen receptors α and β. *Endocrinology*, *138*(3), 863–870.

Lin, L. Z., & Harnly, J. M. (2007). A screening method for the identification of glycosylated flavonoids and other phenolic compounds using a standard analytical approach for all plant materials. *Journal of Agricultural and Food Chemistry*, *55*(4), 1084–1096.

Lin, L. Z., & Harnly, J. M. (2012). Quantitation of flavanols, proanthocyanidins, isoflavones, flavanones, dihydrochalcones, stilbenes, benzoic acid derivatives using ultraviolet absorbance after identification by liquid chromatography–mass spectrometry. *Journal of Agricultural and Food Chemistry*, *60*(23), 5832–5840.

Lin, L.-C., Wang, M.-N., Tseng, T.-Y., Sung, J., & Tsai, T.-H. (2007). Pharmacokinetics of (−)-epigallocatechin-3-gallate in conscious and freely moving rats and its brain regional distribution. *Journal of Agricultural and Food Chemistry*, *55*(4), 1517–1524.

Ma, Y., Tanaka, N., Vaniya, A., Kind, T., & Fiehn, O. (2016). Ultrafast polyphenol metabolomics of red wines using MicroLC-MS/MS. *Journal of Agricultural and Food Chemistry*, *64*(2), 505–512.

Määttä-Riihinen, K. R., Kamal-Eldin, A., & Törrönen, A. R. (2004). Identification and quantification of phenolic compounds in berries of Fragaria and Rubus species (family Rosaceae). *Journal of Agricultural and Food Chemistry*, *52*(20), 6178–6187.

Man, Z., Xin, C.-T., Qingrong, & Huang. (2018). Chemistry and health effect of tea polyphenol (−)-epigallocatechin 3- O-(3- O-methyl)gallate. *Journal of Agricultural and Food Chemistry*, *67*(19), 5374–5378.

Marie, Z., Jean-Paul, M., Emmanuelle, M., Christine, L., & Guernevé. (2018). New flavanol O-glycosides in grape and wine. *Food Chemistry*, *266*, 441–448.

Martins, N. C., Dorneles, G. P., Blembeel, A. S., Marinho, J. P., Proença, I. C. T., da Cunha Goulart, M. J. V., et al. (2020). Effects of grape juice consumption on oxidative stress and inflammation in male volleyball players: A randomized, double-blind, placebo-controlled clinical trial. *Complementary Therapies in Medicine*, *54*, 102570.

Mathias, S., Garland, C., Barrett-Connor, E., & Wingard, D. L. (1985). Coffee, plasma cholesterol, and lipoproteins. A population study in an adult community. *American Journal of Epidemiology*, *121*(6), 896–905.

Mattila, P., Hellström, J., & Törrönen, R. (2006). Phenolic acids in berries, fruits, and beverages. *Journal of Agricultural and Food Chemistry*, *54*(19), 7193–7199.

Milder, I. E. J., Arts, I. C. W., Putte, B. V. D., Venema, D. P., & Hollman, P. C. H. (2005). Lignan contents of Dutch plant foods: A database including lariciresinol, pinoresinol, secoisolariciresinol and matairesinol. *British Journal of Nutrition*, *93*(03), 393–402.

Naveed, M., Hejazi, V., Abbas, M., Kamboh, A. A., Khan, G. J., Shumzaid, M., et al. (2018). Chlorogenic acid (CGA): A pharmacological review and call for further research. *Biomedicine & Pharmacotherapy*, *97*, 67–74.

Neetha, M. C., Panchaksharappa, M. G., Pattabhiramasastry, S., Shivaprasad, N. V., & Venkatesh, U. G. (2020). Chemopreventive synergism between green tea extract and curcumin in patients with potentially malignant oral disorders: A double-blind, randomized preliminary study. *The Journal of Contemporary Dental Practice*, *21*(5), 521–531.

Neveu, V., Perez-Jiménez, J., Vos, F., Crespy, V., du Chaffaut, L., Mennen, L., et al. (2010). Phenol-explorer: An online comprehensive database on polyphenol contents in foods. *Database (Oxford)*, *2010*, bap024.

Nonaka, G.-I., Miwa, N., & Nishioka, I. (1982). Stilbene glycoside gallates and proanthocyanidins from Polygonum multiflorum. *Phytochemistry*, *21*(2), 429–432.

Ogo, Y., Ozawa, K., Ishimaru, T., Murayama, T., & Takaiwa, F. (2013). Transgenic rice seed synthesizing diverse flavonoids at high levels: A new platform for flavonoid production with associated health benefits. *Plant Biotechnology Journal*, *11*(6), 734–746.

Oh, J. K., Sandin, S., Ström, P., Löf, M., Adami, H. O., & Weiderpass, E. (2015). Prospective study of breast cancer in relation to coffee, tea and caffeine in Sweden. *International Journal of Cancer*, *137*(8), 1979–1989.

Okamoto, T., Hashimoto, Y., Kobayashi, R., Nakazato, K., & Willems, M. E. T. (2020). Effects of blackcurrant extract on arterial functions in older adults: A randomized, double-blind, placebo-controlled, crossover trial. *Clinical and Experimental Hypertension*, *42*(7), 640–647.

Onakpoya, I. J., Spencer, E. A., Thompson, M. J., & Heneghan, C. J. (2015). The effect of chlorogenic acid on blood pressure: A systematic review and meta-analysis of randomized clinical trials. *Journal of Human Hypertension*, *29*(2), 77–81.

Osakabe, N. (2013). Flavan 3-ols improve metabolic syndrome risk factors: Evidence and mechanisms. *Journal of Clinical Biochemistry and Nutrition*, *52*(3), 186–192.

Panahande, S. B., Maghbooli, Z., Hossein-Nezhad, A., Qorbani, M., Moeini-Nodeh, S., Haghi-Aminjan, H., et al. (2019). Effects of French maritime pine bark extract (Oligopin®) supplementation on bone remodeling markers in postmenopausal osteopenic women: A randomized clinical trial. *Phytotherapy Research*, *33*(4), 1233–1240.

Park, S. J., Ahmad, F., Philp, A., Baar, K., Williams, T., Luo, H., et al. (2012). Resveratrol ameliorates aging-related metabolic phenotypes by inhibiting cAMP phosphodiesterases. *Cell*, *148*(3), 421–433.

Pastor, R. F., Restani, P., Di Lorenzo, C., Orgiu, F., Teissedre, P. L., Stockley, C., et al. (2017). Resveratrol, human health and winemaking perspectives. *Critical Reviews in Food Science and Nutrition*, *59*(8), 1237–1255.

Peñalvo, J. L., Haajanen, K. M., Botting, N., & Adlercreutz, H. (2005). Quantification of lignans in food using isotope dilution gas chromatography/mass spectrometry. *Journal of Agricultural and Food Chemistry*, *53*(24), 9342.

Penumathsa, S. V., Koneru, S., Zhan, L., John, S., Menon, V. P., Prasad, K., et al. (2008). Secoisolariciresinol diglucoside induces neovascularization-mediated cardioprotection against ischemia-reperfusion injury in hypercholesterolemic myocardium. *Journal of Molecular and Cellular Cardiology*, *44*(1), 170–179.

Power, A. C., Chapman, J., Chandra, S., & Cozzolino, D. (2019). 6—Ultraviolet-visible spectroscopy for food quality analysis. In J. Zhong, & X. Wang (Eds.), *Evaluation technologies for food quality* (pp. 91–104). Woodhead Publishing.

Prasad, K. (2005). Hypocholesterolemic and antiatherosclerotic effect of flax lignan complex isolated from flaxseed. *Atherosclerosis*, *179*(2), 269–275.

Prasad, K. (2007). A study on regression of hypercholesterolemic atherosclerosis in rabbits by flax lignan complex. *Journal of Cardiovascular Pharmacology and Therapeutics*, *12*(4), 304.
Puel, C., Quintin, A., Mathey, J., Obled, C., Davicco, M. J., Lebecque, P., et al. (2005). Prevention of bone loss by phloridzin, an apple polyphenol, in ovariectomized rats under inflammation conditions. *Calcified Tissue International*, 77(5), 311–318.
Qi, H., & Li, S. (2014). Dose-response meta-analysis on coffee, tea and caffeine consumption with risk of Parkinson's disease. *Geriatrics & Gerontology International*, *14*(2), 430–439.
Quezada-Fernández, P., Trujillo-Quiros, J., Pascoe-González, S., Trujillo-Rangel, W. A., Cardona-Müller, D., Ramos-Becerra, C. G., et al. (2019). Effect of green tea extract on arterial stiffness, lipid profile and sRAGE in patients with type 2 diabetes mellitus: A randomised, double-blind, placebo-controlled trial. *International Journal of Food Sciences and Nutrition*, *70*(8), 977–985.
Quirós-Fernández, R., López-Plaza, B., Bermejo, L. M., Palma-Milla, S., & Gómez-Candela, C. (2019). Supplementation with hydroxytyrosol and punicalagin improves early atherosclerosis markers involved in the asymptomatic phase of atherosclerosis in the adult population: A randomized, placebo-controlled, crossover trial. *Nutrients*, *11*(3), 640.
Rensburg, H. V., Steynberg, P. J., Burger, J. F. W., Heerden, P. S. V., & Ferreira, D. (1999). Circular dichroic properties of Flavan-3-ols. *Journal of Chemical Research, Synopses*, (7), 450–451.
Rothwell, J. A., Mireia, U. S., Maria, B.-O., Craig, K., Rafael, L., Roman, E., et al. (2012). Phenol-explorer 2.0: A major update of the phenol-explorer database integrating data on polyphenol metabolism and pharmacokinetics in humans and experimental animals. *Database*, *2012*, bas031.
Saitou, K., Ochiai, R., Kozuma, K., Sato, H., Koikeda, T., Osaki, N., et al. (2018). Effect of chlorogenic acids on cognitive function: A randomized, double-blind, placebo-controlled trial. *Nutrients*, *10*(10), 1337.
Sajadi Hezaveh, Z., Azarkeivan, A., Janani, L., Hosseini, S., & Shidfar, F. (2019). The effect of quercetin on iron overload and inflammation in β-thalassemia major patients: A double-blind randomized clinical trial. *Complementary Therapies in Medicine*, *46*, 24–28.
Schulze, C., Bangert, A., Kottra, G., Geillinger, K. E., Schwanck, B., Vollert, H., et al. (2014). Inhibition of the intestinal sodium-coupled glucose transporter 1 (SGLT1) by extracts and polyphenols from apple reduces postprandial blood glucose levels in mice and humans. *Molecular Nutrition & Food Research*, *58*(9), 1795–1808.
Senanayake, S. P. J. N. (2013). Green tea extract: Chemistry, antioxidant properties and food applications—A review. *Journal of Functional Foods*, *5*(4), 1529–1541.
Shen, Y., Xu, Z., & Sheng, Z. (2017). Ability of resveratrol to inhibit advanced glycation end product formation and carbohydrate-hydrolyzing enzyme activity, and to conjugate methylglyoxal. *Food Chemistry*, *216*, 153–160.
Shoji, T., Yamada, M., Miura, T., Nagashima, K., Ogura, K., Inagaki, N., et al. (2017). Chronic administration of apple polyphenols ameliorates hyperglycaemia in high-normal and borderline subjects: A randomised, placebo-controlled trial. *Diabetes Research and Clinical Practice*, *129*, 43–51.
Singh, N. A., Mandal, A. K. A., & Khan, Z. A. (2016). Potential neuroprotective properties of epigallocatechin-3-gallate (EGCG). *Nutrition Journal*, *15*(1), 60.
Stark, T., & Hofmann, T. (2005). Isolation, structure determination, synthesis, and sensory activity of N-phenylpropenoyl-L-amino acids from cocoa (Theobroma cacao). *Journal of Agricultural and Food Chemistry*, *53*(13), 5419–5428.
Tabatabaie, M., Abdollahi, S., Salehi-Abargouei, A., Clark, C. C. T., Karimi-Nazari, E., Fallahzadeh, H., et al. (2020). The effect of resveratrol supplementation on serum levels of asymmetric de-methyl-arginine and paraoxonase 1 activity in patients with type 2 diabetes: A randomized, double-blind controlled trial. *Phytotherapy Research*, *34*(8), 2023–2031.

Tisserant, L. P., Hubert, J., Lequart, M., Borie, N., & Courot, E. (2016). 13C NMR and LC-MS profiling of stilbenes from elicited grapevine hairy root cultures. *Journal of Natural Products, 79*(11), 2846–2855.

Tominaga, S., Nishi, K., Nishimoto, S., Akiyama, K., Yamauchi, S., & Sugahara, T. (2012). (−)-Secoisolariciresinol attenuates high-fat diet-induced obesity in C57BL/6 mice. *Food & Function, 3*(1), 76–82.

Turner, R. S., Thomas, R. G., Craft, S., van Dyck, C. H., Mintzer, J., Reynolds, B. A., et al. (2015). A randomized, double-blind, placebo-controlled trial of resveratrol for Alzheimer disease. *Neurology, 85*(16), 1383–1391.

Tylewicz, U., Nowacka, M., Martín-García, B., Wiktor, A., & Gómez Caravaca, A. M. (2018). 5—Target sources of polyphenols in different food products and their processing by-products. In C. M. Galanakis (Ed.), *Polyphenols: Properties, recovery, and applications* (pp. 135–175). Woodhead Publishing.

Valls, R. M., Llauradó, E., Fernández-Castillejo, S., Puiggrós, F., Solà, R., Arola, L., et al. (2016). Effects of low molecular weight procyanidin rich extract from french maritime pine bark on cardiovascular disease risk factors in stage-1 hypertensive subjects: Randomized, double-blind, crossover, placebo-controlled intervention trial. *Phytomedicine, 23*(12), 1451–1461.

van Dusseldorp, M., Katan, M. B., & Demacker, P. N. (1990). Effect of decaffeinated versus regular coffee on serum lipoproteins. A 12-week double-blind trial. *American Journal of Epidemiology, 132*(1), 33–40.

Vitrac, X., Bornet, A., Vanderlinde, R., Valls, J., Richard, T., Delaunay, J. C., et al. (2005). Determination of stilbenes (delta-viniferin, trans-astringin, trans-piceid, cis- and trans-resveratrol, epsilon-viniferin) in Brazilian wines. *Journal of Agricultural and Food Chemistry, 53*(14), 5664–5669.

Vitrac, X., Monti, J.-P., Vercauteren, J., Deffieux, G., & Mérillon, J.-M. (2002). Direct liquid chromatographic analysis of resveratrol derivatives and flavanonols in wines with absorbance and fluorescence detection. *Analytica Chimica Acta, 458*(1), 103–110.

Wang, Y., Catana, F., Yang, Y., Roderick, R., & van Breemen, R. B. (2002). An LC-MS method for analyzing total resveratrol in grape juice, cranberry juice, and in wine. *Journal of Agricultural and Food Chemistry, 50*(3), 431–435.

Wang, Y., Li, M., Xu, X., Song, M., Tao, H., & Bai, Y. (2012). Green tea epigallocatechin-3-gallate (EGCG) promotes neural progenitor cell proliferation and sonic hedgehog pathway activation during adult hippocampal neurogenesis. *Molecular Nutrition & Food Research, 56*(8), 1292–1303.

Wang, X., Yang, D. Y., Yang, L. Q., Zhao, W. Z., Cai, L. Y., & Shi, H. P. (2019). Anthocyanin consumption and risk of colorectal cancer: A meta-analysis of observational studies. *Journal of the American College of Nutrition, 38*(5), 470–477.

Wang, W., Zhang, L., Wang, S., Shi, S., Jiang, Y., Li, N., et al. (2014). 8-C N-ethyl-2-pyrrolidinone substituted flavan-3-ols as the marker compounds of Chinese dark teas formed in the post-fermentation process provide significant antioxidative activity. *Food Chemistry, 152*, 539–545.

Williamson, G. (2013). Possible effects of dietary polyphenols on sugar absorption and digestion. *Molecular Nutrition & Food Research, 57*(1), 48–57.

Wise, J. (2016). Mediterranean diet in UK shows positive effects in study. *British Medical Journal, 354*, i5286.

Wu, J., Huang, J., Xiao, Q., Zhang, S., Xiao, Z., Li, Q., et al. (2004). Complete assignments of 1H and 13C NMR data for 10 phenylethanoid glycosides. *Magnetic Resonance in Chemistry, 42*(7), 659–662.

Yari, Z., Tabibi, H., Najafi, I., Hedayati, M., & Movahedian, M. (2020). Effects of soy isoflavones on serum lipids and lipoprotein (a) in peritoneal dialysis patients. *Nutrition, Metabolism, and Cardiovascular Diseases, 30*(8), 1382–1388.

Ye, Q., Wang, H., & Xu, J. (2015). Chapter 2—Application of NMR spectroscopy for the characterization of dietary polyphenols. In A. Ur-Rahman, & M. I. Choudhary (Eds.), *Applications of NMR spectroscopy: Volume 3* (pp. 37–77). Bentham Science Publishers.

Zamora-Ros, R., Andres-Lacueva, C., Lamuela-Raventós, R. M., Berenguer, T., Jakszyn, P., Martínez, C., et al. (2008). Concentrations of resveratrol and derivatives in foods and estimation of dietary intake in a Spanish population: European prospective investigation into cancer and nutrition (EPIC)-Spain cohort. *British Journal of Nutrition, 100*(1), 188–196.

Zhang, L., Ho, C.-T., Zhou, J., Santos, J. S., Armstrong, L., & Granato, D. (2019). Chemistry and biological activities of processed Camellia sinensis teas: A comprehensive review. *Comprehensive Reviews in Food Science and Food Safety, 18*(5), 1474–1495.

Zhang, L., Li, N., Ma, Z.-Z., & Tu, P.-F. (2011). Comparison of the chemical constituents of aged Pu-erh tea, ripened Pu-erh tea, and other teas using HPLC-DAD-ESI-MSn. *Journal of Agricultural and Food Chemistry, 59*(16), 8754–8760.

Zhang, L., Tai, Y., Wang, Y., Meng, Q., & Wan, X. (2017). The proposed biosynthesis of procyanidins by the comparative chemical analysis of five Camellia species using LC-MS. *Scientific Reports*, 7, 46131.

Zhang, L., Wang, Y., Li, D., Ho, C.-T., Li, J., & Wan, X. (2016). The absorption, distribution, metabolism and excretion of procyanidins. *Food & Function*, 7(3), 1273–1281.

Zhou, J., Zhang, L., Meng, Q., Wang, Y., Long, P., Ho, C.-T., et al. (2018). Roasting improves the hypoglycemic effects of a large-leaf yellow tea infusion by enhancing the levels of epimerized catechins that inhibit α-glucosidase. *Food & Function, 9*(10), 5162–5168.

Zhu, J., & Du, C. (2020). Could grape-based food supplements prevent the development of chronic kidney disease? *Critical Reviews in Food Science and Nutrition, 60*(18), 3054–3062.

CHAPTER TWO

The application of polyphenols in food preservation

Qian Wu[a,*] and Jie Zhou[b]
[a]Hubei University of Technology, Wuhan, China
[b]Northwest Agriculture & Forestry University, Yangling, China
*Corresponding author: e-mail address: wuqian@hbut.edu.cn

Contents

Advances in Food and Nutrition Research, Volume 98
ISSN 1043-4526
https://doi.org/10.1016/bs.afnr.2021.02.005

Abstract

Polyphenols are a kind of complex secondary metabolites in nature, widely exist in the flowers, bark, roots, stems, leaves, and fruits of plants. Numerous studies have shown that plant-derived polyphenols have a variety of bioactivities due to their unique chemical structure, such as antioxidant, antimicrobial, and prevention of chronic diseases, cardiovascular disease, cancer, osteoporosis, and neurodegeneration. With the gradual rise of natural product development, plant polyphenols have gradually become one of the research hotspots in the field of food science due to their wide distribution in the plants, and the diversity of physiological functions. Owing to the extraordinary antioxidant and antibacterial activity of polyphenols, plant-derived polyphenols offer an alternative to chemical additives used in the food industry, such as oil, seafood, meat, beverages, and food package materials. Based on this, this chapter provides an overview of the potential antioxidant and antibacterial mechanisms of plant polyphenols and their application in food preservation, it would be providing a reference for the future development of polyphenols in the food industry.

1. Antimicrobial prosperities of polyphenols

1.1 Antimicrobial activity and definition

Nowadays, there is increasing consumer demand for food products with clean labels. Overall, a clean label product is free of chemically synthesized and other substances which, despite their origin, are not perceived by consumers as beneficial for their health. Hence, certain food additives are not recognized as clean. Thus, consumers' preference for natural food additives and their concern about the carcinogenicity of synthetic compounds has encouraged the food industry to consider natural alternatives. As a result, the clean label ingredient market is continuously growing worldwide. Particular attention has been paid to phenolic compounds present in plant foods. This phenomenon is due to the growing number of scientific studies that signify the benefits of these compounds for human health. A boost to research has also been given by the food industry through the production of supplements or additives.

Dietary polyphenols present structural features of phenolics and constitute one of the most numerous and widely distributed groups of natural compounds in the plant kingdom. So far, over 8000 phenolic structures have been recognized, and more than half of which have been identified as flavonoids (Oliver, Vittorio, Cirillo, & Boyer, 2016). Overall, polyphenols are a highly diverse group of compounds with several subgroups that may widely differ in their stability, bioavailability, and physiological functions. Despite that, they have deserved particular attention due to the roles they play in food (Tomadoni, Viacava, Cassani, Moreira, & Ponce, 2016).

Besides the antioxidant role, they may exhibit an important antimicrobial activity, the mechanisms of which have not been completely recognized yet. The known mechanisms include the capability for modifying cell membranes permeability, the changes in several intracellular functions caused by the phenolics to enzymes binding, or by the loss of the cell wall integrity due to various interactions with the cell membrane. While many plant extracts are rich in polyphenols, their applicability is of particular interest for the development of natural alternatives to synthetic preservatives in food (Bouarab-Chibane et al., 2019).

Hence, this section recognizes rather the antimicrobial role and applications of plant originating mixtures of compounds including polyphenols especially those removed as extracts. Such as grapefruit seed extract, green tea extract, grape extract, pomegranate extract, sea buckthorn extract, Indian borage extract, dropwort extract, cloudberry/black chokeberry extract, clove extract, passion fruit extract, turmeric extract.

1.2 Addition mode

The recent development with natural antimicrobials is their incorporation into packaging materials that offer the advantage of concentrating the antimicrobial at the surface of a food product where microbes reside and thus can be effectively inhibited. Similarly, microencapsulated antimicrobials incorporated in food packaging have been demonstrated to control a range of microorganisms, and thus to better preserve the sensory attributes, safety, and the shelf-life of a product (Hintz, Matthews, & Di, 2015). Applications of phenolic-rich natural extracts in the food industry can be further extended to antimicrobial food packaging. Antimicrobial agents incorporated into the polymer interact with the food surface to prevent foodborne microbial growth. Thus, antimicrobial food packaging has a small negative impact on food, but provides antimicrobial properties that can preserve food for a long time.

1.3 Antibacterial mechanism

The exact mechanism of antibacterial action by polyphenols is still unclear and requires further investigation. Phenolic compounds can have a diverse range of chemical structures, which means that there are many possible mechanisms of antimicrobial activity (Fig. 1). Moreover, most studies have focused on investigating not single compounds but extracts, which are mixtures of polyphenols belonging to different groups. This suggests that phenolic compounds have multidimensional activities. Besides, interactions between polyphenols in mixtures may also influence their mechanisms of action.

1.4 Inhibition of cell envelope synthesis

Peptidoglycan is an essential component of the bacterial cell wall, and the inhibition of its synthesis is a common mechanism of action of conventional antimicrobial drugs and flavonoids. Phenolic compounds interact with bacterial cell walls, leading to disruption of the cell wall and the release of cellular contents. Studies have demonstrated that plant extracts can cause significant increases in intracellular protein leakage in bacterial cultures, depending on the concentration and the length of incubation (Tang, Kang, & Lu, 2016). Polyphenols have also been shown to cause the release of cytoplasmic constituents, such as nucleotides and small cellular molecules e.g., potassium and phosphate ions (Stojkovic et al., 2013).

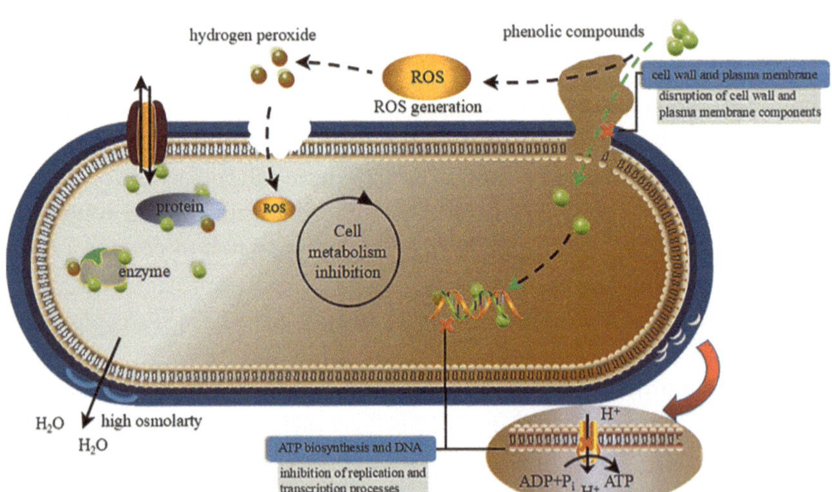

Fig. 1 Diagrammatic representation of action mechanism of polyphenols.

Cell wall damage decreases cell resistance to unfavorable conditions such as high or low osmotic pressure and different external factors. Changes in cell shape, irregular forms, wrinkles on the surfaces of bacteria, and the formation of cell aggregates may be observed. The available data show that Gram-negative bacteria are more resistant to phenolic compounds. This is probably related to the presence of a lipophilic outer that makes the cell wall impermeable to several macromolecules. The interactions of flavonoids with lipid bilayers involve two mechanisms. The first is associated with the partition of the more non-polar compounds in the hydrophobic interior of the membrane, while the second one includes the formation of hydrogen bonds between the polar head groups of lipids and the more hydrophilic flavonoids at the membrane interface. Moreover, nonspecific interactions of flavonoids with phospholipids can induce structural changes in the properties of the membrane (e.g., thickness and fluctuations) and indirectly modulate the distribution/function of membrane proteins, as well as influence pharmacological properties of flavonoids themselves. The high resistance of Gram-negative bacteria to phenolic compounds may also be associated with enzymes in the periplasmic space, which can damage molecules introduced from outside. Nonetheless, other studies have suggested that polyphenols can disintegrate the outer membrane of Gram-negative bacteria, leading to increases in membrane permeability (Plumed-Ferrer et al., 2013). Such as, berry phenolics can reduce outer membrane permeability in a similar manner to EDTA by releasing LPS and chelating divalent cations or by intercalating into the outer membrane and replacing stabilizing cations (Nohynek et al., 2006). Polyphenols also disrupt bacterial cell membranes (the inner membrane), resulting in the loss of chemiosmotic control and leading to cell death. This is probably caused by the ability of polyphenols to bind proteins associated with the cell membrane. Cell wall damage is associated with the prooxidant properties of polyphenols, which cause oxidative stress.

1.5 Inducing oxidative stress and affecting protein biosynthesis in cells

Phenolic compounds can induce endogenous oxidative stress in bacteria cells, by inducing ROS formation, which can lead to cell death. The ability of polyphenols to induce ROS production could be due to their prooxidant properties. When oxidized, phenolic compounds can transform into prooxidant forms and cause H_2O_2 generation via Fenton reactions.

In the presence of transition metals (Cu(I) or Fe (II)) they can cause the generation of ·OH from H_2O_2 (Xiong et al., 2017).

Polyphenols can also influence the biosynthesis of proteins in bacterial cells and thus change their metabolic processes. Metabolic disorders can result in bacteria cell death. Polyphenols could inactivate the essential amino acid synthesis caused by interference in the reactions of phosphoenolpyruvate, erythrose-4-phosphate, and shikimic acid. This favors the production of tryptophan and decreases the production of phenylalanine or tyrosine, modifying the structure of some proteins essential for the formation of fungal appressorium structure.

1.6 Inhibition of nucleic acid synthesis

Polyphenols can also interact with bacteria cells by inhibiting DNA synthesis, which is associated with the inhibition of DNA gyrase. This enzyme is responsible for introducing negative supercoils into DNA. Gyrase is composed of two A (gyrase A) and two B subunits (gyrase B). Subunit A catalyzes DNA breakage, while subunit B catalyzes ATP hydrolysis, which is essential for DNA supercoiling. Both subunits may be targets for antimicrobial agents. However, studies have shown that polyphenols can inhibit enzyme activity by binding to gyrase B (ATP binding site). However, there is no data concerning the interaction of polyphenols with gyrase A.

Furthermore, the hydroxyl group has a key role in the inactivation of microbial enzymes such as ATPase, histidine decarboxylase, amylase, and protease. Inhibition of ATPase may be important for cell death due to disturbed cellular respiration. Polyphenols can also inhibit the electron transport chain and ATP synthesis. A decline in the overall bacterial metabolism can lead to the indirect arrest of the biofilm formation, as well.

1.7 Affect the formation of biofilm

Bacterial biofilm-based infections constitute a significant amount of all microbial and chronic infections in animals and humans, as well as in food spoilage. Microbial biofilms are composed of one or more microbial species, organized as complex tridimensional structures attaching to a surface and each other through a self-produced matrix of extracellular polymeric substances (EPS) (Bjarnsholt, 2013). The development of microbial biofilms on surface materials is a dynamic process composed of five key steps as follows: (i) initial adhesion, (ii) irreversible attachment, (iii) microcolony formation (maturation I), (iv) maturation II and (v) dispersion. When bacteria form biofilms, they present different phenotypic characteristics from planktonic cells as well as greater genetic exchange capacity, thus favoring their survival in hostile environments and getting new virulence genes. The biofilm removal is quite difficult because the EPS matrix provides adhesion to the surface and acts as a physical protection barrier against antimicrobials, acids, high temperature, ultraviolet radiation, and other adverse environmental challenges (Merino, Procura, Trejo, Bueno, & Golowczyc, 2019). For this reason, the inhibition of initial microbial adhesion to surface materials is the most efficient manner to avoid contamination through bacterial biofilms (Satpathy, Sen, Pattanaik, & Raut, 2016). Polyphenols also influence biofilm formation, but the effect is ambiguous. They can stimulate or inhibit biofilm formation depending on the concentration of polyphenolic compounds. Some studies demonstrated that at lower concentrations polyphenols have a stimulatory effect, while at higher concentrations they show an inhibitory effect. Quorum sensing, motility and adherence structures play important roles in biofilm formation and can be also be influenced by polyphenols. Polyphenols can affect biofilm formation by influencing the quorum sensing mechanism, pigment production and bacteria swarming motility, or by changing the biofilm structure. The quorum sensing mechanism also depends on the concentration of polyphenols. The production of signal molecules is inhibited at higher concentrations but stimulated at lower concentrations.

1.8 Influence of structure on antibacterial performance

The antibacterial activity of phenolic compounds is strictly related to their chemical structure, including the position and number of hydroxyl or methoxyl groups, as well as to the presence of galloyl groups in the catechin structure. Other studies have demonstrated that the antibacterial activity of catechins is strongly related to their hydrophobicity. And differences in the number and distribution of hydroxyl groups, the polymerization degree, as well as the presence of methoxy groups in the C ring, can influence the type of interactions that occur between different flavonoids and lipid bilayers. Moreover, flavonoids lacking hydroxyl groups on their B Rings are more active against microbial membranes than those with the –OH groups. This is due to the negative correlation between the relative hydrophobicity of flavonoids and the number of the hydroxyl group present. Furthermore, lipophilic flavonoids which are highly hydroxylated can be more disruptive for membrane structure. It is worth noting that bacterial membrane damage by catechins and other flavonoids may also result in an inability of the bacteria to secrete toxins.

There is a relationship between the activity of flavonoids and their structure, such as the number and position of hydroxyl groups or methoxyl groups. Such as hydroxyl groups at C-5 in the A ring and C-4′ in the B ring as well as methoxyl groups at C-3 and C-8 in the A ring of flavonoids increase their inhibitory effects. However, hydroxyl groups at C-6 in the A ring, at C-3′ and C-5′ in the B ring and at C-3 in the C ring, or methoxyl groups at C-3′ in the B ring contribute to reducing the activity of flavonoids. This effect contributes to strengthen antibacterial activity. A quantitative structure-activity relationship study showed that the activity of flavonoids is associated with molecular hydrophobicity and OH groups situated at position 3 in the C-ring. It has been shown that phenolic compounds act at the membrane level by modifying the polar heads of the lipid molecules. Some results demonstrated that their antimicrobial activity is associated with the presence and position of a free hydroxyl group bonded directly to a C6 aromatic ring as a system for electron delocalization, which favors their ability to modify the microbial cell membrane integrity.

In general, polyphenols may provoke a physical, chemical, or biochemical change in the microorganisms, and different constituents may operate by different mechanisms and may target different kinds of microbes, such as Gram-positive and Gram-negative bacteria, yeasts, or molds, because they differ in the composition of their cell membranes. Therefore, it is difficult to predict how susceptible pathogens are and why the susceptibility varies from strain to strain after applying these compounds. Polyphenols to be capable of inhibiting the growth of spoilage bacteria and fungi, and have suggested their utility in the food industry. Regarding the action mechanisms of polyphenols as antimicrobial agents, it is important to consider that their effectiveness will depend not only on the concentration and chemical nature, but also on the susceptibility and concentration of the pathogen, and even of the microbial strain characteristics. This capacity will vary depending on the composition and nature of the microbial cell surface. Furthermore, it is important to know and understand the mechanisms of action of pathogens, depending on the characteristics of the food and environmental conditions.

1.9 Photodynamic inactivation (PDI) and synergistic antibacterial mechanism with curcumin

Curcumin is a natural polyphenolic compound extracted from the rhizome of the perennial herb *Curcuma longa*. Curcumin has several advantageous properties, including low intestinal absorption, faster metabolism, and low toxicity at high doses in humans (up to 12 g/day) which emphasize its potential applications as an antibacterial agent. Indeed, curcumin exhibits bactericidal activity and can act as a photosensitizer or can be used in combination with other substances to further enhance its antibacterial properties (Shakeri & Sahebkar, 2016).

CUR exerts bioactive properties, i.e., anticancer, anti-inflammatory, antioxidant and antimicrobial. The most recent development is the use of CUR as a natural photosensitizer (PS) in combination with visible light in a technology called photodynamic inactivation (PDI). CUR is an example of natural PS with highly antimicrobial potential–CUR activity is enhanced by the irradiation of visible light at the wavelength of 400–500 nm (Penha et al., 2017). Gram-positive bacteria are known to be more easily killed by anionic, cationic, and nonionic photosensitizers than Gram-negative bacteria because Gram-positive bacteria have a more porous structure in their cell membrane and are more easily penetrated by photosensitizers (Ghate, Zhou, & Yuk, 2019). More importantly, the cell walls of Gram-positive bacteria are composed of a cytoplasmic membrane and a thick layer of peptidoglycan. The attachment effect of lipid-soluble curcumin (hydrophobic) on peptidoglycan is poor. Thus, the Gram-positive cytoplasmic components of Lactobacillus monocytogenes might be vulnerable sites for curcumin-mediated photodynamic inactivation (PDI) with high inactivation efficiency.

The photosensitizer adsorbs photons and induces a variety of reactions including the formation of radicals and reactive oxygen species (ROS) that cause a toxic effect to the cells by the reaction with cell membranes or proteins. PDI efficacy, however, depends on several factors, such as the power output, the type of light source–continuous or pulsed, the length of pre-irradiation and irradiation times or the beam diameter (Dovigo et al., 2013; Jiang, Leung, Hua, Rao, & Xu, 2014; Santezi, Reina, & Dovigo, 2018).

Despite that, there is much interest in research with the emphasis on the combination of natural photosensitizers and PDI, both with LED and UV light derivatives to eliminate foodborne pathogens, as an alternative to thermal processes in food systems. Besides extending the shelf life, the PDI improved the sensory quality, without compromising texture and flavor. CUR-PDI could also effectively delay the browning rate and maintain the luminosity, firmness, water content, and soluble solids content. Another combination of the low concentration of CUR is with UV-A.

This technique is used to treat fruits and vegetables a high bacterial reduction was achieved even under high chemical oxygen demand/dirty conditions. However, lowering the pH value enhanced the antimicrobial photo-activity of CUR (De Oliveira, Tikekar, & Nitin, 2018). The application of CUR-PDI is also considered for the replacement of commonly used fungicides and pesticides. This method proved efficient in the inactivation of *A. flavus*, and it also reduced the production of aflatoxin B1 of contaminated maize kernels (Temba et al., 2019). Moreover, CUR can prevent microbial contamination of ready-to-eat (RTE) meat products and can be used as an edible coating for sausages.

When curcumin is used in combination with different substances (including traditional antibiotics and bacteriostatic agents, strong metal complex bacteriostatic agents, and biological coupling agents), it has been shown to have better synergistic bacteriostatic effects, which increases the sensitivity and uptake capacity of bacteria to other bacteriostatic agents. These changes can lead to increased sensitivity in gram-negative bacteria, in contrast to its effects when used alone. Thus, the synergistic effects of curcumin not only reduce the effective amounts of bacteriostatic agents and expand the range of bacteriostatic species but also improve the bacteriostatic ability of the compound, thereby expanding its potential applications. Curcumin has demonstrated safety and efficacy as a drug, but its pharmaceutical role is restricted as a result of extremely low aqueous solubility, rapid systemic elimination, inadequate tissue absorption and degradation at alkaline pH; properties that severely curtail its bioavailability. The bacteriostatic mechanisms of curcumin when used alone as a photosensitizer and when combined with a synergistic substance (Table 1).

1.10 Antibacterial advantage

Examples of plant extracts selected here exert a broad-spectrum antimicrobial activity, including against food-borne pathogens. This is important since concerns have been raised about the growing number of foodborne outbreaks caused by pathogens associated with the high rate of antibiotic resistance. In contrast, plant extracts are less likely to generate antimicrobial resistance owing to the wide variety of their active compounds, therefore they could reverse the antibiotic resistance and minimize the exposure of humans to resistant bacteria (Gupta & Birdi, 2017). Moreover, foodborne bacteria are frequently found in food processing environments mainly because of their ability to attach and form biofilms on different biotic or abiotic surfaces. Once they adhere and form biofilms, they are very difficult to eradicate even with the best antimicrobial agents. Thus, the antibiofilm activity of plant extracts is also of great importance when studying their potent use in the food industry environment.

Table 1 Phototoxicity mechanism and synergistic antibacterial mechanism of curcumin.

Mechanism	**Material**
Reduction of biofilm adhesion	Curcumin
Destruction of cytoplasmic DNA	Curcumin
Downregulation of virulence gene expression	Curcumin
Inhibition of population effect gene expression	Curcumin
Inhibition of biofilm formation and reduction of FICI	Curcumin and cinnamaldehyde
Enhanced bacterial sensitivity to antibiotics	Curcumin and antibiotics
Enhanced film polarity	Curcumin and chitosan-polyvinyl alcohol
Increased cellular uptake	Curcumin and biological coupling agents
Increased activity against *Escherichia coli*	Curcumin and strong metal complexes

FICI, partial bacteriostatic concentration.

Furthermore, synergistic or additive effect in the antimicrobial activity has been demonstrated by the mixture of natural products like plant extracts and with other food additives owing to the ability to affect multiple targets. Studies have demonstrated that mixtures of polyphenols, such as plant extracts, have a stronger influence on the growth of microorganisms than individual compounds. Therefore, it appears that a positive synergetic effect may occur between the constituents of plant extracts, resulting in strong antimicrobial activity. Current studies are therefore focusing on the antimicrobial properties of polyphenolic mixtures, rather than on individual compounds. Extracts can be obtained from several parts of plants, such as the fruits, skin, leaves, and flowers, each of which has a different composition and activity. Polyphenols may show synergism with other antimicrobials like antibiotics and be, therefore, considered for enhancing the antibiotic efficacy, lowering the antibiotic dosage, and reducing antibiotic side effects.

However, there is limited information about the use of specific combinations regarding their interactions with different components in food. Therefore, the antimicrobial mechanisms of crucial combinations need to be further recognized to be applied consistently in food preservation. Similarly, in vitro results do not necessarily guarantee the success of different combinations in vivo. Nevertheless, this approach offers a cost-efficient alternative to the food industry, at the same time adhering to the hurdle technology in inhibiting the proliferation of foodborne pathogens. It is an attractive alternative not only to improve their preservative effects but also to help decrease their sensory effect in food.

1.11 Trends and challenges of polyphenol as a natural antibacterial agent

Although plant extracts exhibit antimicrobial activity against spoilage microorganisms and foodborne pathogens, the discrepancy in effectiveness of these compounds against microorganisms in laboratory conditions and in food systems is a major concern in their utility.

1.11.1 Effect of food substrate and processing technology on antimicrobial activity

Plant polyphenol-rich extracts have high antimicrobial properties ' But many reasons difficult their incorporation into the biological formulations: fast release, low solubility, low bioactivity, and lability against environment stresses (temperature, moisture, pH, oxygen light, etc.). The oxidative degradation may deteriorate these compounds leading to the generation of free radicals and the development of unpleasant tastes and off-odors. It results in an undesirable effect on shelf stability, sensory characteristics, and consumer acceptability of the products. Thermal stability of polyphenols is thus of prime importance during different stages of food processing (Shishir, Xie, Sun, Zheng, & Chen, 2018).

Substances that have shown antimicrobial activity in vitro may have little or no effect in food matrices due to their interaction with food components such as carbohydrates, lipids, and proteins (Ayala-Zavala, González-Aguilar, & Roura, 2015). For example, polyphenols are known to interact with proteins mainly by hydrogen and hydrophobic bonds, reducing polyphenols' bioavailability and bioactivity. What is more, although some plant compounds deserved the generally recognized as safe (GRAS) status, typical toxicological information of their use in food is not there. Thus, the improvement in toxicological information of these compounds would be beneficial in their use, especially as food preservatives.

An important hurdle in the use of plant extracts in food is the incapacity of the reproducibility of their activity. Although they contain a diversity of natural chemical compounds, their antimicrobial effectiveness is affected by different qualitative and quantitative fluctuations in the content of these compounds. Moreover, mixtures are prone to contain toxic constituents, and they must be carefully investigated and standardized when adopted on an industrial scale. Also, the addition of natural preservatives such as plant extracts often raises the cost of a food product, since they need to be subjected to extraction and refinement procedures that make natural preservatives more expensive than the artificial ones (Olszewska, Gedas, & Simoes, 2020).

The occurrence of dangerous and antibiotic-resistant bacteria has been observed to increase in frequency over the past several decades. Hence, there is a pressing need for finding new antimicrobial drugs. Using plant polyphenol-rich extracts as alternatives to synthetic antimicrobials is a promising trend in the food industry, because of their high antimicrobial properties and increasing consumer demand for all-natural products. Thus, together with the efforts undertaken to search for new antimicrobials, further research on appropriate technological adaptations as well as the development of new, such as PDI, are needed to overcome obstacles associated with losses in polyphenol content.

2. Scavenging AGEs produced in food storage

The consumption of highly processed foods has increased exponentially over the past decades. Changes in processing technologies and diversity of raw materials have provided us an unparalleled variety of food products, but also some known and unknown health-hazardous 'contaminants'. The contents of advanced glycation end-products (AGEs) in foods are much higher than those formed in vivo because of the application of thermal treatments during food processing and home cooking. Once being digested and absorbed across the intestine, dietary advanced glycation end-products (dAGEs) can be partially transformed to fuel the AGEs pool in vivo. Preliminary studies have implied that the consumption of high-AGE foods is associated with oxidative stress, inflammation, and cardiovascular risks in both healthy individuals and patients with chronic diseases. Therefore, we pay more and more attention to the AGEs removal during food storage by adding polyphenols.

2.1 Advanced glycation end-products (AGEs)

2.1.1 Definition and classification

AGEs are a heterogeneous group of irreversible products resulting from nonenzymatic glycation and oxidation of proteins, nucleic acids, and lipids. Glycation produces AGEs compounds with toxic properties associated with inflammation and oxidative stress. AGEs accumulate in the extracellular matrix of various tissues and critically contribute to chronic diseases. Furthermore, AGEs have been shown to become antigenic, and thus, to induce immune responses. AGEs also have adverse effects on cellular functions by cross-linking intracellular and extracellular proteins, which trigger diverse diseases, such as arthritis, neurological disorders, cancers, diabetes, and cardiovascular disorders.

Based on differences in their origins, AGEs primarily include two classes: biologically derived AGEs (biological-AGEs) and food-derived AGEs (food-AGEs or dietary AGEs, d-AGEs). Biological-AGEs have been reported to be toxicants closely associated with various chronic diseases, in particular, diabetes and kidney disorders. Numerous studies have aimed to alleviate the health risk associated with AGEs. Restricting d-AGE intake or consuming a low-calorie diet can alleviate the burden of health risks associated with AGEs. On the other hand, Lys-derived AGEs mainly include *N*(ε)-carboxymethyl-lysine (CML), pyrraline, *N*(ε)-carboxyethyl-lysine (CEL), methylglyoxal-lysine dimer (MOLD), and glyoxal-lysine dimer (GOLD). Arg-derived AGEs mainly include carboxymethylarginine (CMA), hydroimidazolone derived from glyoxal (G-H), and methylglyoxal (MG-H). Additionally, some AGEs such as pentosidine and glucosepane, consist of both the Lys and Arg skeletons. Recent studies also classified cysteine (Cys)-derived compounds, such as *S*-carboxymethylcysteine (CMC), as another source of AGEs (Ahmad et al., 2018).

2.1.2 Formation of AGEs

AGEs are formed by glycation, which is a nonenzymatic reaction between ketones or aldehydes and the amino groups of various proteins and contributes to protein aging. Protein glycation occurs in both normal and hyperglycemic conditions and forms irreversible AGEs. This process begins with the conversion of reversible Schiff base adducts to more stable, covalently bound Amadori rearrangement products and finally the Maillard reaction. AGEs are also formed in the absence of hyperglycemia by homeostatic imbalances, such as redox imbalances, aging, kidney disease, or autoimmune diseases. Also, AGEs are derived from foods (e.g., milk, meat, coffee, cheese), especially those prepared under high-temperature conditions, stored for long periods, or with food additives (Fig. 2) (Liang, Chen, Li, Li, & Yang, 2020).

2.1.3 Toxic properties of AGEs

AGEs have long been considered potentially toxic molecules that promote host cell death and contributing to organ damage in man. AGEs can induce

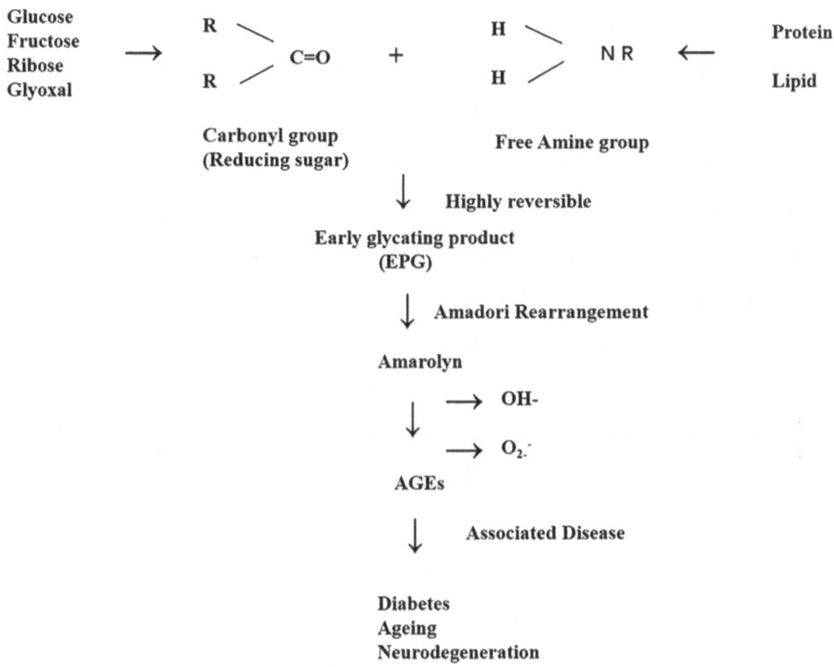

Fig. 2 Association of AGEs and human diseases.

the development or progression of not only diabetic complications but the pathophysiologies of many diseases, including cardiovascular diseases (CVDs) and neurodegenerative diseases like Alzheimer's disease (AD), Parkinson's disease (PD), and alcoholic brain damage. AGEs bind with their receptor RAGE (receptor for advanced glycation end products, which is highly expressed during host cell death through reactive oxygen species (ROS) and strong activations of extracellular signal-regulated kinase 1/2 (ERK1/2) ERK1/2 and nuclear factor kappa B (NF-kB) pathways in many diseases (Fig. 3).

2.2 AGEs in food processing

2.2.1 Contents of AGEs in food processing

AGEs are virtually ubiquitous in foods and can be formed during a serious of heat-related industrial procedures including processing, sterilization, packaging, and long-term storage of food products. Home-cooking steps produce AGEs as well, such as boiling, grilling, frying, and reheating of ready-to-eat foods. Several studies have been dedicated to comparing the AGEs contents in a wide range of food items including those processed in the factory and those prepared at home. Table 2 resents the levels of

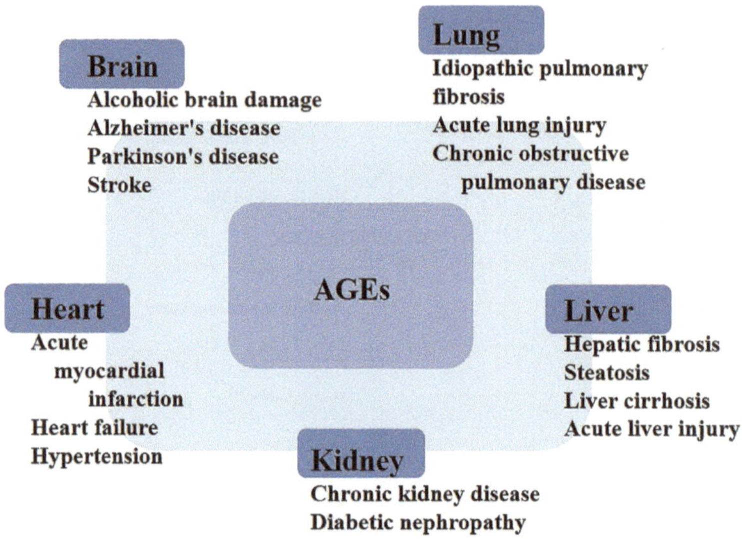

Fig. 3 Schematic representation of probable pathway of macromolecules reacting with reducing sugars to form AGEs.

AGEs in some selected factory-produced and home-made foodstuffs. CML, a well-recognized marker of AGEs, was used for comparison across different categories. It should be noted that several studies expressed the AGEs content in units by ELISA, thus not compatible to be included in the figure. Meat, milk, vegetables, and fruits contained the level of CML are 210.1 (mg/kg protein), 205 (mg/kg protein), 76.7 (mg/kg protein). The differences in sample collection, preparation, and analytical methods constitute an explanation for the paradoxical results. Several other types of AGEs, such as MOLD and GLOD, have been also quantified in a few processed foods to expand the knowledge of dAGEs values. Some studies showed the distribution of 1,2-dicarbonyl compounds (precursors of AGEs) in 173 food items and summarized that 3DG was the most abundant compound while MG was the only precursor minorly detected in all food products.

Therefore, a validated dAGEs database with thorough and national information is of great interest to evaluate the long-term impacts of dAGEs on human health in cohort studies (Zhang, Wang, & Fu, 2020).

2.2.2 The impact of AGEs generation in food processing

A series of chemical reactions between reducing sugars and amino compounds occurring during the production and storage of foods can be summarized as Maillard reaction. The main variables affecting the extent of the

Table 2 Contents of AGEs in selected factory-produced and home-made foodstuffs.

Food categories	AGEs contents (mg/kg protein)	Food categories	AGEs contents (mg/kg protein)
Raw cow milk	1.8	Tomato	10.2
Skimmed milk	4.7	Raw beef	3.9
Infant formula	76	Boiled beef	27.3
Condensed milk	205	Fried beef	61.1
Yoghurt	89.8	Chew sweets	195.8
Whole egg	56.3	Chocolate	502.1
Apple	39.3	Ice cream (dairy)	128.1
Banana	49.1	Butter	60.5
Orange	76.7	Ketchup	12.9

Maillard reaction are temperature and time which depend on processing conditions as well as pH, water activity and type and availability of the reactants which are based on product properties but may be changed as a result of the processing of food and raw materials.

Massive adverse side effects in clinical trials resulted in the abandonment of aminoguanidine, and since then, the discovery of other Maillard inhibitors has been an ongoing quest. It was shown that epicatechin (a polyphenol compound found in plants such as green tea, grapes, and cocoa) traps α-dicarbonyls. When epicatechin was added to UHT-processed milk, it resulted in reduced off-flavor formation. The effects of polyphenols from different plant sources as Maillard inhibitors have gained increasing focus in food systems because these are considered natural compounds and are therefore more accepted as food ingredients than synthetically manufactured compounds. Other naturally occurring compounds such as vitamins and amino acid and peptide derivatives have also been found to inhibit Maillard reactions by targeting reactive sites, intermediates, or products (Jaeger, Janositz, & Knorr, 2010).

Eventually, extensive interests were aroused in exploring natural products due to their relative safety for human consumption, and it was fortunate to find out that a considerable amount of plant extracts were capable of impeding the formation pathway of AGEs. These plants included some fruits (e.g., pomegranate, blackberry, and peach, etc.), vegetables (e.g., okra, tomato, soybean, etc.), spices (e.g., chili pepper, ginger, garlic, etc.), teas (e.g., green tea, black tea, and oolong tea, etc.), flowers (e.g., sweet osmanthus, rose, chrysanthemum, etc.), and medicinal herbs (e.g., *Radix Scutellariae, Rhizoma Anemarrhena, Radix Paeoniae Rubra*, etc.), while the active components were by and large identified as polyphenols, particularly, flavonoids.

2.3 Inhibition AGEs by polyphenols

2.3.1 Inhibition mechanism of AGEs by polyphenols

2.3.1.1 Trapping of α-dicarbonyls

α-Dicarbonyls, such as glyoxal (GO), methylglyoxal (MGO), and deoxyosones, are reactive intermediates that accelerate Maillard reactions due to their higher reactivity compared to glucose. Many of the pharmaceuticals and food ingredients that inhibit Maillard reactions have been found to trap α-dicarbonyls, with most studies being conducted with GO and MGO. Reported trapping agents include creatine, hydroxytyrosol, pyridoxamine, low molecular mass thiols such as cysteine, homocysteine, and glutathione, and phenolic compounds such as flavonoids and phenylpropanoids. It is therefore believed that α-dicarbonyl trapping is one of the major mechanisms by which these compounds inhibit the accumulation of Maillard reaction products.

2.3.1.2 Scavenging of Maillard-derived radicals by polyphenols

Maillard reactions also generate stable radicals, which have been characterized in model systems and foods. In model systems, the radical has been identified as a pyrazinium radical cation, which may be formed by the reaction of an amine with either α-dicarbonyls or carbohydrates, and its formation is accompanied by the development of browning. Addition of catechins [epigallocatechin gallate and (+)-catechin] to a model system generating pyrazinium radicals resulted in increased radical signal intensities as detected by electron paramagnetic resonance (EPR) spectroscopy at concentration ratios up to 10:1 (GO/catechin). Higher concentration ratios of catechins to GO decreased the radical signal. It was concluded that catechins influence the formation of pyrazinium radicals, but it was not possible to determine if catechins catalyzed or quenched pyrazinium radical generation. Other Maillard-derived radicals have been detected by the use of spin traps, but the identity of the radicals was not characterized. Nevertheless, these studies showed that flavonoids and 4-methylcatechol inhibited radical formation (Lund & Ray, 2017).

2.3.1.3 Chelation transition metal ions

Participation of metals in the Maillard reaction and their incorporation into intermediate products is closely related to the biological availability of mineral elements and the chemical damage caused to some essential amino acids. According to several studies, the amounts of soluble and dialyzable Zn and Fe in model solutions or thermally treated products decreased significantly due to the interaction of reducing sugars and amino acids. Some studies show that Maillard reaction products could inhibit the absorbance of Zn, Fe and Cu in vitro and in vivo (in suckling rats). Metals were bound especially by a crude glucose-glutamate Maillard reaction product as compared to the Amadori compound fructosyl-glycine. The strength of metal-binding occurred in the order: $Mg^{2+} > Cu^{2+} = Ca^{2+} > Zn^{2+}$. This order differs somewhat from the Irving-Williams order of transition metals complex stability: $Mn^{2+} < Fe^{2+} < Co^{2+} < Ni^{2+} < Cu^{2+} > Zn^{2+}$, which might suggest the presence of more than one type of ligand. The antioxidant or pro-oxidant activity of melanoidins as chelating agents of metal ions was also described. The metal chelating ability of Maillard reaction products, the stability of the complexes formed and the accumulation of chromophores in the reaction systems depend on the type of cations and amino compounds, and the reaction conditions. It has also been shown that Amadori compounds combine easily with Cu^{2+} to form unstable complexes, which by degradation give hydroxyl radicals and dicarbonyl compounds (Ramonaitytė, Keršienė, Adams, Tehrani, & Kimpe, 2009).

2.3.2 Inhibition AGEs by polyphenols in diverse foods

2.3.2.1 Baked goods

In some studies, cookies were incorporated with three polyphenols at two concentrations (0.02% and 0.2%) and baked under three temperatures (150, 170, and 190 °C) to observe their interaction. They didn't significantly change the cookie texture, but influence the color and pH. The temperature was a significant determinant to affect polyphenols' stability and accordingly influenced cookies' antioxidant capacity. 73.4%, 26.7% and 9.9% of rosmarinic acid, resveratrol, and epicatechin remained after baking at 190 °C for 10 min. Two addition levels of polyphenols decreased fluorescent advanced glycation endproducts (AGEs) by 18.58% to 40.29% and 28.60% to 62.05% respectively, while only resveratrol and epicatechin could reduce the level of glyoxal and methylglyoxal in cookies. Besides, thermal processing could enhance the cytotoxicity of rosmarinic acid against cancer cells.

Thus polyphenols were suggested as beneficial additives for enhancing antioxidant intake and lowering total fluorescent AGEs (Ou, Teng, El-Nezami, & Wang, 2018).

2.3.2.2 Meat

Several studies investigated the effects of irradiation on *N*-epsilon-carboxymethyl-lysine (CML) and *N*-epsilon-carboxyethyl-lysine (CEL) formation in cooked red and white meats during storage. The results showed that irradiation did not affect CML/CEL formation (0 weeks). After 6 weeks, CML/CEL contents in the irradiated samples exhibited a higher growth rate than the non-irradiated samples, especially the red meat. The results of electron spin resonance spectrometry and 2-thiobarbituric acid-reactive substances suggested irradiation had induced free-radical reactions and accelerated lipid oxidation during storage. A linear correlation ($r=0.810$–0.906, $P < 0.01$) was found between the loss of polyunsaturated fatty acids content and the increase of CML/CEL content in the irradiated samples after 0 and 6 weeks of storage. The results indicate that irradiation-induced lipid oxidation promotes CML/CEL formation, and CML/CEL formation by the lipid oxidation pathways may be an important pathway for CML/CEL accumulation in irradiated meat products during storage (Yu et al., 2016).

2.3.2.3 Dairy products

A comprehensive quantitative characterization of Maillard reaction products was carried out for conventional (CON) and lactose-hydrolyzed (LH) ultra-high temperature (UHT) milk during storage at 20, 30, and 40 °C for 1 year. The accumulation of 3-deoxyglucosone (3-DG) and 3-deoxygalactosone (3-DGal) in LH-UHT milk ranged from 20-fold (at 20 °C) to 44-fold (at 40 °C) higher than that in CON-UHT milk. High-temperature storage (40 °C) significantly accelerated the accumulation of 3-DG, 3-DGal, and 5-hydroxymethyl furfural but not the majority of the analyzed advanced glycation endproducts (AGEs). The concentrations of major AGEs including *N*-epsilon-carboxymethyllysine (CML), *N*-epsilon-carboxyethyllysine (CEL), methylglyoxal-hydroimidazolone isomers (MG-H1/H3), glyoxal-hydroimidazolone isomers (G-H1/H3), and G-H2 detected in CON milk during storage were in the range 12–700, 1–14, 8–45, 4–13, and 1–30 μM, respectively, while they were 30–570, 2–88, 17–150, 9–20, and 5–34 μM, respectively, in LH milk. Pyrraline, 5-(carboxymethyl) cysteine (CMC), and glyoxal-lysine dimer were detected in lower levels, while MG-H2, methylglyoxal-lysine dimer, argpyrimidine, glyoxal-lysine-amide, glycolic acid-lysine-amide, and pentosidine were not detected in any of the milk samples. This work demonstrates for the first time that five of the analyzed AGEs (CML, CEL, MG-H1/H3, G-H1/H3, and G-H2) could be selected as markers for evaluation of the extent of the Maillard reaction in UHT milk. These results contribute to a better understanding of how Maillard reactions progress during the storage of UHT milk and can be used to develop strategies to inhibit Maillard reactions in LH milk (van Lieshout, Lambers, Bragt, & Hettinga, 2020).

2.3.2.4 Aquatic products

The purpose of some study was to investigate the formation of advanced glycation end-products in raw and heat-treated fish muscle as affected by fish freshness. The amounts of free and protein-bound *N*-epsilon-carboxymethyllysine (CML) and *NN* epsilon-carboxyethyllysine (CEL) in raw and heat-treated (100 °C, 5, 30 min) white muscle from cultivated grass carp (*Ctenophatyngodon idellus*) ($n = 12$) and catfish (*Clarias leather*) ($n = 24$) steaks previously stored at 0 °C for 0–21 days were quantified with an HPLC-MS/MS method. The levels of free and protein bound CML and CEL in raw fish muscle did not change significantly during 21 days of ice storage. However, the longer the fish muscle was stored at 0 °C, the more protein bound CML and CEL were formed upon heating. The heat-induced formation of protein-bound CML in catfish muscle previously-stored at 0 °C for 21 days increased by 107% (5 min heating) or 172% (30 min heating), and CEL increased by 448% (5 min heating) or 191% (30 min heating), as compared with that in heat-treated freshly killed catfish muscle. In grass carp muscle, the increases were from 144% (5 min) to 131% (30 min) for protein-bound CML, and 135% (5 min) to 95% (30 min) for protein-bound CEL (Niu et al., 2017).

Thus, the AGEs-scavenging by polyphenols should be more emphasized in future studies from our perspective of view. Fruits, vegetables, and beverages are important dietary sources of natural polyphenols. These polyphenolics are of current research interest due to their important biological and pharmacological properties, which can be attributed to their antioxidant capacities. It is hoped that the development and investigation of AGE inhibitors, especially natural anti-AGE agents with fewer side-effects, may provide a therapeutic approach for delaying and preventing premature aging and diabetic complications. It was quite important that a reasonable selection of food processing and storage methods for the content of AGEs by polyphenol.

3. Inhibition on chemical oxidation during oil storage

3.1 Oxidation of oil during storage

Oils and fats from edible oils are constituents of many foods and play important organoleptic and nutritional roles. It is constituted of triacylglycerol molecules, mainly formed by unsaturated and saturated fatty acids esterified to glycerol units (Fig. 4). They are usually containing three ester bonds, which can be hydrolyzed to form glycerol and three fatty acids. (Fig. 4). The major fatty acids present in triacylglycerols from edible vegetable oils are the unsaturated oleic, linoleic, and linolenic acids, followed by the saturated acids, mainly myristic, palmitic, and stearic acids (Fig. 4). As the composition of fatty acids of the triacylglycerols in edible vegetable oils changes from plant to plant, each vegetable oil has a characteristic composition that allows its botanical origin to be identified.

3.1.1 Oil oxidation type

One of the major problems in fats and oil is the development of rancidity (oxidative) and deterioration of its quality due to multiple environmental and storage conditions. The oxidation of lipids reduces the sensorial quality and shelf life of many fatty foods (Cui, Cho, McClements, Decker, & Park, 2016). Oxidation occurring at the double bond sites of the fatty acid molecule is the primary route of deterioration of oil quality. Of course, the more unsaturation (i.e., the more double bonds in the molecules), as determined by the types of fatty acids making up the triglyceride structure, the more susceptible fats and oils are to oxidative breakdown. The deterioration process can be broadly classified into two types, viz. autoxidation and photosensitive oxidation.

3.1.1.1 Oil autooxidation

Autoxidation of oils, free radical chain reaction, includes initiation, propagation, and termination steps (Fig. 5). Autoxidation of oils requires fatty

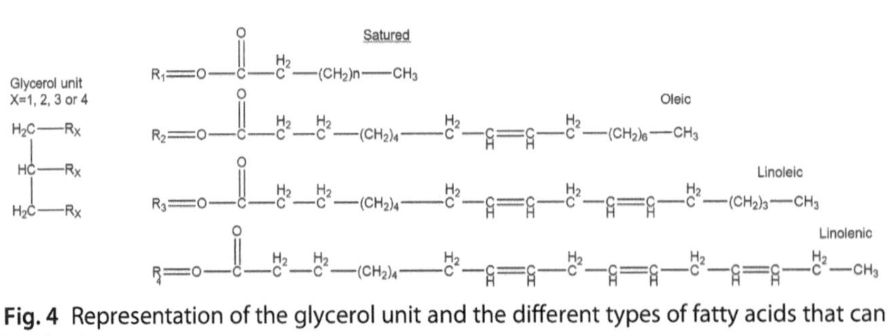

Fig. 4 Representation of the glycerol unit and the different types of fatty acids that can be the R-groups.

Initiation	RH	$\longrightarrow$	R·+H·
Propagation	R·+O_2	$\longrightarrow$	ROO·
	ROO·+RH	$\longrightarrow$	ROOH+R ·
Termination	ROO·+R·	$\longrightarrow$	ROOH
	R·+R·	$\longrightarrow$	RR
	ROO·+ROO·	$\longrightarrow$	ROOR+ O_2

(R : lipid alkyl)

Fig. 5 The autooxidation process of oils.

acids or acylglycerols to be in radical forms. The initiating step is the formation of a fatty free radical when hydrogen (actually a proton) departs from the methylene carbon in the unsaturated fatty acid group of the fat molecule (RH). The energy required to remove hydrogen from fatty acids or acylglycerols is dependent on the hydrogen position in the molecules. The hydrogen atom adjacent to the double bond, especially hydrogen attached to the carbon between the 2 double bonds, is removed easily. The resultant free radical(R·) becomes very susceptible to attack by atmospheric oxygen to form an unstable peroxide free radical (ROO·). In the final or terminating stage of glyceride oxidation, following the induction period, hydroperoxides that form in ever-increasing quantity split or decompose into short-chain aromatic organic compounds (mainly aldehydes, ketones, alcohols, and acids), which cause the rancidity condition that ultimately destroys acceptability and usefulness of fats and oils. Autoxidation of linoleic and linolenic acids produces only conjugated products. The hydroperoxide positional isomers formed in the autoxidation of oleic, linoleic and linolenic acids are shown in Table 3.

3.1.1.2 Oil photosensitive oxidation

Photosensitizers (such as chlorophyll, heme, myoglobin, etc.) can transfer the absorbed energy to oxygen molecules in the air under the excitation of light energy so that it can react with fatty acids or esters after activation to form hydroperoxides. This reaction is called photosensitivity. The hydroperoxides generated by photosensitive oxidation are easily decomposed into various free radicals, especially in the case of metals or heat, these free radicals act as inducers of the auto-oxidation reaction induction period to initiate or induce auto-oxidation reactions.

Various methods are currently used to evaluate the oxidation of lipids, which can be assessed by monitoring organoleptic properties or different chemical and physical changes (Frankel, 1993; Gray, 1978; Ho, 1996).

Table 3 Hydroperoxides of fatty acids by autoxidation.

Fatty acid	Hydroperoxides at	Relative amount (%)
Linolenic acid	C9	28∼35
	C12	8∼13
	C13	10∼13
	C16	28–35
Linoleic acid	C9	48∼53
	C13	48∼53
Oleic acid	C8	26∼28
	C9	22∼25
	C10	22∼24
	C11	26∼28

The oxidative products can be monitored through p-anisidine value (AV), peroxide value (PV), thiobarbituric acid test, weight gain, conjugated diene value, chromatographic techniques that analyze specific oxidation compounds, loss of unsaturated fatty acids, and others.

3.2 Influencing factors

Oxidative stability signifies the degree of resistance offered by the oil against oxidative deterioration. The oxidation of oil is influenced by the fatty acid composition of the oil, oil processing, energy of heat or light, the concentration and type of oxygen, and free fatty acids, mono- and diacylglycerols, transition metals, peroxides, thermally oxidized compounds, pigments, and antioxidants. These factors interactively affect the oxidation of oil and it is not easy to differentiate the individual effect of the factors.

3.3 Suppression method

The edible oil oxidation can be inhibited or delayed by adding a specific additive such as antioxidants (Mohammadi, Jafari, Esfanjani, & Akhavan, 2016). A variety of effective synthetic antioxidants have been widely used such as BHT, butylated hydroxy anisole (BHA), tertiary butyl hydroquinone (TBHQ), and propyl gallate (PG). Currently, synthetic antioxidants (TBHQ) are used for deferring lipid oxidation in the industry. The synthetic antioxidant is mostly used by the edible oil industry as a food additive for restraining oxidation of edible oil, due to its low cost and high antioxidation. But due to the bad nutritional and carcinogenic effects of these constituents, as well as consumers' tendency toward using natural compounds, the use of natural antioxidants has received far more attention.

In recent years, a great deal of research has been devoted to the replacement of chemical materials with natural ones to eliminate or reduce chemical and synthetic constituents in food. Natural antioxidants are considered a good alternative to synthetic antioxidants, due to their potential health benefits and safety. Hence, many efforts have been made to find natural antioxidants from herbal sources. Natural antioxidants such as tocopherols, green tea polyphenols, vitamin C, carnosic acid, rutin, and dihydrocaffeic acid have been the subject of well-publicized studies. It has been found that most of the natural antioxidants possess more thermal stability and anti-oxidation in comparison to synthetic ones in different edible oils (Taghvaei & Jafari, 2015). Quinones antioxidants can form a more stable free radical to terminate auto-oxidation chain reaction. Green tea polyphenols are primarily attributed to the neutralization of phenolic hydroxyl groups with lipid-free radicals. Phenolic antioxidants can donate hydrogen-atoms to lipid radicals to produce more stable lipid derivatives and antioxidant radicals.

3.4 Effect of polyphenols on edible oil oxidation

Bouaziz et al. used Chemlali olive leaves and its hydrolysate extract in refined and husk olive oils. It was concluded that hydrolysate extracts at 400 ppm had the highest protective effect against oil oxidation (even more than BHT) (Bouaziz, Fki, Jemai, Ayadi, & Sayadi, 2008). Koketsu and Satoh (1997) studied the antioxidant activity of green tea extracts in lard and soybean oil and compared them with a-tocopherol and BHT (Koketsu & Satoh, 1997). They reported stronger antioxidative activity of tea polyphenols in soybean oil compared with a-tocopherol. Myricetin is effective owing to strong radical scavenging and metal-chelating properties, whereas quercetin has weaker radical scavenging activity, although it is also active by metal-chelation (Roedig-Penman & Gordon, 1998). However, myricetin appears to be less effective than TBHQ (Olsen, Vogt, Saarem, Greibrokk, & Nilsson, 2005). Rafiee et al. found that methanol extract of Cronaiky at 1000 ppm could control the oxidation of sunflower oil comparable to BHA and BHT at 100 and 200 ppm levels (Rafiee, Jafari, Alami, & Khomeiri, 2018). In one study, myricetin, quercetin, and rutin, as well as (−) epicatechin (flavan-3-ols), exhibited superior oxidative stability than synthetic antioxidants tested. Commercial rosemary extract and its active constituents (carnosic acid, rosmarinic acid and carnosol) have been extensively studied for their antioxidant properties in oil (Table 4). Carnosic acid

Table 4 Effect of phenolic compounds and their components on oxidative stability of edible oils.

Concentration	Oil	Conditions	Conclusions	Authors
Carnosic acid (100, 200, 300 ppm), tocopherols (300 ppm), TBHQ (300 ppm)	n-3 PUFA rich fish oil	4 and 30 °C for 66 days	Dose-dependent enhancement of oxidative stability and the antioxidant activity of 300 ppm of carnosic acid was stronger than that of tocopherols but still weaker than that of TBHQ	Wang et al. (2011)
Rosemary essential oil (REO) fractions obtained by molecular distillation (MD)	Sunflower oil	In a dark place at 23 °C for 115 days	The residue fractions of rosemary had lower IC_{50} for REO and better antioxidativity than REO and distillate fractions. Hence, it could be utilized as natural antioxidants for the sunflower oil	Mezza et al. (2018)
Carnosic acid (100 ppm and 1000 ppm)	Virgin olive oil	Accelerated oxidation conditions at 60 °C	Dose-dependent inhibition in the formation of primary and secondary oxidation products	Zunin, Leardi, Bisio, Boggia, and Romussi (2010)
Mulberry leaves extract and mulberry powder	Rice bran oil (RBO)	Subjected to accelerated oxidative treatment at 100 °C for 5 days	Mulberry leaves methanolic extract (200 ppm) and mulberry powder (500 ppm) were found to be more effective in maintaining oxidative stability of RBO in comparison to 200 ppm BHT, under accelerated oxidation condition (100 °C/5 days) and heat treatment (180 °C/1 h)	Roy, Arabshahi-Delouee, and Urooj (2010)

100 ppm of catechin, quercetin, rutin, ferulic acid, caffeic acid and sinapic acid in	Refined n-3 PUFA rich sardine oil	Stored at 37 °C in the dark in contact with atmospheric air for 14 days	Quercetin and rutin followed by caffeic acid were effective in restraining sardine oil oxidation	Chandrasekar, Belur, and Regupathi (2016)
Potato peel and sugar beet pulp extract	Sunflower and soybean oils	subjected to accelerated oxidation in the dark in an oven at 70 °C for 72 h	Potato peels extract (100 mg/kg) and sugar beet pulp extract (200 mg/kg) respectively had sunflower and soybean stabilization efficiency under accelerated condition (70 °C/72 h), equivalent to BHT and BHA at their legal limit i.e. 200 mg/kg, but less effective than the TBHQ at 200 mg/kg	Mohdaly, Sarhan, Mahmoud, Ramadan, and Smetanska (2010)
200 ppm of various phenolic compounds and synthetic antioxidants (tannic acid, ferulic acid, caffeic acid, sinapic acid, catechin, quercetin, rutin, sesamol, TBHQ, BHT, and BHA)	Cold-pressed pecan oil	Aerated condition in dark at 60 °C	Caffeic acid inhibited oxidation of pecan oil effectively, and its performance was superior to BHT and BHA and was close to TBHQ	Zhang et al. (2018)
Commercial rosemary extract and its active constituents (carnosol, carnosic acid, and rosmarinic acid)	Tocopherol-stripped corn oil	Kept in shaker oven at 60 °C	Best antioxidant activity was obtained with 50 ppm of rosmarinic acid, 50 ppm of carnosic acid and 500 ppm of rosemary extract	Frankel, Huang, Aeschbach, and Prior (1996)

Continued

Table 4 Effect of phenolic compounds and their components on oxidative stability of edible oils.—cont'd

Concentration	Oil	Conditions	Conclusions	Authors
Grape seed extract	Sunflower oil	Samples were separately heated at 200 °C for 240 min	Grape seed extract (600–800 ppm) delayed the oxidation of sunflower oil under microwave heating (240 min) similar to BHT at 200 ppm. At concentration over 800 ppm, grape seed extract had more stabilization efficiency than BHT	Poiana (2012)
200 ppm of natural antioxidants [apigenin, chrysin, quercetin, myricetin, morin, kaempferol, rutin, naringin, naringenin, taxifolin and (−) epicatechin] and 200 ppm of synthetic antioxidants [BHA, BHT]	Purified canola oil	Accelerated oxidation at 65 °C	Myricetin, quercetin, and rutin, as well as (−) epicatechin exhibited superior oxidative stability than synthetic antioxidants	Shahidi and Wanasundara (1995)
Myricetin compared with α-tocopherol and TBHQ (100, 500 and 1000 ppm)	Soy bean oil methyl esters	At room temperature under darkness for 90 days	Myricetin (100 ppm) possessed greater antioxidant activity than α-tocopherol, but was inferior to TBHQ at 500 ppm	Moser (2008)
Quercetin (120 ppm) and rutin (244 ppm)	Purified soy bean oils containing α-tocopherol and α-tocopherol-stripped oil	Stored in sealed bottle with headspace, incubated at 25 °C for 45 days	Both acted as antioxidant in α-tocopherol-stripped soy bean oils and as a pro-oxidant in oil containing α-tocopherol. Rutin fared better compared with quercetin in exhibiting antioxidant effects in the bulk oil	Lee, Gim, Kim, and Lee (2017)

reportedly exhibited a dose-dependent inhibition of primary and secondary oxidation product formation and radical scavenging activity (Wang et al., 2011; Zunin et al., 2010). Carnosic acid exhibited a superior antioxidant effect compared with tocopherols but still weaker than TBHQ (Yang et al., 2016). The superior performance of canola meal extract (rich in phenolic compounds) was found compared with synthetic antioxidants (BHA, BHT and BHA/BHT/monoglyceride citrate) in purified canola oil in another study (Su et al., 2004).

In conclusion, myricetin, quercetin, rutin, carnosic acid, caffeic acid, catechins and theaflavins exhibited superior antioxidant properties in purified bulk oils during storage. In general, the antioxidant activity is determined by the number of hydroxyl groups and other aromatic ring substituents present in the phenolic compounds (Olszowy, 2019). Their performance was comparable or better than some of the synthetic antioxidants (BHA, BHT and TBHQ) tested in those studies. Several studies indicated the positive correlation between antioxidant concern.

4. Polyphenols used in food packaging materials

Food packaging is a dispensable part of the modern food industry system, predominantly protects foods from the damage of biological, chemical, and physical factors from field or factory to consumers, and it is also convenient for the consumption of food. The main function of packaging is extending shelf-life and maintaining the sensory and nutritional quality of more stable food products. With the improvement of living standards and consciousness in food quality and safety, people pay more attention to bioactive, biodegradable, functional, and edible packaging to avoid the immigration of chemicals or additives from traditional packaging materials (Jakobek, 2019). Polyphenols are widely present in various parts of vegetables, fruits, herbs, and other plants, more than 8000 polyphenolic compounds have been isolated and identified (de Araujo, de Paulo Farias, Neri-Numa, & Pastore, 2021). It can be classified into three main types, flavonoids, phenolic acids, and stilbenoids (Papuc, Goran, Predescu, Nicorescu, & Stefan, 2017). Owing to their extraordinary antioxidant and antibacterial activities, and accessibility, polyphenols have been applied in food packaging materials to improve the physical and functional properties.

In the latest decades, many researchers have attempted to graft natural polyphenols or phenolic-rich extracts with synthetic polymers or biopolymers to improve the packaging properties of materials, replace the chemical additives and reduce potential migrants in food. As shown in Fig. 6, the development of polyphenol-rich packaging materials generally consists of

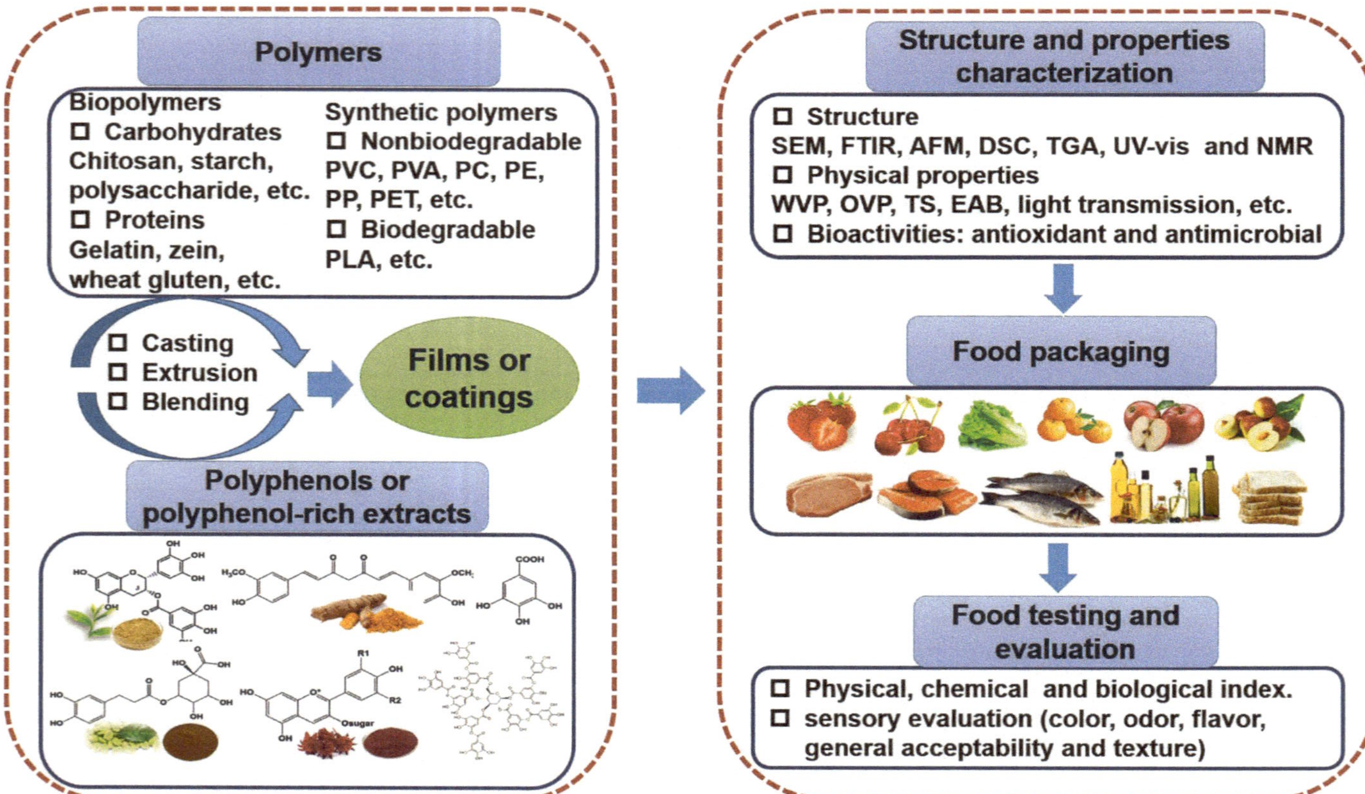

Fig. 6 Schematic diagram of the general procedure of active film or coating for food packaging.

several steps, including incorporating polyphenols or phenolic-rich extracts into polymers, structure characterization by various modern instruments, testing barrier and mechanical properties, and bioactivities of materials, then application in food packaging, finally, food evaluation by sensory, physical, chemical and biological index.

Numerous researches have studied the combination of polyphenols with polymers for packaging. Among these studies, most of them were merely focused on the film's preparation, characterization properties with minor in vitro antioxidant and antimicrobial test, while neglected the applications of these films or coatings in food packaging. Therefore, more attention should be paid to their application in food packaging.

4.1 Polyphenols grafted with synthetic polymers for food packaging

Since the application of plastics in packaging, plastic packaging materials have been maintained rapid growth in production and consumption. It possesses many advantages, such as convenience, cheapness, lightweight, toughness, and multifunction, which were extensively used for food packaging (Majhi, Tyagi, & Mishra, 2020). They play a critical role in keeping food quality, safe, and nutritional. The matrix of plastics mainly is petroleum-chemical-based and nonbiodegradable polymers, commonly are polyvinyl chloride (PVC), polyvinyl alcohol (PVA), polyamide (PA), polypropylene (PP), polystyrene (PS), polyethylene (PE) and its oxide (PEO), polyethylene terephthalate (PET), and ethylene-vinyl alcohol copolymer (EVOH) (Asgher, Qamar, Bilal, & Iqbal, 2020). Although plastic materials facilitate our life, they have caused serious environmental pollution because of hardly degradable and recycle, especially the single-use containers were extensively used for food package (Nesic, Castillo, Castaño, Cabrera-Barjas, & Serrano, 2020). As a hot topic in recent years, microplastic contamination threatened oceans and wildlife health. What's more, long-term exposure to microplastic and organic pollutants released from plastic may lead to metabolic disorders, such as metabolic syndrome, neurodegenerative diseases, and carcinogenesis (Zwierello et al., 2020).

In general, chemical synthetic antioxidants, are mixed with these polymers to protect from oxidation and aging caused by heat, oxygen, UV-light, free radicals, and other factors, which possesses potential risk for humans (Topuz & Uyar, 2020). To replace synthetic plastic antioxidants, some studies have attempted to graft natural polyphenols or phenolic-rich extracts with conventional chemical synthetic polymers to replace the chemical additives and reduce potential migrants in food. In Table 5, polyphenols

Table 5 Polyphenols grafted with synthetic polymers for food packaging.

Matrix	Polyphenols	Applications	Evaluation index	Reference
EVOH	Ferulic acid, quercetin, green tea extract	Film, against lipid oxidation of brined sardines	Thermogravimetric analysis (TGA), water vapor permeability (WVP), oxygen permeability (OP), antioxidant release tests, peroxide index and malondialdehyde (MDA) content	López-de-Dicastillo, Gómez-Estaca, Catalá, Gavara, and Hernández-Muñoz (2012)
PE	Green tea extract	Film, extend the shelf-life of minced pork meat for 3 days	Metmyoglobin (MetMb) content, organoleptic assay (visual and aroma), color measurement (L*, a*, b*), migration rate of catechins	Wrona, Nerín, Alfonso, and Caballero (2017)
PPT, PP	Olive leaf extract	Film, against lipid oxidation of sunflower seeds	Migration rate of polyphenols, peroxide value (POV)	Cejudo Bastante, Casas Cardoso, Fernández Ponce, Mantell Serrano, and Martínez de la Ossa-Fernández (2018)
PET-PP	Mango polyphenols	Film, extend shelf-life of lettuce and tangerine	Scanning electron microscopy (SEM), 1,1-diphenyl-2-picrylhydrazyl (DPPH), preservation perishable food	Belizón, Fernández-Ponce, Casas, Mantell, and Martínez de la Ossa-Fernández (2018)
PVA	Tea polyphenols	Film, extend shelf-life of strawberries	FTIR, tensile strength (TS), OP, water resistance, biodegradation, antimicrobial properties, fruit's weight loss, decay rate, firmness, pH, titratable acidity, and total soluble solids.	Lan, Zhang, Ahmed, Qin, and Liu (2019)

PLA-chitosan	Tea polyphenols	Film, extend shelf-life of cherries	SEM, FTIR, density, WVP, tensile strength (TS), breaking elongation rate, biodegradability, cherries' physical indexes (rotting rate and weight loss), hardness, soluble solids, titratable acid, vitamin C content	Ye, Wang, Lan, Qin, and Liu (2018)
PVA-chitosan, Starch-PVA	Roselle anthocyanins	Film, monitor the freshness of pork	FTIR, SEM, thickness, mechanical properties, moisture content, color stability, DPPH, response to ammonia, color, total volatile basic nitrogen (TVB-N)	Zhang et al. (2019)
HPMC-PEO	Gallic acid	Nanofiber, decrease the oxidation of walnut	SEM, loading efficiency, DPPH, TGA, DSC, FTIR, antioxidant analysis (POV, TBARS, p-anisidine value, totox value)	Aydogdu, Sumnu, and Sahin (2019)
Starch-PVA	Roselle anthocyanins	Film, monitor the real-time fish freshness of silver carp	Fourier transform infrared spectroscopy (FTIR), X-ray diffraction (XRD), SEM, thickness, moisture content (MC), TS, elongation at break (EAB), color stability (RGB), sensitive to volatile ammonia, fish spoilage trial, TVB-N	Zhai et al. (2017)
Zein-PLA	Carvacrol	Fiber, extend shelf-life of wheat bread	FTIR, SEM, TGA, release study, DPPH, bread preservation (growth inhibition rate of aerobic bacteria and mold and yeast)	Altan, Aytac, and Uyar (2018)

or polyphenol-rich extracts such as ferulic acid, quercetin, olive leaf extract, mango and tea polyphenols, were grafted with chemical synthetic polymers, like EVOH, PET, PP, and PVA, the formed films exhibited good packaging properties and bioactivities (Belizón et al., 2018; Cejudo Bastante et al., 2018; Lan et al., 2019; López-de-Dicastillo et al., 2012). For instance, Lan et al. developed various volumetric ratios of PVA/PA films by tap-casting method, and characterized their properties by serials of approaches, and used for strawberries packaging. The result indicated that the incorporation of tea polyphenols effectively improved the mechanical properties (tensile strength) and physical properties (oxygen permeability and water resistance), as well as antimicrobial activity. Samples packed in PVA/PT(8/2) film had a lower decay rate than in pure PVA film (10/0), meanwhile could significantly extend the shelf-life of strawberries (Lan et al., 2019). Besides, some studies also produced synthetic polymer-biopolymer-polyphenols-based films to enhance the physical and mechanical performance of the film. These polyphenols-containing were used for packaging cherries (Ye et al., 2018), pork (Zhang et al., 2019), walnut (Aydogdu et al., 2019), silver carp (Zhai et al., 2017) and wheat bread (Altan et al., 2018), to extend the shelf-life or monitor the real-time freshness of food (Table 5).

4.2 Polyphenols incorporated with biopolymers for food packaging

To address the issue of plastic pollution and chemicals immigration, more and more attention has been paid to eco-friendly packaging materials with biodegradable and sustainable. Consequently, incorporating natural polyphenols or polyphenol-rich extracts into bio-based materials has become a new trend. These biopolymers were used for food package general are edible carbohydrates, e.g., chitosan, cellulose, carboxymethyl cellulose (CMC), starch, pectin, agar, hydroxypropyl methylcellulose (HPMC) and proteins (e.g., gelatin, zein, wheat gluten), commonly extracted from the crustacean, plants, and bioengineer (Jakobek, 2019; Shankar & Rhim, 2017). Due to the excellent film-formation property of these biopolymers and the extraordinary antioxidative, antimicrobial activities of polyphenols, these polyphenols-biopolymers' conjugates were commonly made into active, edible, and biodegradable films or coatings for fruits, vegetables, oil, and meat packaging.

As shown in Table 6, tea polyphenols and its monomer are the most common polyphenols applied in the development of food packaging materials, besides curcumin, quercetin, chlorogenic acid, and anthocyanins occasionally used. These newly made materials for food packaging are

Table 6 The applications of polyphenols in food packaging.

Biopolymer	Compounds	Application	Evaluation index	Reference
Pectin-chitosan	Tea polyphenols	Film, inhibit the degeneration of color fresh beef	Edible, slow-release, antioxidant and antiseptic activities	Gao, He, Sun, He, and Zeng, (2019)
Chitosan	Green tea extract	Film, extend shelf-life of pork sausages	Antioxidant and antimicrobial properties	Siripatrawan and Noipha (2012)
Gelatin-chitosan	Tea polyphenols	Film, retard the oxidation of fish oil	Antioxidant and radical-scavenging activity	Bao, Xu, and Wang (2009)
Chitosan	Tea polyphenols, rosemary extract	Coating, extend shelf-life of large yellow croaker	TVC, pH, K value, PV, TBARS, sensory evaluation (eyes, gill, skin, body surface slime and color)	Li, Hu, et al. (2012), Li, Li, et al. (2012)
Chitosan	Tea polyphenols, grape seed extract	Coating, extend shelf-life of red drum by 6–8 days	TVC, pH, TVB-N, K-value, TBARS, hardness and sensory evaluation (color, odor, flavor, general acceptability and texture)	Li, Li, Hu, and Li (2013)
Gelatin	Tea polyphenols	Film, inhibit oxidation and rancidity the sunflower oil	SEM, atomic force microscope (AFM), visual appearance, light transmission, FRAP, DPPH, Oil's POV and TBARS	Liu et al. (2015)
Alginate	Tea polyphenols	Coating, maintain the quality of fresh winter jujube	Respiration intensity, flesh firmness, electrolyte leakage, red indices, ascorbic acid, MDA, total phenol, total chlorophylls content, SOD, POD and CAT activity	Zhang, Li, Dong, Zhi, and Zong (2016)

Continued

Table 6 The applications of polyphenols in food packaging.—cont'd

Biopolymer	Compounds	Application	Evaluation index	Reference
Gelatin-carrageenan	Curcumin, gallic acid, quercetin	Film, extend shelf-life of raw broiler meat	TS, EAB, EM, MC, water solubility, WVP, Hunter color values (L*, a*, b*), antioxidant and antibacterial activity, FTIR, TVC, pH	Khan, Sadiq, and Mehmood (2020)
Chitosan	Tea polyphenols	Film, extend shelf-life of pork meat patties	Microbiological, physicochemical (pH, TBARS values, MetMb), and sensory qualities (color and odor)	Qin et al. (2013)
Gelatin	EGCG	Film, extend shelf-life of tilapia fillets	Texture (springiness and hardness), TBARS, TVB-N, K-value, color (L*, a*, b*), biogenic amines, microbiological analysis	Cao et al. (2020)
Sodium alginate and CMC	EGCG	Coating, extend shelf-life of fresh pork	Weight loss, pH, color (L*, a*, b*), TVC, TBARS, TVB-N, sensory evaluation	Ruan et al. (2019)
Curdlan-chitosan	Tea polyphenols	Coating, extend shelf-life of chilled meat for 3–5 days	Mechanical properties (TS, EAB), WVP, MC, DPPH, SEM, antibacterial activity, TVB-N, TBARS, color (L*, a*, b*), pH	Zhou et al. (2019)
Zein	Curcumin	Coating, limit postharvest decay on coated apples	ATR-FTIR, DSC, SEM, encapsulation efficiency and releasing kinetics, antifungal properties (*Botrytis cinerea* and *Penicillium expansum*), lesion diameter	Yilmaz et al. (2016)

Gelatin	Tea polyphenols	Coating, prevent the spoilage of fish fillet during cold storage	Physiochemical index (color, weight loss), nanostructure and compositions of myofibrillar proteins, volatile organic compounds, aerobic mesophilic count (AMC), aerobic psychrotrophic count (APC) and yeasts and molds	Feng, Ng, Mikš-Krajnik, and Yang (2016)
Chitosan	Young apple polyphenols	Film, extend shelf life of grass carp fillets	Microbial counts, POV, TBARs, TVB-N, physical properties (pH, color, water holding capacity), soluble myofibrillar protein, SEM, texture properties, amino acids, sensory evaluation (odor, color, tissue elasticity, surface character and general acceptability)	Sun et al. (2018)
Pullulan-CMC sodium	Tea polyphenols	Nanofiber, decrease weight loss and maintain firmness of strawberries	SEM, FTIR, Differential scanning calorimeter (DSC), fruits' weight loss, titratable acidity, firmness	Shao, Niu, Chen, and Sun (2018)
Chitosan	Chlorogenic acid	Coating, extend the preservative time of peach fruit	Antioxidant (ABTS, DPPH, FRAP, CUPRAC), UV–vis, FTIR, ^{1}H-NMR, peach fruit' weight loss, decay rate, respiration rate, firmness, soluble solids, and L-ascorbic acid contents, titratable acidity, reactive oxygen species, SEM	Jiao et al. (2019)

Continued

Table 6 The applications of polyphenols in food packaging.—cont'd

Biopolymer	Compounds	Application	Evaluation index	Reference
Chitosan-oxidized chitin	Black rice bran anthocyanins	Film, monitor the spoiling of pomfret meat and shrimp	Color (L*, a*, b*), FTIR, X-ray diffraction (XRD), SEM, AFM, optical transmittance, TS, EAB, OP, DPPH, TVB-N	Wu et al. (2019)
κ-Carrageenan	Mulberry polyphenolic extract	Film, monitor freshness of milk	Color, thickness, MC,WVP, UV–vis light transmittance, TS, EAB, TGA, SEM, FTIR, XRD, DPPH, pH-sensitive property, color change, pH, acidity of milk	Liu et al. (2019)
Gelatin-chitosan	Tea polyphenols	Film, prolong the stability of soybean oil	FTIR, SEM, TS, EAB, WVP, OP, light transmission and transparency, TP-release, properties of films in simulants for fats, DPPH, POV	Wang et al. (2020)
Pomelo peel flour and sodium alginate	Tea polyphenols	Film, delay soybean oil oxidation during storage	FTIR, SEM, thickness, color (L*, a*, b*), light transmission, MC, WVP, TS, EAB, DPPH, antimicrobial activity, anti-permeate ability for soybean oil, POV, weight loss	Wu et al. (2019)

Abbreviation: cupric (Cu^{2+}) reducing antioxidant capability (CUPRAC), thermogravimetric (TGA), ferric reducing antioxidant power (FRAP).

multifunction, such as antioxidant, antimicrobial, biodegradable, and anti-UV. Some of them still have excellent packaging and barrier properties, such as thermal and water stability, water vapor and oxygen permeability, pH, and oxygen sensitivity.

4.2.1 Vegetables and fruits

Raw or unpackaged vegetables and fruits have high water content, carbohydrates, and respiration rate that makes them susceptible to ofloss water and nutrients, infection by various fungus, bacteria, and mold. Currently, with the growing concept of environment protection and food safety and quality, various polyphenols incorporated in biopolymers or synthetic polymers have been applied for extending shelf-life, keeping nutritional value and organoleptic qualities of vegetables and fruits. The predominant packaging approaches are film or coating to forming the film. Films were generally made of active compounds, biopolymers, or synthetic polymers by casting, extrusion, and coating procedure (Dominguez et al., 2018). The optimization packaging way is the coating on the surface of vegetables and fruits. The blended coating solutions of alginate-TP, chitosan-chlorogenic acid, and zein-curcumin were respectively uniform-covered the surface of winter jujube (Zhang et al., 2016), peach fruit (Jiao et al., 2019), and apples (Yilmaz et al., 2016). Polyphenol-coatings effectively extended the shelf-life by reducing weight loss, respiration rate, decay rate, and flesh firmness, the nutritional or flavor indicators like soluble solids contents, titratable acidity, and L-ascorbic acid contents were also improved (Jiao et al., 2019; Shao et al., 2018). The inherent antioxidant of fruits and were enhanced by initiating the endogenous antioxidant enzyme, such as SOD, POD, and CAT, to decrease the MDA and ROS levels (Zhang et al., 2016). Meanwhile, some common pathogens of fruits were also inhibited in relatively lower counts compared with the untreated group and polyphenol-free group (Yilmaz et al., 2016).

4.2.2 Oils and nuts

Oils and nuts are rich in unsaturated fatty acids, which are easily oxidized to forming peroxide substances, triggered by oxygen, light, heat, and other factors. When the storage environment is improper, the fat will deteriorate due to oxidation. Long-term intake causes adverse effects on people's health. It was reported that the bags made of gelatin-TP and the gelatin-chitosan-TP film could inhibit the oxidation and rancidity of fish oil (Bao et al., 2009) and sunflower oil (Liu et al., 2015). PPF (pomelo peel flour)-sodium alginate-TP and the gelatin-chitosan-TP film also effectively maintain the stability and delay the oxidation during storage of soybean oil (Wang et al., 2020; Wu, Lei, et al., 2019). The morphological characterization of these films was analyzed by scanning electron microscopy (SEM), Fourier transform infrared spectroscopy (FTIR), and atomic force microscope (AFM), and showed that polyphenols were well dispersed and grafted with biopolymers, thereby the physical and functional properties were also improved, such as water vapor and oxygen permeability, light transmission and transparency, tensile strength (TS), elastic modulus (EM) and elongation at break (EAB), antioxidant (DPPH) and antimicrobial activities (Aydogdu et al., 2019; Bao et al., 2009; Wang et al., 2020; Wu, Lei, et al., 2019). Besides, incorporated tea polyphenols could sustainably release or immigrate from films to foods, consequently, the peroxide value (POV) of oil or nuts was significantly reduced (Cejudo Bastante et al., 2018; Liu et al., 2015).

4.2.3 Meat and aquatic products

Meat and seafood products contain a high level of fats and proteins, are highly perishable by microbial and must be chilled or frozen in the refrigerator. So, microbial contamination, lipid oxidation, protein oxidation, and nitration are the major issues for the quality and safety of meat, seafood, and their related products. As above mentioned, polyphenols present in films or coatings play a protective role in the antimicrobial and antioxidant effects of foods. These active materials were extensively applied in the packaging of meat and fish products, which could effectively extend the shelf life of fresh beef and pork (Gao et al., 2019; Ruan et al., 2019), pork sausages and meat patties (Qin et al., 2013; Siripatrawan & Noipha, 2012), large yellow croaker (Li, Hu, et al., 2012; Li, Li, et al., 2012), red drum (Li et al., 2013), raw broiler meat (Khan et al., 2020), chilled meat (Zhou et al., 2019) and fish fillet (Cao et al., 2020; Feng et al., 2016; Sun et al., 2018; Wu, Sun, et al., 2019). Bioactive films were improved the sensory quality or indicator during the storage of meat and fish products, including smell, texture, and color. The biochemical evaluation demonstrated that the presence of polyphenols notably decreased total viable counts (TVC), K value, POV, thiobarbituric acid reactive substances (TBARS) and metmyoglobin (MetMb). Especially, the emission of biogenic amines form spoiled meat and milk could disturb the balance of the internal gas-environment of packaging, consequently, the color of films contained anthocyanins sensitively changed to respond to the changes of internal packaging conditions and product quality (Liu et al., 2019; Wu,

Sun, et al., 2019; Yousefi et al., 2019). Thereby, anthocyanins-rich films were developed to monitor the real-time freshness of meat, fish and dairy products.

Numerous studies have developed polyphenol contained films or coatings and applied both in vitro and in vivo for food packaging. However, there was a lack of a comparison test between active films/coatings and commercial or conventional packaging materials in the real food system, it leads to the discrepancy of laboratory results with reality. For example, some antimicrobial films merely tested the antimicrobial effect in vitro by bacteriostatic zone, neglected the complexity of the food matrix. Besides, there still have some limitations of polyphenols-biopolymers in the application of food packaging, such as higher cost, heavy metals/pesticides or herbicides, reproducibility, sensitivity, toxicological information (Olszewska et al., 2020). Therefore, widely and large-scale application tests should be carried to overcome the limitations and narrow the gap between ideal and in practice.

5. Preserving meat and aquatic products

Meat and aquatic products are abundant in fats, proteins, and other micro-nutrients, which are the main source of high-quality protein for the demand of the human body. However, fresh meat and aquatic products are highly spoilage during the transport, trade, processing, and storage, which leads to sensory quality deterioration, nutrients loss and shelf-life shorten, even the potential toxicants formation (Papuc et al., 2017). Various modern measurements were taken to address these issues, mainly through cold chain storage, preservatives or antioxidants addition, modified atmosphere packaging, irradiation, sterilization, and so on (Villalobos-Delgado et al., 2019). The addition of preservatives is the most effective and convenient way among them. Therefore, chemical preservatives were extensively used to inhibit lipids and proteins oxidation, suppress discoloration and harmful microbial growth, to extend the shelf-life of meat and aquatic products, such as nitrates, benzoic acid, sorbic acid and its salts, butylhydroxytoluene (BHT), butylhydroxyanisole (BHA), *tert*-butylhydroquinone (TBHQ) (Efenberger-Szmechtyk, Nowak, & Czyzowska, 2021; Maqsood, Benjakul, & Shahidi, 2013). However, long-term intake of chemical preservatives with a potential risk for human health. With the ever-growing demand for (free or fewer chemicals in meat) healthy and functional meat, much attention has been paid to the development and utilization of natural preservatives and antioxidants, which were commonly extracted from plants, animals, and microorganisms, which has the advantages of non-toxic, biodegradable, and has certain nutritional value and health function, especially plant-derived polyphenols (Baptista, Horita, & Sant'Ana, 2020). As the previous section mentioned, natural polyphenols in addition to applying for food packaging against the potential hazardous from physical and biochemical factors and can also use for the preservation of meat and aquatic products as an antioxidant, antimicrobial, colorant and cryoprotectant to maintain or improve their sensory quality, nutritional value and safety.

When animals are slaughtered and fished, they are rich in lipids and proteins, with a high level of water activity, and easy to be oxidized to produce acidity, off-flavor, and discoloration, thus loss of edible and commercial value. Apart from microbial contamination, the oxidation of proteins, lipids, and myoglobin, is the biggest contributor to sensory and nutritional quality deterioration, which is induced by oxygen, metal ion, light, heat, the breaking of the self-antioxidant defense system, and other factors (Villalobos-Delgado et al., 2019).

The nonenzymatic or autoxidation is the predominant pathway of lipid oxidation, especially the fatty acids. Peroxides are the main product of automatic oxidation of lipids, extremely unstable and easy decomposition to alkoxy radicals, and can be further cracked into low molecular weight compounds, such as carbonyls (aldehydes and ketones), alcohols, acids, and dienes, which contribute to the main off-odor and off-flavor to oxidized meat and aquatic products (Maqsood, Benjakul, Abushelaibi, & Alam, 2014; Papuc et al., 2017). Lipid oxidation commonly accompanies protein oxidation, the oxidation products of lipids, free radicals, peroxides, and its degradation products can trigger protein oxidation, leading the changes of physicochemical and nutritional properties, such as solubility, gelation, emulsification, digestibility, and bioavailability of proteins (Villalobos-Delgado et al., 2019). Myoglobin is the main heme protein responsible for the meat red color, which consists of globin and heme. The state of the heme group determines the meat color. Myoglobin and oxymyoglobin naturally oxidize into metmyoglobin converting the meat color from bright red to brown (Lorenzo et al., 2018). Biochemical indicators are used to evaluate the oxidation degree of meat and aquatic products commonly are POV, TBARS, conjugated dienes (CD), carbonyls and free fatty acids, MetMb, physicochemical and nutritional properties of the protein (Papuc et al., 2017; Villalobos-Delgado et al., 2019).

Meat and aquatic products are protein-rich foods, the spoilage is mainly characterized by protein degradation. Proteins are first hydrolysates into peptides, and further cleavage to free amino acids under the action of exogenous enzymes secreted by microorganisms. Amino acids are subsequently decomposed into the corresponding ammonia, amines, organic acids, and various hydrocarbons by dehydrogenation, deaminization, and desulfurization, etc., showing the spoilage characteristics of meat and aquatic products (Maqsood et al., 2013). Methods used to judge the quality and safety of meat and aquatic products induced by microbial, ranging from sensory evaluations to microbiological and chemical methods. The sensory analysis mainly focused on color, odor, and texture by sensory panelists. For the chemical and microbiological analysis, total volatile basic nitrogen (TVB-N), K and pH value, total viable count, and pathogens are widely used indicators for assessing the degree of freshness (Papuc et al., 2017).

As shown in the Table 7, beef, pork, chicken meat, and various aquatic products treated with a different source of polyphenols alone or combination by direct coating on the surface or blending with mixtures, dipping and blending, were able to extend the shelf life and maintain sensory quality compared with the control group (no polyphenols) during the cold storage.

5.1 Tea polyphenols applied in meat and aquatic products preservation

Tea polyphenols (TP) are the most widely studied and applied plant polyphenols, which have been used for meat and preservation by replacing or decreasing conventional chemical antioxidants in meat and aquatic products. Silver carp dipped with 0.2% tea polyphenols (TP) solution and packed in plastic trays with polyvinyl dichloride film, then stored at −3 °C, could significantly inhibit TVC, decrease pH, TVB-N, TBARS, *K* value (nucleotide breakdown), and retain sensory qualities for longer compared with TP-free (distilled water) group (Fan et al., 2008). The latter study sliced silver carp into fillets and immersed in higher concentrations (0.5% and 1%) of TP solution for 30 min at 4 °C, which showed that TP treatment also reduced the formation of putrescine, cadaverine, and hypoxanthine (Hx), retarded the degradation of inosine monophosphate (IMP) and hypoxanthine ribonucleotide (HxR), and modified the composition of the microbiota of fillets, even changed the spoilage pattern and process of fish fillets (Jia et al., 2018). For the meat and its products, the addition of 0.3% TP in the basic formulation of pork sausages showed lower variations in TVC, TBARS values, Hunter color of L* and b*, and sensory quality than control samples (Wenjiao et al., 2014). Similarly, compared with the same levels of grape seed extract (GSE) and α-tocopherol, blending 0.03% TP with fresh minced

Table 7 Polyphenols applied in meat and aquatic products preservation.

Polyphenols	Food species	Additional treatments	Main outcomes	References
Quince polyphenolic extract (10%, 20%)	Mackerel (*Scomber scombrus*) fillets	Storage at 4 °C or −80 °C	↓ Lipid oxidation (PV, TBARS), the growth of *Vibrio fluvialis* (in vitro)	Fattouch, Sadok, Raboudi-Fattouch, and Slama (2008)
Tea polyphenols (0.2%), dip	Silver carp (*Hypophthalmicthys molitrix*)	Packed with plastic trays and polyvinyl dichloride film, storage at −3 °C	↓ TVC, pH, TVB-N, TBARS, K-value ↑ Sensory scores	Fan, Chi, and Zhang, (2008)
Apple polyphenols (300, 500, 1000 mg/kg)	Sliced cooked cured beef and pork hams	Vacuum packed and storage at (8 ± 1) °C	↓ Lipid oxidation (TBARS), discoloration	Sun, Zhang, Zhou, Xu, and Peng (2010)
Tea polyphenols (0.5, 1.0%)	Silver carp (*Hypophthalmichthys molitrix*)	Packed in PVC bags and storage at (4 ± 1) °C	↓ TVC, TBARS, TVB-N, K value, Hx and putrescine, cadaverine Delay the degradation of IMP and HxR, sensory deterioration	Jia et al. (2018)
Catechin, tea polyphenols, gallic acid, Na-citrate (2.0%)	Grass carp (*Ctenopharyngodon idella*) fillets	Storage at ice (0.2 ± 0.1) °C and super chilling at (−1.5 ± 0.2) °C	↓ Collagenase activity Retard the decrease of total collagen content and fillets softening	Xu et al. (2017)
Sea buckthorn fruits' polyphenolic extract (250 mg/kg)	Carp muscle (*Cyprinus carpio*)	Packed in PE foil and storage at −18 °C	↓ Lipid oxidation (POV, TBARS, CD), MetMb concentration, carbonyl levels, proteins cross-linking process ↑ Myoglobin and soluble proteins concentration	Papuc, Crivineanu, Nicorescu, Papuc, and Predescu (2012)

Hamamelis virginiana phenolic fraction (30 ppm)	Atlantic mackerel (*Scombrus scombrus*) muscle	Storage at 4 °C	↓ Lipid oxidation (POV and TBARS)	Gonzalez, Torres, and Medina (2010)
Tannic acid (200 mg/kg)	Ground beef	Modified atmosphere packaging and storage at 4 °C	↓ Lipid oxidation (POV and TBARS), non-heme iron content, protein degradation, psychrophilic bacterial count (PBC)	Maqsood and Benjakul (2010a, 2010b)
Tannic acid, catechin, caffeic acid and gallic acid (200 mg/kg)	Ground camel meat	Packed in polystyrene tray and wrapped with plastic film, storage at 4 °C	↓ The increase in POV and TBARS, mesosphilic bacterial count (MBC) and psychrophilic bacterial count (PBC)	Maqsood, Abushelaibi, Manheem, Al Rashedi, and Kadim (2015)
			↑ Sensory scores	
Tannic acid, catechin (200 mg/kg), date seed extract, green tea extract (400 mg/kg)	Camel meat sausages	Packed in a sealable PE bag and storage at 4 °C	↓ The increase in POV and TBARS, total plate counts (TPC) and PBC	Maqsood, Manheem, Abushelaibi, and Kadim (2016)
			↑ Sensory scores and textural properties	
Tea polyphenols (0.1–0.3 g/kg)	*Collichthys* fish ball	Vacuum packed and storage at 0 °C	↓ TVC, TVB-N, TBARS	Yi et al. (2011)
			Extend shelf-life and retain excellent fish ball sensory quality	

Continued

Table 7 Polyphenols applied in meat and aquatic products preservation.—cont'd

Polyphenols	Food species	Additional treatments	Main outcomes	References
Tannic acid (100, 200 mg/kg)	Striped catfish (*Pangasius hypophthalmus*) slice	Modified atmosphere packaging and storage at 4 °C	↓Lipid oxidation (POV and TBARS), non-haem iron content, protein degradation, TVC and PBC	Maqsood and Benjakul (2010a, 2010b)
			Retard the formation of primary oxidation products and free fatty acids	
Tea polyphenols or rosemary extract (0.2%)	Crucian carp (*Carassius auratus*)	Packed in air-proof PE pouches and storage at (4 ± 1) °C	↓ The increase of pH, TVB-N, K-value and TBARS value	Li, Hu, et al. (2012); Li, Li, et al. (2012)
			↑ Shelf-life and sensory quality	
Tea polyphenols (0.2%)	Black sea bream (*Sparus macrocephalus*)	Storage at 4 °C	↓Nucleotide breakdown (K value), lipid oxidation (POV, TBARS), protein decomposition (TVB-N and —TMA-N), and microbial growth (TVC)	Feng, Jiang, Wang, and Li (2012)
			Keep texture, color and sensory characteristics within acceptable limits, extend the shelf-life of fish	
Barley husk phenolic extracts (0.5%, 1.0%)	Coho salmon (*Oncorhynchus kisutch*)	Vacuum-packed in PE bag, storage at −18 °C	↓ Peroxide and anisidine values, α-, γ-tocopherol losses, rancid odor	Rodríguez, Cruz, Paseiro-Losada, and Aubourg (2011)
Tea polyphenols (0.15%)	Dried-seasoned squid (*Dosidicus gigas*)	Air-packed in PE bag and storage at 25 °C	↓ Moisture loss, oxidation of lipids (TBARS), TMAO and TVB-N accumulation	Dong, Zhu, Li, and Li (2013)

Tea polyphenols (300 mg/kg), α-tocopherol (700 mg/kg)	Fresh minced pork	Packaged in a HiOx atmosphere and storage at (2 ± 1) °C	↓ Lipid oxidation (TBARS) and protein oxidation (carbonyls, sulfhydryls and free amines), *N*-nitrosodiethylamine (NDEA)	Yang et al. (2013)
Tea polyphenol (0.03%)	Pork sausages	Storage at 20 °C	↓ Changes in TVC, pH, TBARS value, color, and sensory characteristics	Wenjiao, Yunchuan, Junxiu, and Yongkui (2014)
GTP, GSE, α-tocopherol (300 mg/kg), GTP-GSE (150 + 150 mg/kg)	Bacon belly pieces	Vacuum-packaged and storage at 4 °C	↓ pH, lipid oxidation (TBARS) and residual nitrite content, aerobic plate and enterobacteriaceae counts (APC), biogenic amines and *N*-nitroso-dimethylamine (NDMA)	Wang et al. (2015)
			↑ Sensory quality	
Tea polyphenols, 6-gingerol (0.3%), Tea polyphenols (0.15%) + 6-gingerol (0.15%)	Shrimp paste	Storage at 25 °C	↓ Oxidation of protein (TVB-N) and lipids (TBARS), microorganism counts, and biogenic amines	Cai et al. (2015)
			Maintain sensory quality and total amino acids	
Polyphenolic extracts of cherry leaves (0.5%) or blackcurrant leaves (1.0%)	Pork sausage	Vacuum packed and storage at (4 ± 2) °C	↓ Lipid oxidation (TBARS)	Nowak, Czyzowska, Efenberger, and Krala (2016)
			↑ Shelf-life and sensory stability	
Grape seeds polyphenolic extracts (0.02%, 0.04%)	Beef sausage	Storage at −18 °C	↓ Lipid oxidation (TBARS) and total bacterial counts (TBC)	El-Zainy, Morsy, Sedki, and Mosa (2016)

Continued

Table 7 Polyphenols applied in meat and aquatic products preservation.—cont'd

Polyphenols	Food species	Additional treatments	Main outcomes	References
Tannic acid (2.5, 5, 10 ppm)	Raw and cooked chicken breast meat	Packaged in PE bags and storage at 4 °C	↓ Lipid and protein oxidation (TBARS, lipid oxidation-related volatiles and total carbonyl)	Al-Hijazeen, Lee, Mendonca, and Ahn (2016)
Apple peel polyphenol extracts (0.1, 0.15, 0.3%)	Pan fried beef patties	–	↓ Heterocyclic amines (MeIQx, PhIP and 4,8-DiMeIQx)	Sabally, Sleno, Jauffrit, Iskandar, and Kubow (2016)
Thinned young apple polyphenols (0.05%, 0.1%)	Grass carp (*Ctenopharyngodon idellus*) surimi	Sealed in preservative bags and storage at 4 °C	↓ Lipid oxidation (POV, TBARS), TVB-N increase, soluble myofibrillar protein (SMP) degradation and changes of color	Sun, Sun, Thavaraj, Yang, and Guo (2017)
			↑ SMPs' functional properties, gel strength and textural properties	
Green tea (3 ppm) + carvacrol (3, 10 ppm)	Lamb burger meat	Packaged in PS-EVA-PET tray with PVC film, storage at 4 °C	↓ Lipid oxidation (TBARS), microbial counts and discoloration	Belles, Alonso, Roncales, and Beltran (2019)
Cabbage leaves, banana peels polyphenolic extracts (0.5, 1, 1.5%)	Rohu (*Labeo rohita*) meat ball	Vacuum sealed in plastic bags, storage at 4 °C and −18 °C	↓ POV and free fatty acids	Ali et al. (2019)
Kelp polyphenol extracts (150 μmol KPE/g protein)	Mackerel (*Scomberomorus niphonius*) myofibrillar protein gel	Storage at 4 °C	↓ Total sulfhydryl and free amino groups, endogenous tryptophan content, free radicals production	He et al. (2020)
			↑ Gel properties (mechanical and rheological and microstructure), surface hydrophobicity, cross-linking between proteins	

pork or bacon belly pieces exhibited the most effective in the reduction of lipid oxidation (TBARS) and protein oxidation (carbonyls and sulfhydryls), aerobic plate count (APC), *Enterobacteriaceae*, as well as in suppressing the formation of free biogenic amines (putrescine, cadaverine, histamine, tyramine, and spermine) and *N*-nitrosodiethylamine (NDMA) (Wang et al., 2015; Yang et al., 2013).

Except for the individual use, TP combined with other antioxidants, such as GSE and gingerol, also showed the synergistic effect to inhibit the oxidation of lipids and proteins, the growth of microorganisms, and the sensory quality and shelf-life of meat and aquatic products (Cai et al., 2015; Wang et al., 2015). Also, grass carp fillets were respectively dipped with 2% of catechin, TP, gallic acid, and Na-citrate solution, catechin showed more effective in the inhibition of collagenase activity, delay the decrease of total collagen content and the softening of fillets, followed by TP, gallic acid and Na-citrate (Xu et al., 2017). These two suggested that using monomeric TP (pure phenolics) or combination with other natural products might be more promising than TP alone in keeping the sensory quality and extending the shelf-life of meat and aquatic products.

5.2 Fruits polyphenols applied in meat and aquatic products preservation

Polyphenols are also abundant in fruits, such as apple, grape, and pomegranate, mainly distribute in fruit peel and seeds and consist of non-gallated catechins, anthocyanidins, tannins, etc., which is the critical secondary metabolites responsible for the formation of fruits quality. Several studies have reported that polyphenol-rich fruits (extracts or polyphenolic extracts) present an excellent effect on the preservation of meat and aquatic meat. The addition of apple polyphenols (AP, 0.03–0.1%) in sliced cooked cured beef and pork hams could delay discoloration and the increase of TBARS value (Sun et al., 2010). While, the equivalent of AP treated with grass carp surimi, packed in preservative bags and stored at cold condition, which also retarded variations in L*, a* and b* value, the decrease of POV, TBARS and TVB-N value, the degradation of the soluble myofibrillar protein (SMP), meanwhile, AP protected the SMP functional properties (surface hydrophobicity, emulsifying stability, and activity), gel strength and textural properties of surimi, including hardness, springiness, cohesiveness, gumminess, and chewiness (Sun et al., 2017). Besides, surface treatment of apple peels polyphenolic extract (0.1–0.3%) with beef patties before frying was able to minimize the production of carcinogenic heterocyclic aromatic amines (HAs) in pan-fried beef patties (Sabally et al., 2016). This result was consistent with green tea marinated beef with lower levels of PhIP and AαC after the pan-frying process, compared with the un-marinated beef samples (Quelhas et al., 2010).

5.3 Tannic acid applied in meat and aquatic products preservation

Tannins are water-soluble polyphenolic compounds, which can be classified into condensed and hydrolysable tannins, the latter one is generally used tannic acid (TA) in some food products, such as beverages, dairy, and meat products. Maqsood and colleagues found that addition tannic acid with striped catfish slices and ground beef, combined with modified atmosphere packaging and stored at 4 °C, showed a synergistic effect on decreasing POV and TBARS value, non-haem iron content, protein degradation, TVC and psychrophilic bacterial count (PBC), retarding the formation of primary oxidation products and free fatty acids (Maqsood & Benjakul, 2010a, 2010b). Subsequently, they explored the use of tannic acid in camel meat and camel meat sausages preservation, TA exhibited more effectiveness in keeping sensory quality, delaying lipid oxidation and microbial growth compared with simple phenolic acids (gallic acid and caffeic acid) and extracts (date seed extract and green tea extract) during the cold storage (Maqsood et al., 2015, 2016). Al-Hijazeen et.al also studied the antioxidant effects and quality improvement of tannic acid (2.5–10 mg/kg) on the raw and cooked chicken breast meat during refrigerate storage. Meat treated with 5 and 10 mg/kg of tannic acid could effectively inhibit lipids and protein oxidation, decrease the formation of off-aroma volatiles, maintain color stability compared with 5 mg/kg of BHA (Al-Hijazeen et al., 2016). Besides, they continually combined tannic acid with oregano oil to assess the antioxidant effects on chicken breast and thigh meats. The results indicated that the combination was before the individuals in reducing TBARS, total carbonyls, and off-aroma volatiles production, and was an excellent replacement for the chemical antioxidant (Al-Hijazeen, Mendonca, Lee, & Ahn, 2018).

5.4 Other polyphenols applied in meat and aquatic products preservation

Except for the above-mentioned commonly used polyphenols, other plant-derived polyphenolic extracts, such as quince polyphenolic extract (Fattouch et al., 2008), sea buckthorn fruits' polyphenols (Papuc et al.,

2012), *Hamamelis virginiana* (Gonzalez et al., 2010), barley husk (Rodríguez et al., 2011), cherry leaves and blackcurrant leaves (Nowak et al., 2016), kelp (He et al., 2020), even the argo-waste of cabbage leaves and banana peels (Ali et al., 2019), were used to decrease the lipid and protein oxidation, microbial spoilage and foodborne growth, retain sensory quality, nutritional value and extend the shelf-life of meat and aquatic products. It should be noted that the extraction of these polyphenolic extracts was time-consume, high-cost, and inefficient, meanwhile, polyphenolic extracts were complex and lack of specific composition and toxicological information, these factors would limit their future application in the meat and aquatic industry.

6. Application of polyphenols in beverage

Owing to the extensive distribution in nature and multifunction in sensory quality and health, polyphenols becomes one of the most studied, applied, and consumed bioactive compounds for improving human health and shelf-life of foods. Therefore, with the ever-growing demand for polyphenol-rich foods, the global market for polyphenols continues to rapidly elevate. It was estimated that both the global industrial production and monetary value of extracted polyphenols will reach 33,880 tons and 1.33 billion by the end of 2024. The predominant application of polyphenols is the functional beverages industry with 44%, followed by the functional foods sector with 33% (Adebooye, Alashi, & Aluko, 2018). Beverages became a daily-consuming product during the past decades. However, in recent years, the quality and the potential health problem for humans were received much attention due to their high concentrations of sugar and preservatives. Polyphenols are plentiful in our daily diet includes fruits, vegetables, and whole grains, so, the direct application of polyphenols is generally intake by directly eating as foods or drinking as beverages form and strengthens the beverages' health benefits to address the health issue. What's more, it is also used for improving the sensory quality and extending the shelf-life of products by stabilizing various ingredients.

The addition of tea polyphenols to beverages has a positive effect on the stability of beverage quality and the extension of shelf-life. Zhong et al. compared the preservation efficacy of fruit juices between tea polyphenols and chemical preservatives (potassium sorbate and sodium benzoate). Tea polyphenols could effectively keep the flavor stable and extend the shelf-life of juices for 7–10 days, compared with preservatives free juices. Three common types of juices including apple, orange, and pear juice, were added

100 mg/kg of tea polyphenols showed the equivalent preservative effect with 500 mg/kg of potassium sorbate and sodium benzoate and had a better inhibitory effect on *E. coli* and *Staphylococci* than them in juices (Zhong & Li, 2011). Proper addition of tea polyphenols (50–100 mg/kg) can effectively inhibit the growth of bacteria, maintain the original nutrients and flavor of the juice, and extend the shelf life, without bitter and astringent taste.

6.1 Improving the stability of the ingredients of beverages

During the process and storage of beverages, the degradation of nutrients, off-flavor formation, and discoloration have negative effects on the consumer acceptability and shelf-life of products. The degradation of nutrients, pigments, aroma, and flavor compounds is commonly affected by temperature, oxygen, light, pH, metal, microorganisms, enzymes, and ingredient interactions during processing and storage. Based on these factors, many studies have explored the utilization of polyphenols in enhancing the stability of beverages' ingredients to guarantee a minimum loss of beverage quality.

Ascorbic acid as a common ingredient used to evaluate the nutritional quality of vegetables and fruits, has been widely applied for fortification nutritional value of beverages to meet the market demands. Anthocyanins and anthocyanins-rich extracts are widely used as colorant and functional ingredients in the beverage industry (Morata, López, Tesfaye, González, & Escott, 2019). However, the presence of ascorbic acid could accelerate discoloration and degradation of anthocyanins in the model beverage system in both the solution and powder forms, and this effect was irrelevant to anthocyanin structure (Poei-Langston & Wrolstad, 1981; West & Mauer, 2013). It was reported that the addition of 0.2% epigallocatechin gallate (EGCG) and green tea extract in the model beverage system showed the most potent inhibition effect on color degradation of anthocyanin induced by ascorbic acid, compared with the concentration of vanillin, quillaja saponin, and protocatechualdehyde. The addition of EGCG or GTE (containing ~50% EGCG) extended the half-lives ($t_{1/2}$) of anthocyanin more than two folds under the accelerated storage conditions. EGCG or GTE might protect anthocyanin from the condensation with ascorbic acid by hydrophobic interaction with anthocyanins (Chung, Rojanasasithara, Mutilangi, & McClements, 2016). Gallic acid was used as a co-pigment to facilitate anthocyanin stabilities and color characteristics of blueberry juice or ascorbic acid fortified cranberry juice (Roidoung, Dolan, & Siddiq, 2016; Zhang et al., 2020).

Apart from catechins and gallic acid, other co-pigments such as flavonoid C-glycosides (rutin, vitexin, and orientin), phenolic acids (chlorogenic acid, tannic acid, caffeic acid, and ferulic acid), and rosemary polyphenolic extract were also added to berry's juice as a potent color enhancer and anthocyanins stabilizer (Hernandez-Herrero & Frutos, 2015; Kopjar, JakŠIĆ, & PiliŽOta, 2012; Pan et al., 2014; Sari, 2016). These flavan-3-ols, flavonoid C-glycosides, and phenolic acids are merely colorless compounds, which were attracted by anthocyanins formed intermolecular co-pigmentation through hydrophobic and hydrogen bond interactions, contributing to the increase or stability in color intensity of anthocyanins (Pan et al., 2014). The other two studies also proved that the addition of GT and GTE into blackberry juice and jams could against thermal degradation to improve the stability of anthocyanin through co-pigmentation (Kopjar, Bilić, & Piližota, 2014; Piližota, Kopjar, Županić, & Balija, 2012). Besides, the addition of polyphenolic extract from passion fruits seeds inhibited the formation of lipid oxidation products (conjugated dienes and aldehydes) of dairy beverages fortified with linseed oil (rich in omega-3 fatty acids) during storage and simulated digestion (Rotta et al., 2020).

As we all know, the stability of polyphenols has troubled the food industry and limited its utilization. Some research groups have developed stable polyphenols for beverages by microencapsulation. Blueberry-derived anthocyanins extract was encapsulated with chitosan nanoparticles and used in a model beverage. The stability of anthocyanins was significant increased during the storage. At the same time, chitosan encapsulated anthocyanin nanoparticles showed a decreased degradation rate in simulated gastrointestinal fluid (He et al., 2017). Muhammad et al. prepared the colloidal cinnamon nanoparticles by shellac and xanthan gum using anti-solvent precipitation and were added in the chocolate beverage. Incorporation of cinnamon-loaded nanoparticles in chocolate beverages improved total polyphenols content, antioxidant activity, and suspension stability through colloidal network decreasing the sedimentation index, while without any significant effects on pH and color of chocolate beverages (Muhammad et al., 2019). Feng et al. successfully encapsulated resveratrol liposomes, EGCG liposomes, and their combination liposomes uncoated and coated with low-methoxyl pectin, and characterized their properties that include encapsulation efficiency, dispersion index, particle size, zeta potential, morphology, and antioxidant capacity, and evaluated pasteurization and orange juice on their stability. Compared with the other three liposomes, low-methoxyl pectin coated complex liposomes showed the optimal physical properties and higher stability after pasteurization and in orange juice (Feng et al., 2019).

6.2 Removing the undesirable odor and improving the sensory quality

In ultra-high temperature (UHT) processed beverages, Maillard reactions are obvious owing to the presence of high levels of proteins and reducing sugars, especially in lactose-hydrolyzed milk, milk-based beverages, and plant-protein-based beverages. As a result, some of Maillard reaction products such as furfurals, aldehydes, and α-dicarbonyls, play a negative role in nutritional value, color, aroma, flavor, and shelf-life during storage. In particular, α-dicarbonyls, such as glyoxal and methylglyoxal, can further react with free amino acids leading to the formation of Strecker aldehydes, which contributes to the off-flavor for UHT processed milk. Tea polyphenols and monomeric catechins were able to trap α-dicarbonyls forming catechins-carbonyl conjugates to terminate their further reaction with other nutrients of milk and beverages (Jansson et al., 2017). The A ring of catechins can form highly nucleophilic centers at C-6 and C-8 positions, reactive carbonyl species were trapped at these two positions by aromatic electrophilic substitution (Meng et al., 2019). A similar reaction was observed in the processing of bottled tea beverages, leading to the formation of ascorbyl adducts with epigallocatechin gallate and gallocatechin gallate (Hung et al., 2018). Also, tea polyphenols could bind with amino acids and proteins to inhibit proteolysis during milk storage (Jansson et al., 2017). Therefore, Aromatic electrophilic substitution and polyphenol-protein binding are deemed as the main mechanisms that polyphenols inhibit Maillard reactions in milk and beverages.

Apple and grape seed polyphenols were supplemented with a milk-based beverage to investigate the impact of these two types of polyphenols on the flavor profile, samples were evaluated by a trained sensory panel through descriptive analysis with vocabulary and scale intensities. Compared with control, samples added with polyphenols had a robust inhibitory effect on the formation of milk UHT odor and flavor but were perceived high intensity of bitterness, particularly in apple extracts (Axten, Wohlers, & Wegrzyn, 2008).

6.3 Fortification health benefits of beverages

Polyphenols widely exist secondary metabolites found in various plants, including fruits, vegetables, herbs, cereals, and beverages such as coffee, cocoa, and tea. Myriad of epidemiological and experimental studies have reported multi-health-promoting effects of these plant-derived polyphenols, including the well-known antioxidant properties, and the preventive efficacy on cancer, metabolic syndromes, neurodegenerative diseases, etc. (Perez-Jimenez, Neveu, Vos, & Scalbert, 2010). It was reported that polyphenol consumption is tightly linked with the low intake of vegetables, fruits, and whole grains of the American population in the past 10 years, while foods and beverages contributed 99.8% of total polyphenol intake, among them coffee, beans, and tea are the major dietary contributors for US adults, accounting for 39.6%, 9.8% and 7.6%, respectively (Huang, Braffett, Simmens, Young, & Ogden, 2020). It suggested that beverages are one of the important routes to intake polyphenols, more polyphenol-rich beverages need to intake to fulfill the demand for health.

Ahmad and others developed green tea catechins based functional beverage, called *Green cool*, could effectively improve the antioxidant status of Sprague-Dawley rats fed with high cholesterol and high sucrose diets (Ahmad, Butt, Huma, & Sultan, 2013). Lamothe et al. found that the combination of milk with polyphenol-rich beverages, green tea, and grape juice, could increase the antioxidant activity and synergistically reduces the formation of lipid oxidation related toxic substances during simulated digestion of linseed oil emulsions (PUFA-rich) (Lamothe, Guerette, Dion, Sabik, & Britten, 2019). Different types of tea extracts were supplemented during the fermentation of soymilk to promoting the anti-radical activities and the functional properties of fermented soymilk (Zhao & Shah, 2014).

Banana peel, a common fruit biowaste, is rich in polyphenols. It was respectively supplemented to freshly squeezed and concentrated orange juices, which showed a more stable antioxidant effect and sensory qualities of orange juice during the pasteurization process and cold storage, compared with the untreated orange juice (Ortiz et al., 2017). As a by-product of plum juice or pulp process, plum skins are plentiful in flavan-3-ols, anthocyanins, and flavonols, which can be extracted and applied as a functional enhancer for beverages (de Beer, Steyn, Joubert, & Muller, 2012). Supplementation of the plum skin extracts to plum nectar could fortify the color, total polyphenol contents, and antioxidant activity, while there observed some negative impact on sensory characteristics, including the intensified acidity and astringency, plant-like flavor and aroma with the increased plum skin extracts of addition (de Beer et al., 2012). It suggested that the application of these polyphenol extracts needs to balance the relationship between function and sensory by feasible adjusting the addition amount or formulation, to elevate the health benefits and acceptability of beverages. Olive leaf extract has abundant oleuropein and hydroxytyrosol, with a bitter taste and antioxidant capacity, was used to develop low bitterness perception and functional fruit smoothies by addition sodium cyclamate and sucrose (Kranz, Braun, Schulze, & Kunz, 2010). Besides, olive vegetation water, one of the concomitant products of olive oil extraction, is rich in phenolic compounds. The extracted polyphenols were added in milk beverages to fortify its functional properties and did not disturb the fermentation process, activities of lactic acid bacteria (*Lactobacilli*), and volatiles compositions (Servili et al., 2011). Therefore, it is suggested that polyphenols can successfully enhance the function of probiotics beverages without affecting the survival and fermentation of probiotic bacterium.

References

Adebooye, O. C., Alashi, A. M., & Aluko, R. E. (2018). A brief review on emerging trends in global polyphenol research. *Journal of Food Biochemistry*, *42*(4), e12519.

Ahmad, R. S., Butt, M. S., Huma, N., & Sultan, M. T. (2013). Green tea catechins based functional drink (Green cool) improves the antioxidant status of SD rats fed on high cholesterol and sucrose diets. *Pakistan Journal of Pharmaceutical Sciences*, *26*(4), 721–726.

Ahmad, S., Khan, H., Siddiqui, Z., Khan, M. Y., Rehman, S., Shahab, et al. (2018). AGEs, RAGEs and s-RAGE; friend or foe for cancer. *Seminars in Cancer Biology*, *49*, 44–55.

Al-Hijazeen, M., Lee, E. J., Mendonca, A., & Ahn, D. U. (2016). Effects of tannic acid on lipid and protein oxidation, color, and volatiles of raw and cooked chicken breast meat during storage. *Antioxidants (Basel)*, *5*(2), 19.

Al-Hijazeen, M., Mendonca, A., Lee, E. J., & Ahn, D. U. (2018). Effect of oregano oil and tannic acid combinations on the quality and sensory characteristics of cooked chicken meat. *Poultry Science*, *97*(2), 676–683.

Ali, M., Imran, M., Nadeem, M., Khan, M. K., Sohaib, M., Suleria, H. A. R., et al. (2019). Oxidative stability and sensoric acceptability of functional fish meat product supplemented with plant-based polyphenolic optimal extracts. *Lipids in Health and Disease*, *18*(1), 35.

Altan, A., Aytac, Z., & Uyar, T. (2018). Carvacrol loaded electrospun fibrous films from zein and poly (lactic acid) for active food packaging. *Food Hydrocolloids*, *81*, 48–59.

Asgher, M., Qamar, S. A., Bilal, M., & Iqbal, H. M. N. (2020). Bio-based active food packaging materials: Sustainable alternative to conventional petrochemical-based packaging materials. *Food Research International*, *137*, 109625.

Axten, L. G., Wohlers, M. W., & Wegrzyn, T. (2008). Using phytochemicals to enhance health benefits of milk: Impact of polyphenols on flavor profile. *Journal of Food Science*, *73*(6), H122–H126.

Ayala-Zavala, J. F., González-Aguilar, G. A., & Roura, S. I. (2015). Using natural antimicrobials to enhance the safety and quality of fresh and processed fruits and vegetables. In *Handbook of natural antimicrobials for food safety and quality* (pp. 287–313). Woodhead Publishing Series in Food Science, Technology and Nutrition.

Aydogdu, A., Sumnu, G., & Sahin, S. (2019). Fabrication of gallic acid loaded hydroxypropyl methylcellulose nanofibers by electrospinning technique as active packaging material. *Carbohydrate Polymers*, *208*, 241–250.

Bao, S., Xu, S., & Wang, Z. (2009). Antioxidant activity and properties of gelatin films incorporated with tea polyphenol-loaded chitosan nanoparticles. *Journal of the Science of Food and Agriculture, 89*(15), 2692–2700.

Baptista, R. C., Horita, C. N., & Sant'Ana, A. S. (2020). Natural products with preservative properties for enhancing the microbiological safety and extending the shelf-life of seafood: A review. *Food Research International, 127*, 108762.

Belizón, M., Fernández-Ponce, M. T., Casas, L., Mantell, C., & Martínez de la Ossa-Fernández, E. J. (2018). Supercritical impregnation of antioxidant mango polyphenols into a multilayer PET/PP food-grade film. *Journal of CO_2 Utilization, 25*, 56–67.

Belles, M., Alonso, V., Roncales, P., & Beltran, J. A. (2019). Sulfite-free lamb burger meat: Antimicrobial and antioxidant properties of green tea and carvacrol. *Journal of the Science of Food and Agriculture, 99*(1), 464–472.

Bjarnsholt, T. (2013). The role of bacterial biofilms in chronic infections. *APMIS, 121*, 1–58.

Bouarab-Chibane, L., Forquet, V., Lanteri, P., Clement, Y., Leonard-Akkari, L., Oulahal, N., et al. (2019). Antibacterial properties of polyphenols: Characterization and QSAR (quantitative structure activity relationship) models. *Frontiers in Microbiology, 10*, 829.

Bouaziz, M., Fki, I., Jemai, H., Ayadi, M., & Sayadi, S. (2008). Effect of storage on refined and husk olive oils composition: Stabilization by addition of natural antioxidants from Chemlali olive leaves. *Food Chemistry, 108*(1), 253–262.

Cai, L., Liu, S., Sun, L., Wang, Y., Ji, H., & Li, J. (2015). Application of tea polyphenols in combination with 6-gingerol on shrimp paste of during storage: Biogenic amines formation and quality determination. *Frontiers in Microbiology, 6*, 981.

Cao, J., Wang, Q., Ma, T., Bao, K., Yu, X., Duan, Z., et al. (2020). Effect of EGCG-gelatin biofilm on the quality and microbial composition of tilapia fillets during chilled storage. *Food Chemistry, 305*, 125454.

Cejudo Bastante, C., Casas Cardoso, L., Fernández Ponce, M. T., Mantell Serrano, C., & Martínez de la Ossa-Fernández, E. J. (2018). Characterization of olive leaf extract polyphenols loaded by supercritical solvent impregnation into PET/PP food packaging films. *The Journal of Supercritical Fluids, 140*, 196–206.

Chandrasekar, V., Belur, P. D., & Regupathi, I. (2016). Effect of hydroxybenzoic acids antioxidants on the oxidative stability of sardine oil. *Resource-Efficient Technologies, 2*, S114–S118.

Chung, C., Rojanasasithara, T., Mutilangi, W., & McClements, D. J. (2016). Stabilization of natural colors and nutraceuticals: Inhibition of anthocyanin degradation in model beverages using polyphenols. *Food Chemistry, 212*, 596–603.

Cui, L., Cho, H. T., McClements, D. J., Decker, E. A., & Park, Y. (2016). Effects of salts on oxidative stability of lipids in Tween-20 stabilized oil-in-water emulsions. *Food Chemistry, 197*(15Pt. B), 1130–1135.

De Araujo, F. F., de Paulo Farias, D., Neri-Numa, I. A., & Pastore, G. M. (2021). Polyphenols and their applications: An approach in food chemistry and innovation potential. *Food Chemistry, 338*, 127535.

De Beer, D., Steyn, N., Joubert, E., & Muller, N. (2012). Enhancing the polyphenol content of a red-fleshed Japanese plum (*Prunus salicina* Lindl.) nectar by incorporating a polyphenol-rich extract from the skins. *Journal of the Science of Food and Agriculture, 92*(13), 2741–2750.

De Oliveira, E. F., Tikekar, R., & Nitin, N. (2018). Combination of aerosolized curcumin and UV-A light for the inactivation of bacteria on fresh produce surfaces. *Food Research International, 114*, 133–139.

Dominguez, R., Barba, F. J., Gomez, B., Putnik, P., Bursac Kovacevic, D., Pateiro, M., et al. (2018). Active packaging films with natural antioxidants to be used in meat industry: A review. *Food Research International, 113*, 93–101.

Dong, L., Zhu, J., Li, X., & Li, J. (2013). Effect of tea polyphenols on the physical and chemical characteristics of dried-seasoned squid (*Dosidicus gigas*) during storage. *Food Control, 31*(2), 586–592.
Dovigo, L. N., Carmello, J. C., de Souza Costa, C. A., Vergani, C. E., Brunetti, I. L., Bagnato, V. S., et al. (2013). Curcumin-mediated photodynamic inactivation of *Candida albicans* in a murine model of oral candidiasis. *Medical Mycology, 51*(3), 243–251.
Efenberger-Szmechtyk, M., Nowak, A., & Czyzowska, A. (2021). Plant extracts rich in polyphenols: Antibacterial agents and natural preservatives for meat and meat products. *Critical Reviews in Food Science and Nutrition, 61*(1), 149–178.
El-Zainy, A. R., Morsy, A. E., Sedki, A. G., & Mosa, N. M. (2016). Polyphenols grape seeds extract as antioxidant and antimicrobial in beef sausage. *International Journal of Current Science, 19*(2), 10.
Fan, W., Chi, Y., & Zhang, S. (2008). The use of a tea polyphenol dip to extend the shelf life of silver carp (*Hypophthalmicthys molitrix*) during storage in ice. *Food Chemistry, 108*(1), 148–153.
Fattouch, S., Sadok, S., Raboudi-Fattouch, F., & Slama, M. B. (2008). Damage inhibition during refrigerated storage of mackerel (*Scomber scombrus*) fillets by a presoaking in quince (*Cydonia oblonga*) polyphenolic extract. *International Journal of Food Science & Technology, 43*(11), 2056–2064.
Feng, L., Jiang, T., Wang, Y., & Li, J. (2012). Effects of tea polyphenol coating combined with ozone water washing on the storage quality of black sea bream (*Sparus macrocephalus*). *Food Chemistry, 135*(4), 2915–2921.
Feng, X., Ng, V. K., Mikš-Krajnik, M., & Yang, H. (2016). Effects of fish gelatin and tea polyphenol coating on the spoilage and degradation of myofibril in fish fillet during cold storage. *Food and Bioprocess Technology, 10*(1), 89–102.
Feng, S., Sun, Y., Wang, P., Sun, P., Ritzoulis, C., & Shao, P. (2019). Co-encapsulation of resveratrol and epigallocatechin gallate in low methoxyl pectin-coated liposomes with great stability in orange juice. *International Journal of Food Science & Technology, 55*(5), 1872–1880.
Frankel, E. N. (1993). In search of better methods to evaluate natural antioxidants and oxidative stability in food lipids. *Trends in Food Science & Technology, 4*(7), 220–225.
Frankel, E. N., Huang, S. W., Aeschbach, R., & Prior, E. (1996). Antioxidant activity of a rosemary extract and its constituents, carnosic acid, carnosol, and rosmarinic acid, in bulk oil and oil-in-water emulsion. *Journal of Agricultural and Food Chemistry, 44*(1), 213–220.
Gao, H. X., He, Z., Sun, Q., He, Q., & Zeng, W. C. (2019). A functional polysaccharide film forming by pectin, chitosan, and tea polyphenols. *Carbohydrate Polymers, 215*, 1–7.
Ghate, V. S., Zhou, W. B., & Yuk, H. G. (2019). Perspectives and trends in the application of photodynamic inactivation for microbiological food safety. *Comprehensive Reviews in Food Science and Food Safety, 18*(2), 402–424.
Gonzalez, M. J., Torres, J. L., & Medina, I. (2010). Impact of thermal processing on the activity of gallotannins and condensed tannins from *Hamamelis virginiana* used as functional ingredients in seafood. *Journal of Agricultural and Food Chemistry, 58*(7), 4274–4283.
Gray, J. I. (1978). Measurement of lipid oxidation: A review. *Journal of the American Oil Chemists' Society, 55*(6), 539–546.
Gupta, P. D., & Birdi, T. J. (2017). Development of botanicals to combat antibiotic resistance. *Journal of Ayurveda and Integrative Medicine, 8*(4), 266–275.
He, B., Ge, J., Yue, P., Yue, X., Fu, R., Liang, J., et al. (2017). Loading of anthocyanins on chitosan nanoparticles influences anthocyanin degradation in gastrointestinal fluids and stability in a beverage. *Food Chemistry, 221*, 1671–1677.
He, B., Ming, Y., Pu, Y., Sun, Y., Jin, M., Yu, C., et al. (2020). The dual effects of riboflavin and kelp polyphenol extracts on the gel properties of myofibrillar protein from *Scomberomorus Niphonius* under UVA irradiation. *Food Chemistry, 332*, 127373.

Hernandez-Herrero, J. A., & Frutos, M. J. (2015). Influence of rutin and ascorbic acid in colour, plum anthocyanins and antioxidant capacity stability in model juices. *Food Chemistry*, *173*, 495–500.

Hintz, T., Matthews, K. K., & Di, R. (2015). The use of plant antimicrobial compounds for food preservation. *Biomed Research International*, *2015*, 246264.

Ho, C. T. (1996). Methods to assess quality and stability of oils and fat-containing foods. *Trends in Food Science & Technology*, 7(5), 176.

Huang, Q., Braffett, B. H., Simmens, S. J., Young, H. A., & Ogden, C. L. (2020). Dietary polyphenol intake in US adults and 10-year trends: 2007–2016. *Journal of the Academy of Nutrition and Dietetics*, *120*(11), 1821–1833.

Hung, W. L., Wang, S., Sang, S., Wan, X., Wang, Y., & Ho, C. T. (2018). Quantification of ascorbyl adducts of epigallocatechin gallate and gallocatechin gallate in bottled tea beverages. *Food Chemistry*, *261*, 246–252.

Jaeger, H., Janositz, A., & Knorr, D. (2010). The Maillard reaction and its control during food processing. The potential of emerging technologies. *Pathologie Biologie (Paris)*, *58*(3), 207–213.

Jakobek, L. (2019). Food packaging materials with polyphenols as active compounds. *Meso: prvi hrvatski časopis o mesu*, *21*(5), 469–474.

Jansson, T., Rauh, V., Danielsen, B. P., Poojary, M. M., Waehrens, S. S., Bredie, W. L. P., et al. (2017). Green tea polyphenols decrease Strecker aldehydes and bind to proteins in lactose-hydrolyzed UHT milk. *Journal of Agricultural and Food Chemistry*, *65*(48), 10550–10561.

Jia, S., Huang, Z., Lei, Y., Zhang, L., Li, Y., & Luo, Y. (2018). Application of Illumina-MiSeq high throughput sequencing and culture-dependent techniques for the identification of microbiota of silver carp (*Hypophthalmichthys molitrix*) treated by tea polyphenols. *Food Microbiology*, *76*, 52–61.

Jiang, Y., Leung, A. W., Hua, H., Rao, X., & Xu, C. (2014). Photodynamic action of LED-activated curcumin against *Staphylococcus aureus* involving intracellular ROS increase and membrane damage. *International Journal of Photoenergy*, *2014*, 637601.

Jiao, W., Shu, C., Li, X., Cao, J., Fan, X., & Jiang, W. (2019). Preparation of a chitosan-chlorogenic acid conjugate and its application as edible coating in postharvest preservation of peach fruit. *Postharvest Biology and Technology*, *154*, 129–136.

Khan, M. R., Sadiq, M. B., & Mehmood, Z. (2020). Development of edible gelatin composite films enriched with polyphenol loaded nanoemulsions as chicken meat packaging material. *CyTA Journal of Food*, *18*(1), 137–146.

Koketsu, M., & Satoh, Y. I. (1997). Antioxidative activity of green tea polyphenols in edible oils. *Journal of Food Lipids*, *4*(1), 1–9.

Kopjar, M., Bilić, B., & Piližota, V. (2014). Anthocyanins, phenols, and antioxidant activity in blackberry juice with plant extracts addition during heating. *Acta Alimentaria*, *43*(2), 333–343.

Kopjar, M., JakŠIĆ, K., & PiliŽOta, V. (2012). Influence of sugars and chlorogenic acid addition on anthocyanin content, antioxidant activity and color of blackberry juice during storage. *Journal of Food Processing and Preservation*, *36*(6), 545–552.

Kranz, P., Braun, N., Schulze, N., & Kunz, B. (2010). Sensory quality of functional beverages: Bitterness perception and bitter masking of olive leaf extract fortified fruit smoothies. *Journal of Food Science*, *75*(6), S308–S311.

Lamothe, S., Guerette, C., Dion, F., Sabik, H., & Britten, M. (2019). Antioxidant activity of milk and polyphenol-rich beverages during simulated gastrointestinal digestion of linseed oil emulsions. *Food Research International*, *122*, 149–156.

Lan, W., Zhang, R., Ahmed, S., Qin, W., & Liu, Y. (2019). Effects of various antimicrobial polyvinyl alcohol/tea polyphenol composite films on the shelf life of packaged strawberries. *LWT—Food Science and Technology*, *113*, 108297.

Lee, C. K., Gim, S. Y., Kim, M. J., & Lee, J. H. (2017). Effects of quercetin or rutin on the oxidative stability of stripped or non-stripped soybean oils containing α-tocopherol. *European Journal of Lipid Science and Technology, 119*(7), 1600329.

Li, T., Hu, W., Li, J., Zhang, X., Zhu, J., & Li, X. (2012). Coating effects of tea polyphenol and rosemary extract combined with chitosan on the storage quality of large yellow croaker (*Pseudosciaena crocea*). *Food Control, 25*(1), 101–106.

Li, T., Li, J., Hu, W., & Li, X. (2013). Quality enhancement in refrigerated red drum (*Sciaenops ocellatus*) fillets using chitosan coatings containing natural preservatives. *Food Chemistry, 138*(2–3), 821–826.

Li, T., Li, J., Hu, W., Zhang, X., Li, X., & Zhao, J. (2012). Shelf-life extension of crucian carp (*Carassius auratus*) using natural preservatives during chilled storage. *Food Chemistry, 135*(1), 140–145.

Liang, Z. L., Chen, X., Li, L., Li, B., & Yang, Z. (2020). The fate of dietary advanced glycation end products in the body: From oral intake to excretion. *Critical Reviews in Food Science and Nutrition, 60*(20), 3475–3491.

Liu, F., Antoniou, J., Li, Y., Yi, J., Yokoyama, W., Ma, J., et al. (2015). Preparation of gelatin films incorporated with tea polyphenol nanoparticles for enhancing controlled-release antioxidant properties. *Journal of Agricultural and Food Chemistry, 63*(15), 3987–3995.

Liu, Y., Qin, Y., Bai, R., Zhang, X., Yuan, L., & Liu, J. (2019). Preparation of pH-sensitive and antioxidant packaging films based on kappa-carrageenan and mulberry polyphenolic extract. *International Journal of Biological Macromolecules, 134*, 993–1001.

López-de-Dicastillo, C., Gómez-Estaca, J., Catalá, R., Gavara, R., & Hernández-Muñoz, P. (2012). Active antioxidant packaging films: Development and effect on lipid stability of brined sardines. *Food Chemistry, 131*(4), 1376–1384.

Lorenzo, J. M., Pateiro, M., Dominguez, R., Barba, F. J., Putnik, P., Kovacevic, D. B., et al. (2018). Berries extracts as natural antioxidants in meat products: A review. *Food Research International, 106*, 1095–1104.

Lund, M. N., & Ray, C. A. (2017). Control of Maillard reactions in foods: Strategies and chemical mechanisms. *Journal of Agricultural and Food Chemistry, 65*(23), 4537–4552.

Majhi, S., Tyagi, A., & Mishra, A. (2020). Bio-polymeric packaging material for packaging of raw food. In *Reference module in materials science and materials engineering* (pp. 44–51).

Maqsood, S., Abushelaibi, A., Manheem, K., Al Rashedi, A., & Kadim, I. T. (2015). Lipid oxidation, protein degradation, microbial and sensorial quality of camel meat as influenced by phenolic compounds. *LWT- Food Science and Technology, 63*(2), 953–959.

Maqsood, S., & Benjakul, S. (2010a). Preventive effect of tannic acid in combination with modified atmospheric packaging on the quality losses of the refrigerated ground beef. *Food Control, 21*(9), 1282–1290.

Maqsood, S., & Benjakul, S. (2010b). Synergistic effect of tannic acid and modified atmospheric packaging on the prevention of lipid oxidation and quality losses of refrigerated striped catfish slices. *Food Chemistry, 121*(1), 29–38.

Maqsood, S., Benjakul, S., Abushelaibi, A., & Alam, A. (2014). Phenolic compounds and plant phenolic extracts as natural antioxidants in prevention of lipid oxidation in seafood: A detailed review. *Comprehensive Reviews in Food Science and Food Safety, 13*(6), 1125–1140.

Maqsood, S., Benjakul, S., & Shahidi, F. (2013). Emerging role of phenolic compounds as natural food additives in fish and fish products. *Critical Reviews in Food Science and Nutrition, 53*(2), 162–179.

Maqsood, S., Manheem, K., Abushelaibi, A., & Kadim, I. T. (2016). Retardation of quality changes in camel meat sausages by phenolic compounds and phenolic extracts. *Animal Science Journal, 87*(11), 1433–1442.

Meng, Q., Li, S., Huang, J., Wei, C. C., Wan, X., Sang, S., et al. (2019). Importance of the nucleophilic property of tea polyphenols. *Journal of Agricultural and Food Chemistry, 67*(19), 5379–5383.

Merino, L., Procura, F., Trejo, F. M., Bueno, D. J., & Golowczyc, M. A. (2019). Biofilm formation by *Salmonella* sp. in the poultry industry: Detection, control and eradication strategies. *Food Research International*, *119*, 530–540.
Mezza, G. N., Borgarello, A. V., Grosso, N. R., Fernandez, H., Pramparo, M. C., & Gayol, M. F. (2018). Antioxidant activity of rosemary essential oil fractions obtained by molecular distillation and their effect on oxidative stability of sunflower oil. *Food Chemistry*, *242*, 9–15.
Mohammadi, A., Jafari, S. M., Esfanjani, A. F., & Akhavan, S. (2016). Application of nano-encapsulated olive leaf extract in controlling the oxidative stability of soybean oil. *Food Chemistry*, *190*, 513–519.
Mohdaly, A. A. A., Sarhan, M. A., Mahmoud, A., Ramadan, M. F., & Smetanska, I. (2010). Antioxidant efficacy of potato peels and sugar beet pulp extracts in vegetable oils protection. *Food Chemistry*, *123*(4), 1019–1026.
Morata, A., López, C., Tesfaye, W., González, C., & Escott, C. (2019). Anthocyanins as natural pigments in beverages. In *Vol. 14. Value-added ingredients and enrichments of beverages* (pp. 383–428). Academic Press.
Moser, B. R. (2008). Efficacy of myricetin as an antioxidant in methyl esters of soybean oil. *European Journal of Lipid Science and Technology*, *110*(12), 1167–1174.
Muhammad, D. R. A., Gonzalez, C. G., Sedaghat Doost, A., Van de Walle, D., Van der Meeren, P., & Dewettinck, K. (2019). Improvement of antioxidant activity and physical stability of chocolate beverage using colloidal cinnamon nanoparticles. *Food and Bioprocess Technology*, *12*(6), 976–989.
Nesic, A., Castillo, C., Castaño, P., Cabrera-Barjas, G., & Serrano, J. (2020). Bio-based packaging materials. In *Biobased products and industries* (pp. 279–309). Elsevier.
Niu, L. H., Sun, X. H., Tang, J. M., Wang, J., Rasco, B. A., Lai, K. Q., et al. (2017). Formation of advanced glycation end-products in fish muscle during heating: Relationship with fish freshness. *Journal of Food Composition and Analysis*, *63*, 133–138.
Nohynek, L. J., Alakomi, H. L., Kähkönen, M. P., Heinonen, M., Helander, I. M., Oksman-Caldentey, K.-M., et al. (2006). Berry phenolics: Antimicrobial properties and mechanisms of action against severe human pathogens. *Nutrition and Cancer*, *54*(1), 18–32.
Nowak, A., Czyzowska, A., Efenberger, M., & Krala, L. (2016). Polyphenolic extracts of cherry (*Prunus cerasus* L.) and blackcurrant (*Ribes nigrum* L.) leaves as natural preservatives in meat products. *Food Microbiology*, *59*, 142–149.
Oliver, S., Vittorio, O., Cirillo, G., & Boyer, C. (2016). Enhancing the therapeutic effects of polyphenols with macromolecules. *Polymer Chemistry*, 7(8), 1529–1544.
Olsen, E., Vogt, G., Saarem, K., Greibrokk, T., & Nilsson, A. (2005). Autoxidation of cod liver oil with tocopherol and sscorbyl palmitate. *Journal of the American Oil Chemists Society*, *82*(2), 97–103.
Olszewska, M. A., Gedas, A., & Simoes, M. (2020). Antimicrobial polyphenol-rich extracts: Applications and limitations in the food industry. *Food Research International*, *134*, 109214.
Olszowy, M. (2019). What is responsible for antioxidant properties of polyphenolic compounds from plants? *Plant Physiology and Biochemistry*, *144*, 135–143.
Ortiz, L., Dorta, E., Gloria Lobo, M., Antonio González-Mendoza, L., Díaz, C., & González, M. (2017). Use of banana peel extract to stabilise antioxidant capacity and sensory properties of orange juice during pasteurisation and refrigerated storage. *Food and Bioprocess Technology*, *10*(10), 1883–1891.
Ou, J. Y., Teng, J., El-Nezami, H. S., & Wang, M. F. (2018). Impact of resveratrol, epicatechin and rosmarinic acid on fluorescent AGEs and cytotoxicity of cookies. *Journal of Functional Foods*, *40*, 44–50.
Pan, Y. Z., Guan, Y., Wei, Z. F., Peng, X., Li, T. T., Qi, X. L., et al. (2014). Flavonoid C-glycosides from pigeon pea leaves as color and anthocyanin stabilizing agent in blueberry juice. *Industrial Crops and Products*, *58*, 142–147.

Papuc, C., Crivineanu, M., Nicorescu, V., Papuc, C., & Predescu, C. (2012). Increase of the stability to oxidation of lipids and proteins in carp muscle (*Cyprinus Carpio*) subject to storage by freezing by polyphenols extracted from sea buckthorn fruits (*Hippophae Rhamnoides*). *Revista de Chimie-Bucharest-Original Edition*, *63*(12), 1198–1203.

Papuc, C., Goran, G. V., Predescu, C. N., Nicorescu, V., & Stefan, G. (2017). Plant polyphenols as antioxidant and antibacterial agents for shelf-life extension of meat and meat products: Classification, structures, sources, and action mechanisms. *Comprehensive Reviews in Food Science and Food Safety*, *16*(6), 1243–1268.

Penha, C. B., Bonin, E., da Silva, A. F., Hioka, N., Zanqueta, E. B., Nakamura, T. U., et al. (2017). Photodynamic inactivation of foodborne and food spoilage bacteria by curcumin. *LWT-Food Science and Technology*, *73*, 198–202.

Perez-Jimenez, J., Neveu, V., Vos, F., & Scalbert, A. (2010). Systematic analysis of the content of 502 polyphenols in 452 foods and beverages: An application of the phenol-explorer database. *Journal of Agricultural and Food Chemistry*, *58*(8), 4959–4969.

Piližota, V., Kopjar, M., Županić, N., & Balija, N. (2012). Anthocyanin content and antioxidant activity of reduced-calorie blueberry jams fortified with green tea or pine bark extracts. *Acta Alimentaria*, *41*(4), 424–432.

Plumed-Ferrer, C., Vakevainen, K., Komulainen, H., Rautiainen, M., Smeds, A., Raitanen, J. E., et al. (2013). The antimicrobial effects of wood-associated polyphenols on food pathogens and spoilage organisms (vol 164, pg 99, 2013). *International Journal of Food Microbiology*, *166*(1), 163.

Poei-langston, M. S., & Wrolstad, R. E. (1981). Color degradation in an ascorbic acid-anthocyanin-flavanol model system. *Journal of Food Science*, *46*, 1218–1236.

Poiana, M. A. (2012). Enhancing oxidative stability of sunflower oil during convective and microwave heating using grape seed extract. *International Journal of Molecular Sciences*, *13*(7), 9240–9259.

Qin, Y. Y., Yang, J. Y., Lu, H. B., Wang, S. S., Yang, J., Yang, X. C., et al. (2013). Effect of chitosan film incorporated with tea polyphenol on quality and shelf life of pork meat patties. *International Journal of Biological Macromolecules*, *61*, 312–316.

Quelhas, I., Petisca, C., Viegas, O., Melo, A., Pinho, O., & Ferreira, I. M. P. L. V. O. (2010). Effect of green tea marinades on the formation of heterocyclic aromatic amines and sensory quality of pan-fried beef. *Food Chemistry*, *122*(1), 98–104.

Rafiee, Z., Jafari, S. M., Alami, M., & Khomeiri, M. (2018). Antioxidant effect of microwave-assisted extracts of olive leaves on sunflower oil. *Journal of Agricultural Science and Technology*, *14*(3), 1497–1509.

Ramonaitytė, D. T., Keršienė, M., Adams, A., Tehrani, K. A., & Kimpe, N. D. (2009). The interaction of metal ions with Maillard reaction products in a lactose–glycine model system. *Food Research International*, *42*(3), 331–336.

Rodríguez, A., Cruz, J. M., Paseiro-Losada, P., & Aubourg, S. P. (2011). Effect of a polyphenol–vacuum packaging on lipid deterioration during an 18-month frozen storage of Coho Salmon (*Oncorhynchus kisutch*). *Food and Bioprocess Technology*, *5*(6), 2602–2611.

Roedig-Penman, A., & Gordon, M. H. (1998). Antioxidant properties of myricetin and quercetin in oil and emulsions. *Journal of the American Oil Chemists Society*, *75*(2), 169–180.

Roidoung, S., Dolan, K. D., & Siddiq, M. (2016). Gallic acid as a protective antioxidant against anthocyanin degradation and color loss in vitamin-C fortified cranberry juice. *Food Chemistry*, *210*, 422–427.

Rotta, E. M., Giroux, H. J., Lamothe, S., Bélanger, D., Sabik, H., Visentainer, J. V., et al. (2020). Use of passion fruit seed extract (*Passiflora edulis Sims*) to prevent lipid oxidation in dairy beverages during storage and simulated digestion. *LWT-Food Science and Technology*, *123*, 109088.

Roy, L. G., Arabshahi-Delouee, S., & Urooj, A. (2010). Antioxidant efficacy of mulberry (*Morus indica* L.) Leaves extract and powder in edible oil. *International Journal of Food Properties*, *13*(1), 1–9.

Ruan, C., Zhang, Y., Sun, Y., Gao, X., Xiong, G., & Liang, J. (2019). Effect of sodium alginate and carboxymethyl cellulose edible coating with epigallocatechin gallate on quality and shelf life of fresh pork. *International Journal of Biological Macromolecules*, *141*, 178–184.

Sabally, K., Sleno, L., Jauffrit, J. A., Iskandar, M. M., & Kubow, S. (2016). Inhibitory effects of apple peel polyphenol extract on the formation of heterocyclic amines in pan fried beef patties. *Meat Science*, *117*, 57–62.

Santezi, C., Reina, B. D., & Dovigo, L. N. (2018). Curcumin-mediated photodynamic therapy for the treatment of oral infections—A review. *Photodiagnosis and Photodynamic Therapy*, *21*, 409–415.

Sari, F. (2016). The copigmentation effect of different phenolic acids on berberis crataegina anthocyanins. *Journal of Food Processing and Preservation*, *40*(3), 422–430.

Satpathy, S., Sen, S. K., Pattanaik, S., & Raut, S. (2016). Review on bacterial biofilm: An universal cause of contamination. *Biocatalysis and Agricultural Biotechnology*, *7*, 56–66.

Servili, M., Rizzello, C. G., Taticchi, A., Esposto, S., Urbani, S., Mazzacane, F., et al. (2011). Functional milk beverage fortified with phenolic compounds extracted from olive vegetation water, and fermented with functional lactic acid bacteria. *International Journal of Food Microbiology*, *147*(1), 45–52.

Shahidi, F., & Wanasundara, U. (1995). Effect of natural antioxidants on the stability of canola oil. *Developments in Food Science*, *37*, 469–479.

Shakeri, A., & Sahebkar, A. (2016). Optimized curcumin formulations for the treatment of Alzheimer's disease: A patent evaluation. *Journal of Neuroscience Research*, *94*(2), 111–113.

Shankar, S., & Rhim, J. W. (2017). Bio-nanocomposites for food packaging applications. In *Nanotechnology applications in food* (pp. 363–379). Elsevier.

Shao, P., Niu, B., Chen, H., & Sun, P. (2018). Fabrication and characterization of tea polyphenols loaded pullulan-CMC electrospun nanofiber for fruit preservation. *International Journal of Biological Macromolecules*, *107*(Pt. B), 1908–1914.

Shishir, M. R. I., Xie, L. H., Sun, C. D., Zheng, X. D., & Chen, W. (2018). Advances in micro and nano-encapsulation of bioactive compounds using biopolymer and lipid-based transporters. *Trends in Food Science & Technology*, *78*, 34–60.

Siripatrawan, U., & Noipha, S. (2012). Active film from chitosan incorporating green tea extract for shelf life extension of pork sausages. *Food Hydrocolloids*, *27*(1), 102–108.

Stojkovic, D. S., Zivkovic, J., Sokovic, M., Glamoclija, J., Ferreira, I. C. F. R., Jankovic, T., et al. (2013). Antibacterial activity of *Veronica montana* L. extract and of protocatechuic acid incorporated in a food system. *Food and Chemical Toxicology*, *55*, 209–213.

Su, Y.-L., Xu, J.-Z., Ng, C. H., Leung, L. K., Huang, Y., & Chen, Z.-Y. (2004). Antioxidant activity of tea theaflavins and methylated catechins in canola oil. *Journal of the American Oil Chemists Society*, *81*(3), 269–274.

Sun, L., Sun, J., Liu, D., Fu, M., Yang, X., & Guo, Y. (2018). The preservative effects of chitosan film incorporated with thinned young apple polyphenols on the quality of grass carp (*Ctenopharyngodon idellus*) fillets during cold storage: Correlation between the preservative effects and the active properties of the film. *Food Packaging and Shelf Life*, *17*, 1–10.

Sun, L., Sun, J., Thavaraj, P., Yang, X., & Guo, Y. (2017). Effects of thinned young apple polyphenols on the quality of grass carp (*Ctenopharyngodon idellus*) surimi during cold storage. *Food Chemistry*, *224*, 372–381.

Sun, W. Q., Zhang, Y. J., Zhou, G. H., Xu, X. L., & Peng, Z. Q. (2010). Effect of apple polyphenol on oxidative stability of sliced cooked cured beef and pork hams during chilled storage. *Journal of Muscle Foods*, *21*(4), 722–737.

Taghvaei, M., & Jafari, S. M. (2015). Application and stability of natural antioxidants in edible oils in order to substitute synthetic additives. *Journal of Food Science and Technology*, *52*(3), 1272–1282.

Tang, Q. L., Kang, A. R., & Lu, C. X. (2016). Phytochemical analysis, antibacterial activity and mode of action of the methanolic extract of *Scutellaria barbata* against various clinically important bacterial pathogens. *International Journal of Pharmacology*, *12*(2), 116–125.

Temba, B. A., Fletcher, M. T., Fox, G. P., Harvey, J., Okoth, S. A., & Sultanbawa, Y. (2019). Curcumin-based photosensitization inactivates *Aspergillus flavus* and reduces aflatoxin B1 in maize kernels. *Food Microbiology*, *82*, 82–88.

Tomadoni, B., Viacava, G., Cassani, L., Moreira, M. R., & Ponce, A. (2016). Novel biopreservatives to enhance the safety and quality of strawberry juice. *Journal of Food Science and Technology*, *53*(1), 281–292.

Topuz, F., & Uyar, T. (2020). Antioxidant, antibacterial and antifungal electrospun nanofibers for food packaging applications. *Food Research International*, *130*, 108927.

Van Lieshout, G. A. A., Lambers, T. T., Bragt, M. C. E., & Hettinga, K. A. (2020). How processing may affect milk protein digestion and overall physiological outcomes: A systematic review. *Critical Reviews in Food Science and Nutrition*, *60*(14), 2422–2445.

Villalobos-Delgado, L. H., Mateo, J., Caro, I., Leal Ramos, M.-Y., Mendez, N. G., Cansino, R. G., et al. (2019). Natural antioxidants in fresh and processed meat. In *Sustainable meat production and processing* (pp. 207–236). Academic Press.

Wang, Y., Li, F., Zhuang, H., Chen, X., Li, L., Qiao, W., et al. (2015). Effects of plant polyphenols and α-tocopherol on lipid oxidation, residual nitrites, biogenic amines, and *N*-nitrosamines formation during ripening and storage of dry-cured bacon. *LWT-Food Science and Technology*, *60*(1), 199–206.

Wang, H., Liu, F., Yang, L., Zu, Y., Wang, H., Qu, S., et al. (2011). Oxidative stability of fish oil supplemented with carnosic acid compared with synthetic antioxidants during long-term storage. *Food Chemistry*, *128*(1), 93–99.

Wang, Y., Zhang, R., Qin, W., Dai, J., Zhang, Q., Lee, K., et al. (2020). Physicochemical properties of gelatin films containing tea polyphenol-loaded chitosan nanoparticles generated by electrospray. *Materials & Design*, *185*, 108277.

Wenjiao, F., Yunchuan, C., Junxiu, S., & Yongkui, Z. (2014). Effects of tea polyphenol on quality and shelf life of pork sausages. *Journal of Food Science and Technology*, *51*(1), 191–195.

West, M. E., & Mauer, L. J. (2013). Color and chemical stability of a variety of anthocyanins and ascorbic acid in solution and powder forms. *Journal of Agricultural and Food Chemistry*, *61*(17), 4169–4179.

Wrona, M., Nerín, C., Alfonso, M. J., & Caballero, M.Á. (2017). Antioxidant packaging with encapsulated green tea for fresh minced meat. *Innovative Food Science & Emerging Technologies*, *41*, 307–313.

Wu, H., Lei, Y., Zhu, R., Zhao, M., Lu, J., Xiao, D., et al. (2019). Preparation and characterization of bioactive edible packaging films based on pomelo peel flours incorporating tea polyphenol. *Food Hydrocolloids*, *90*, 41–49.

Wu, C., Sun, J., Zheng, P., Kang, X., Chen, M., Li, Y., et al. (2019). Preparation of an intelligent film based on chitosan/oxidized chitin nanocrystals incorporating black rice bran anthocyanins for seafood spoilage monitoring. *Carbohydrate Polymers*, *222*, 115006.

Xiong, L. G., Chen, Y. J., Tong, J. W., Huang, J. A., Li, J., Gong, Y. S., et al. (2017). Tea polyphenol epigallocatechin gallate inhibits *Escherichia coli* by increasing endogenous oxidative stress. *Food Chemistry*, *217*, 196–204.

Xu, Y., Jiang, X., Ge, L., Zang, J., Xia, W., & Jiang, Q. (2017). Inhibitory effect of edible additives on collagenase activity and softening of chilled grass carp fillets. *Journal of Food Processing and Preservation*, *41*(2), e12836.

Yang, H., Meng, P., Xiong, Y. L., Ma, L., Wang, C., & Zhu, Y. (2013). Oxidation in HiOx-packaged pork Longissimus muscle predisposes myofibrillar and sarcoplasmic proteins to N-nitrosamine formation in nitrite-curing solution. *Meat Science*, *95*(3), 465–471.

Yang, Y., Song, X., Sui, X., Qi, B., Wang, Z., Li, Y., et al. (2016). Rosemary extract can be used as a synthetic antioxidant to improve vegetable oil oxidative stability. *Industrial Crops and Products*, *80*, 141–147.

Ye, J., Wang, S., Lan, W., Qin, W., & Liu, Y. (2018). Preparation and properties of polylactic acid-tea polyphenol-chitosan composite membranes. *International Journal of Biological Macromolecules*, *117*, 632–639.

Yi, S., Li, J., Zhu, J., Lin, Y., Fu, L., Chen, W., et al. (2011). Effect of tea polyphenols on microbiological and biochemical quality of Collichthys fish ball. *Journal of the Science of Food and Agriculture*, *91*(9), 1591–1597.

Yilmaz, A., Bozkurt, F., Cicek, P. K., Dertli, E., Durak, M. Z., & Yilmaz, M. T. (2016). A novel antifungal surface-coating application to limit postharvest decay on coated apples: Molecular, thermal and morphological properties of electrospun zein–nanofiber mats loaded with curcumin. *Innovative Food Science & Emerging Technologies*, *37*, 74–83.

Yousefi, H., Su, H. M., Imani, S. M., Alkhaldi, K., CD, M. F., & Didar, T. F. (2019). Intelligent food packaging: A review of smart sensing technologies for monitoring food quality. *ACS Sensors*, *4*(4), 808–821.

Yu, L. G., Gao, C., Zeng, M. M., He, Z. Y., Wang, L. X., Zhang, S., et al. (2016). Effects of raw meat and process procedure on N-epsilon-carboxymethyllysine and N-epsilon-carboxyethyl-lysine formation in meat products. *Food Science and Biotechnology*, *25*(4), 1163–1168.

Zhai, X., Shi, J., Zou, X., Wang, S., Jiang, C., Zhang, J., et al. (2017). Novel colorimetric films based on starch/polyvinyl alcohol incorporated with roselle anthocyanins for fish freshness monitoring. *Food Hydrocolloids*, *69*, 308–317.

Zhang, L., Li, S., Dong, Y., Zhi, H., & Zong, W. (2016). Tea polyphenols incorporated into alginate-based edible coating for quality maintenance of Chinese winter jujube under ambient temperature. *LWT- Food Science and Technology*, *70*, 155–161.

Zhang, Q. Z., Wang, Y. B., & Fu, L. L. (2020). Dietary advanced glycation end-products: Perspectives linking food processing with health implications. *Comprehensive Reviews in Food Science and Food Safety*, *19*(5), 2559–2587.

Zhang, L., Wang, W., Yue, X., Wu, G., Yue, P., & Gao, X. (2020). Gallic acid as a copigment enhance anthocyanin stabilities and color characteristics in blueberry juice. *Journal of Food Science and Technology*, *57*(4), 1405–1414.

Zhang, Y. Y., Zhang, F., Thakur, K., Ci, A. T., Wang, H., Zhang, J. G., et al. (2018). Effect of natural polyphenol on the oxidative stability of pecan oil. *Food and Chemical Toxicology*, *119*, 489–495.

Zhang, J., Zou, X., Zhai, X., Huang, X., Jiang, C., & Holmes, M. (2019). Preparation of an intelligent pH film based on biodegradable polymers and roselle anthocyanins for monitoring pork freshness. *Food Chemistry*, *272*, 306–312.

Zhao, D., & Shah, N. P. (2014). Antiradical and tea polyphenol-stabilizing ability of functional fermented soymilk-tea beverage. *Food Chemistry*, *158*, 262–269.

Zhong, Y. M., & Li, Y. B. (2011). Preservation effects of tea polyphenols in fresh crude juices. *Modern Food Science and Technology*, *27*(6), 4.

Zhou, Y., Xu, T., Zhang, Y., Zhang, C., Lu, Z., Lu, F., et al. (2019). Effect of tea polyphenols on curdlan/chitosan blending film properties and its application to chilled meat preservation. *Coatings*, *9*(4), 262.

Zunin, P., Leardi, R., Bisio, A., Boggia, R., & Romussi, G. (2010). Oxidative stability of virgin olive oil enriched with carnosic acid. *Food Research International*, *43*(5), 1511–1516.

Zwierello, W., Maruszewska, A., Skorka-Majewicz, M., Goschorska, M., Baranowska-Bosiacka, I., Dec, K., et al. (2020). The influence of polyphenols on metabolic disorders caused by compounds released from plastics-Review. *Chemosphere*, *240*, 124901.

CHAPTER THREE

Technological applications of phenolic-rich extracts for the development of non-dairy foods and beverages

Marina Fidelis[a] and Daniel Granato[b,*]

[a]Food Processing and Quality, Production Systems Unit, Natural Resources Institute Finland (Luke), Helsinki, Finland

[b]Department of Biological Sciences, Faculty of Science and Engineering, University of Limerick, Limerick, Ireland

*Corresponding author: e-mail address: granatod@gmail.com

Contents

Abstract

Fruits and other vegetables are sources of bioactive compounds, especially carotenoids, terpenoids, and phenolic compounds. With the focus on sustainability, these compounds' recovery has become a research trend in the last 20 years. However, the correct use of solvents and the steps required to assess the extracts' suitability to be added in food models have been poorly described. Thus, in this review, we attempt to show the pathways and provide guidance on the tailored-made use of solvents for recovering bioactive polyphenolic compounds from food matrices. Special attention is given to the toxicological safety of polyphenol-rich extracts and also their impacts on bioactivity and sensory acceptance of foods and beverages. Practical examples are described and commented on the applications of polyphenol-rich extracts in non-dairy foods and beverages. In summary, the alliance among food science, food technologies, biochemistry, and pharmacology are required to make the development of non-dairy polyphenol-rich foods feasible.

Advances in Food and Nutrition Research, Volume 98
ISSN 1043-4526
https://doi.org/10.1016/bs.afnr.2021.02.006

1. Introduction

In the last 15 years, food companies worldwide have faced a wide variety of challenges, not only in terms of new processing technologies and nutritional aspects, but also the challenges related to the fulfillment of consumers' demands and needs (Linan, Arroyo, & Carrete, 2019). In this scenario, consumers are not only more aware of the nutritional facts and their relations with overall health but also they pay special attention on the impacts of food production on the environment (e.g., organic versus conventional foods) and animal welfare (e.g., intensive production systems). Additionally, more than ever, a considerable portion of consumers has ideological beliefs and perceptions to purchase foods and beverages (Wang, Worsley, & Cunningham, 2008). However, according to Manohar, Rehman, and Sivakumaran (2021), consumers tend to opt for familiar ingredients over foods/beverages with new ingredients. Thus, when developing new products, food companies need to consider not only the physicochemical composition, commercial value, sensory properties, microbiological stability, nutritional content, packaging, and logistics but also ideological beliefs and impacts on the environment.

When it comes to the alliance between food innovation, nutrition, and consumer trends, non-dairy beverages and foods represent an important segment in the food sector. The global non-dairy food/beverage market is driven by growing popularity and demand because of health-related concerns. In fact, the modern generation is becoming more interested in pursuing a healthy lifestyle and the consumption of non-dairy, vegan foods is linked to healthy diets (Costa et al., 2020). Thus, it is not surprising that the growing trend of vegan, vegetarian, flexitarian, and other new consumption habits among the millennials worldwide has been on the rise (Smith, 2019).

The development of healthy foods is a promising task for food companies that need to pay attention to the taste and overall appearance and, more recently, to add nutritional appealing ingredients into different foods (Ferreira, Fernández, Bilbao, & Fernández, 2019). In line with this strategy, many commercial examples on the development of healthy foods are available in different countries: olive oil added with phenolic compounds from olive leaf, peanut butter added with omega-3 fatty acids, chocolate added with extra antioxidants, and yogurts made with tea extracts that contain phenolic compounds. From the commercial point of view, yogurts, cheeses, milk chocolate, milk-based beverages and other dairy foods represent the main types of products with nutritional and functional claims because they are consumed regularly and are perceived as traditional foods (Talaei, Pan, Yuan, & Koh, 2018). However, as mentioned before, non-dairy foods and beverages have gained spaces in all continents, including developing countries (Freire, Ramos, Souza, Cardoso, & Schwan, 2017; Rocha-Guzmán et al., 2019; Tesfaye, Suarez-Lepe, Loira, Palomero, & Morata, 2019). Non-dairy foods and beverages made or added with natural extracts containing bioactive compounds, especially phenolics, represent suitable alternatives to milk-based products (Rodino & Butu, 2019; Voss, Monteiro, Jauregi, Valente, & Pintado, 2021). In fact, some researches have shown that the manufacture of foods and beverages coming from legumes, cereals, seeds, fruit pulps, fruit leaves, and herbs are the main ingredients (Costa et al., 2020; Ivana & Marija, 2019; Martinović, Polak, Ulrih, & Abramovič, 2020; Santana et al., 2019). These materials represent rich sources of phenolic compounds (e.g., phenolic acids, flavonoids, stilbenes, and tannins) that have been linked to multiple health benefits (e.g., antioxidant, anti-inflammatory, anticancer, and antimicrobial effects) when consumed regularly and as a part of the diet (Granato et al., 2020). The familiarity and acceptance by potential consumers represent some positive characteristics of using these sources as ingredients of non-dairy foods, beverages, natural dyes, biofilms, and other value-added applications (de Araújo, de Paulo Farias, Neri-Numa, & Pastore, 2021).

Allying the importance of legumes, cereals, seeds, fruit pulps, fruit leaves, and herbs in the regular diet and considering the health-promoting effects of polyphenols contained in those matrices in humans, the main objectives of this chapter are to describe the latest developments on the field and to incentivize the use of phenolic-rich extracts in the development of potentially functional food products and beverages.

2. Applications of phenolic compounds in non-dairy foods and beverages

2.1 Natural extracts and ingredients sources of polyphenols used in foods and beverages

Consumers have paid attention to what the food label contains, and the list of ingredients have become the target to decide whether the product will be purchased or not. In practice, most worldwide companies have dedicated part of their activities to the development of more sustainable and healthy foods that contain minimal or no use of synthetic chemical compounds, such as stabilizers, coloring agents, or preservatives (Battacchi, Verkerk, Pellegrini, Fogliano, & Steenbekkers, 2020; Peterson, Minton, Liu, & Bartholomew, 2021). In turn, in academia, the application of natural extracts as full replacements for synthetic preservatives is already a reality for dairy, non-dairy, beverages, and meat-based products. Most of the technological applications are related to extracts from plants (e.g., herbs with traditional use, fruit peels and seeds, encapsulated isolated compounds, leaves and steams, among others). For example (Forte et al., 2018) used an aqueous extract from oregano (*Origanum vulgare*) to supplement the feed of broilers to avoid the oxidative reactions of polyunsaturated fatty acids added in the feeds. Consumers linked the addition of oregano extract in the broiler feed as a positive action. Similarly, Damerau et al. (2020) added supercritical CO_2 extracts from different natural materials (e.g., sage, licorice root, temulawak, clove bud, oregano and ajowan fruit) into Baltic herring mince to avoid lipid oxidation and compared the results with synthetic antioxidants (EDTA, ascorbic acid and tocopherol). Results showed that oxidation markers (peroxide value, the formation of volatile oxidation products, and loss of polyunsaturated fatty acids) of the batch added with natural extracts had greater or similar protective effects compared to the synthetic counterpart. Other examples on the use of natural extracts rich in polyphenols in distinct foods models can be found elsewhere (Bellucci, Munekata, Pateiro, Lorenzo, & Barretto, 2021; Chung, Rojanasasithara, Mutilangi, & McClements, 2016; Foffe et al., 2020; Lamothe, Guérette, Dion, Sabik, & Britten, 2019).

Polyphenols are usually extracted using different solvents, such as ethanol, methanol, acetone, water (acidified or not with a food grade acid), ethyl acetate, or a mixture of these solvents (Soto-Covasich et al., 2020; Zanoni, Primiterra, Angeli, & Zoccatelli, 2020). For screening purposes (e.g., phytochemistry-oriented works), the use of methanol, acetone and ethyl acetate is acceptable because they present unique polarity, enabling the extraction of different polyphenols (and other secondary metabolites) for further qualitative and quantitative analysis. For food and feed applications in food industries, their use should be avoided, as it is tough to guarantee their complete elimination from the final extract (liquid or in a dry form).

For technological applications, one should consider the following steps for using natural extracts containing polyphenols in food applications: (i) select materials that come from traditional use (e.g., in the case of herbs) or use materials free from foodborne pathogens; (ii) consider the use of materials in a dry form based on initial experiments to assess the effects of the drying technique on the stability of polyphenols; (iii) use water, ethanol or binary mixtures, acidified or not with food grade organic acids, to extract polyphenols (e.g., phenolic acids, stilbenes, or flavonoids); (iv) post extraction techniques, such as freeze-drying, solvent evaporation under vacuum, spray-drying, or any other technique should be considered to enhance and/or preserve the chemical stability of polyphenols and preserve their bioactivity; (v) consideration of using the crude extract (no further processing) or using (partial) purification of the polyphenol-rich extract by employing microporous adsorbent resins (e.g., Amberlite XAD-7HP, XAD-4, HPD-450, or XAD-16), especially when anthocyanins are the main target of the technological application; (vi) assessing the toxicological safety of the extracts using different normal and non-normal humans cell lines (e.g., IMR-90, HaCaT, HepG2, and Caco-2); (vii) finally, apply in the food model and measure the effect of the polyphenol-rich extract on the quality traits of the final product (Brezoiu et al., 2019; Escher, Marques, et al., 2020; Silva, Castellanos, & Ottens, 2018). Considering the steps to be followed to recover polyphenols and to preserve their bioactivity for further technological utilization in food models, and taking on that water and ethanol are of low toxicity, below we provide a list of examples in which polyphenol-rich extracts were used for the development of different non-dairy foods and beverages.

While some researchers aim to extract and apply different phenolic-rich extracts in distinct food models, others aim to produce ingredients. For instance, considering that the processing of red chicory (*Chicorium intibus*) and red cabbage (*Brassica oleracea*) generates between 40% and 50% of by-products, which are important sources of polyphenols, Zanoni et al. (2020) aimed to produce spray-dried beads from these sources. For this purpose, water and water/ethanol mixtures were compared to extract the polyphenols from red chicory and red cabbage and spray-dried using corn starch as encapsulating material. Total phenolic content (TPC) ranged from 170 to 220 mg gallic acid equivalent (GAE)/100 g of powder, and the final microparticles had antioxidant activity as formally assayed by the ABTS (2,2-azinobis-(3-ethylbenzothiazoline-6-sulfonic acid)) radical scavenging assay. The ingredients had thermal stability, and at different pH values, the color varied from intense red to yellow/green in solution. Thus, the authors concluded that the microencapsulate powders made from by-products of vegetable companies could be effectively used as potential functional ingredients for the innovative foods/beverages manufacture and packaging materials.

To develop a new polyphenol-rich ingredient from agro-industrial side-streams, Papillo et al. (2019) used cocoa hulls as a source of bioactive compounds. Extraction was performed using water and ethanol (50:50 v/v). The extract was spray-dried and characterized by its phenolic composition and antioxidant activity, measured by the DPPH (2,2-diphenyl-1-picrylhydrazyl) free-radical scavenging assay. The powder extract had roughly 10% of phenolics (93 mg/g) and high antioxidant activity (215 mg Trolox equivalent/g). The most remarkable phenolic compounds responsible for the antioxidant activity were epicatechin, catechin, procyanidins B1 and B2, protocatechuic acid, *p*-hydroxybenzoic and gallic acids. When applied to biscuits, foods that are heat-treated at high temperatures, phenolic compounds from the spray-dried extract were partially degraded, showing that the baking process of foods containing polyphenols may be a technological limitation for the manufacture of such products.

Green tea polyphenols (GTP), such as flavanols, are recognized as antioxidant and anti-inflammatory compounds. Aiming to develop an innovative fruit-based alternative rich in GTP, Zokti, Baharin, Abdulkarim, and Abas (2016) extracted *Camellia sinensis* leaves using supercritical CO_2 extraction and microencapsulated the extract using different carbohydrates (e.g., maltodextrin, gum Arabic, and chitosan). After that, different concentrations of the microcapsules (0.5%, 1.0%, and 2.0%) containing catechin concentrations in the range of 40.65 to 63.42 μg/g were added into the mango juice. The antioxidant activity of the GTP-enriched mango juice was dose-dependent, and the results showed stability of such compounds during refrigerated storage for 30 days. No sensory evaluation of the products was made, but the approach seems to be interesting from the beverage technology point of view.

Anthocyanins, stilbenes, and other phenolic compounds from grape pomace (Cabernet Sauvignon and Feteasca Neagra from the Black Sea region) were extracted using ethanol under a microwave-assisted extraction apparatus. The extracts were encapsulated using CM-41 silica and Zn- and Mg-modified MCM-41 silica (Brezoiu et al., 2019). Encapsulated extracts presented a TPC from 210 to 280 mg/g, in which total anthocyanins ranged from 3.4 to 13.1 mg/g. Besides antioxidant activity measured by the DPPH and ABTS assays, the extracts did not show cytotoxicity in the HaCaT cell line (spontaneously transformed aneuploid immortal keratinocyte). Thus, the encapsulation technique allied to the extraction technology applied can be considered to recover bioactive polyphenols from grape pomace and for future high value-added applications in food and cosmetics.

Consumer demand and advances in science and technology are factors responsible for generating new approaches and food innovations. Additionally, health promotion will remain to be at the forefront of new product development strategies. However, health claim credibility, market access specifications, and lack of product awareness among consumers have shown to be barriers to further business expansion of the non-dairy food category (Dey, 2018).

2.2 Bioactivity and safety assessments of natural extracts containing polyphenols

The pharmacological activity and bioactivities of herbs and extracts are highly linked to the content and profile of different chemical compounds that can produce a beneficial effect or toxic to distinct cells and organs (Luo et al., 2020; Zdunić et al., 2020). Indeed, despite the worldwide use of botanical dietary supplements and herbal preparations, robust safety data are limited and should be better studied to understand how these matrices may affect cells and organs (Waidyanatha et al., 2020). Safety data are of paramount importance because they can be compared across studies and used to evaluate the risk of different materials and extracts. When new ingredients are developed, especially those coming from herbs and other materials that are not usually consumed as foods (e.g., tree barks, some fruit peels, and nutshells), both therapeutic and toxicological indices need to be evaluated. Additionally, it is necessary and mandatory to evaluate their safety, which encompasses genotoxic and cytotoxic assessments to detect if the extracts/ingredients have the ability to induce changes in DNA and if they produce any deleterious effects on cells or cellular mechanisms (Ahmadi, Gandomi, Derakhshandeh, Misaghi, & Noori, 2021; Candela et al., 2020). Overall, these ingredients (e.g., extracts produced from those matrices) need to be safe for consumption, and in vitro tests (e.g., cell-based tests using human cells) and in vivo protocols (e.g., acute and (sub)chronic evaluation in mice/rats) (Granato et al., 2020).

Safety testing should range from chemical test tube-based assays, human cell-based assessments (e.g., healthy and cancerous cell lines), small animals' safety testing (e.g., zebra fish), acute and chronic rodent-based protocols (Al-Ghanim et al., 2020; Kiptisia, Maiyoh, Macharia, & Tuei, 2020; Lu et al., 2020). Similarly, bioactivity testing should involve not only chemical assays (e.g., antioxidant and inhibition of metabolic enzymes) but also cell-based assays and in vivo protocols to unveil the mechanisms of action and to comprehensively understand the impact of chemical composition on the safety indices and bioactivity pathways (Croft et al., 2018; Fidelis et al., 2018; Santos et al., 2020; Wang et al., 2020). Another possible way to predict the bioactivity, toxicological and safety indexes of isolated compounds can be obtained by in silico tests, that is, via computer simulation to assess pharmacokinetics, key effects and the underlying mechanisms and their interconnections to endpoints such as gene mutation, carcinogenicity, reproductive toxicity, and developmental toxicity (Hasselgren et al., 2019; Mugaranja & Kulal, 2020; Myatt et al., 2018). For this purpose, a consortium of toxicologists, computational scientists, and regulatory scientists across several industries to support the implementation and acceptance of the in silico approach is assembled (Hyacienth et al., 2020). Then, standardized protocols when conducting toxicity-related predictions should be developed and implemented to be reproducible, consistent, and well-documented across stakeholders (Myatt et al., 2018). In silico protocols have been successfully used to analyze plant-based extract containing phenolics and isolated compounds (Hyacienth et al., 2020; Locatelli et al., 2018; Ziani et al., 2020).

Another issue to be thoroughly considered is the use of extracts that are produced with nontoxic solvents. For instance, extracts and fractions obtained with methanol should not be used in foods whatsoever because it is toxic to humans. Instead, extracts obtained with water, ethanol, and their binary mixtures should be considered as they are not a toxicological issue (Luo et al., 2020; Schechtel et al., 2019). As an example, Schechtel et al. (2019) extracted bioactive compounds from flaxleaf fleabane leaves (*Conyza bonariensis*) using water and ethanol at different proportions (0:100, 25:75, 50:50, 75:25, and 100:0 v/v). The authors analyzed the phenolic composition, in vitro antioxidant capacity, and toxicological and safety testing on different human cells. Data showed the free-ze dried extract obtained with 50% ethanol and 50% water contained ellagic acid, procyanidin A2, caffeic, rosmarinic, gallic, and 2,5-dihydroxybenzoic acids and extracts had antioxidant capacity by scavenging free radicals, by reducing transition metals, and by inhibiting the lipid oxidation of Wistar rat's brain homogenate. Additionally, the freeze-dried extract protected human red blood cells (i.e., erythrocytes) from mechanical hemolysis and did not present any toxic effect on normal human lung cells (IMR90). Authors concluded that the extract, which also contains natural pigments, can be used as a bioactive food ingredient.

To conclude the toxicological, safety, and bioactivity of plant-based extracts and natural ingredients, one should ponder to use water and/or ethanol to extract bioactive compounds (e.g., phenolics) from raw materials, perform a detailed and comprehensive chemical analysis of these compounds and interlink the chemical data with the bioactivity measured by in vitro tests (e.g. antioxidant activity, inhibition of key enzymes, among others) and in vivo data (e.g. toxicity data in different organs/cells, safety indices obtained using distinct animal models—*Caenorhabditis elegans*, brine shrimp (*Artemia salina*), zebra fish, cat, dog, Guinea pig, mouse, and rat). A focus should be given to the mechanistic studies, where possible, to elucidate the route pathways that can turn the extract/ingredient a bioactive or a health-threatening agent. The potential for applying research findings to human and environmental health issues makes cell-based and animal-based protocols scientific valuable alternative models to assess the safety and toxicological events and the bioactivity of these matrices. Although there are ethical issues and method uncertainties about using animals in safety/toxicological testing, animals can be useful models for predicting toxic effects in humans (Saganuwan, 2017).

Considering toxicological data produced by in vitro and initial in vivo testing and the food sector's current scenario, the development of new non-dairy food products is still challenging from the sensory and technological standpoints because of consumer achievement's expectations for healthy, safe, and tasty products. The market of polyphenol-enriched non-dairy foods and beverages is still marginal, but considering the experience in both academia and the food sector, a tentative step-by-step approach to develop non-dairy products containing polyphenols is shown in Fig. 1.

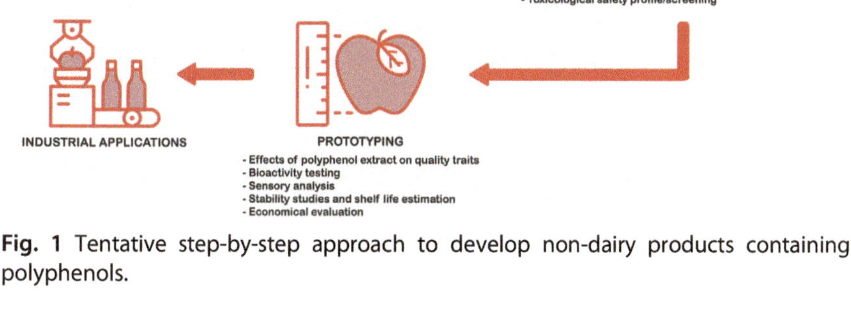

Fig. 1 Tentative step-by-step approach to develop non-dairy products containing polyphenols.

From market trends to industrial applications, some aspects should be considered when developing polyphenol-rich foods and beverages: characterization of the raw material, extraction optimization of polyphenols from the raw material, post-extraction technologies to preserve the bioactivity and chemical stability of polyphenols (encapsulation, lyophilization, etc.), bioactivity and toxicological safety screening of the polyphenol-rich extract, and analysis of its effects on the bioactivity, sensory acceptance, and stability of the food/beverage model.

2.3 Application of polyphenols in non-dairy foods

Nowadays, increasing demand for high-quality products has made the natural products market find healthy sources and substitutes in several food categories. For example, a study proved that it is possible to obtain a low-fat dessert rich in dietary fiber and bioactive substances. Cropotova, Popel, Draganova, Parshakova, and Colesnicenco (2017) replaced dairy fat with inulin, xanthan gum, and sunflower oil in non-dairy creams prepared with white and black seedless grapes varieties. Since earlier studies already proved the metabolic, physiological, and health benefits, this potential non-dairy product could contribute to' individuals who suffer from lactose intolerance, allergy to milk protein, high cholesterol, and obesity. Results concluded that lower sunflower oil concentrations (<15%) and a high amount of grape puree (>80%) positively influence the resistance for creams by raising their apparent viscosity and yield stress. Creams prepared from black seedless grapes presented higher phenolic content measured by Folin-Ciocalteu reducing capacity (FCRC) (1443 versus 1198 mg tannin equivalent/kg) and antioxidant activity (liquid chromatography method coupled with amperometric detection; 0.07 versus 0.16 mg quercetin equivalent QE/g) compared to the white seedless grape. According to the authors, the difference in the phenolic composition is attributable to the grape variety.'

Macho-Gonzalez et al. (2020) included carob fruit extract in meat products, and the results demonstrated an antidiabetic effect in the final product. The authors tested antidiabetic mechanisms of carob fruit extract-enriched meat (CFE-RM) through a late-stage type 2 diabetes mellitus (T2DM) Wistar rats model induced by the action of a high saturated-fat/high-cholesterol diet. It was tested rats fed the CFE-Cholesterol-diet (Chol-diet) since the beginning of the study (ED), and rats fed the CFE-Chol-diet when the diabetic state (DE). Findings revealed that CFE-RM, incorporated into the diet as a preventive procedure (ED) or curative treatment (DE), positively influenced glycemia (283.7 (ED) and 251 mg/dL (DE) versus 326 mg/dL) and enhanced hepatic insulin sensitivity (10.6 (ED) and 8.9 μLU/mL (DE) versus 5.4 μLU/mL) compared to rats fed only the Chol-diet. The study also unveiled improvement in pancreatic β-cell regeneration and an antilipogenic effect by reducing the accumulation of liver lipids and de novo lipogenesis (DNL). In conclusion, CFE-RM can be considered functional food from a meat matrix and a suitable nutritional strategy to prevent or manage late-stage T2DM.

An innovative vegan Camembert cheese analogue from flaxseed oil cake (FOC) was investigated using *Penicillium camemberti* (PC) and *Geotrichum candidum* (GC) as a starter culture. The antioxidant activity, oxidative stability, physicochemical and microbial modifications of a non-dairy alternative product were accessed (Lopusiewicz et al., 2020). The study revealed that combining two starter cultures (*P. camemberti* and *G. candidum*) was responsible for increasing the final vegan product's biological activities and physicochemical properties. During the storage and ripening process, FOC-based products remained stable toward oxidation. The highest TPC was detected in sample PC + GC on day 28 (35.78 mg GAE/g), while the highest TFC was observed in sample PC + GC on day 14 (12.52 mg QE/g). The authors suggested that the fermentation process can influence the enhancement of TPC and TFC content. Both PC and PC + GC samples showed an increase in antioxidant activity at the end of the storage period. The antioxidant activity (DPPH assay), during ripening, ranged from 77.46% to 81.55% for PC and 76.23% to 80.36% for PC + GC samples. At the same time, ABTS ranged from 42.84% to 57.63% for PC and 45.18% to 59.04% for PC + GC. Finally, polyphenols in the flaxseed composition may explain the antioxidant capacity and, consequently, the vegan product's oxidative stability. FOC biotransformation into bioactive non-dairy cheese is a promising technological alternative to develop novel potential functional non-dairy foods due to the bioactivity. Complementary sensory analysis should be conducted to assess the product's acceptability by potential consumers and improve the industrial process.

As non-dairy probiotic products represent good food alternatives for vegetarian and lactose intolerant individuals, Saadiya, Jan, and Sharma (2018) highlighted some non-dairy probiotic products containing different polyphenols developed recently. Even though beverages are the majority of products found in the literature, the authors cited various other food options. Fermented banana, probiotic banana puree, dried fruits, and olive were some examples of previously studied products within the fruit and vegetable-based category. Furthermore, it is possible to find several products using cereal-based, non-fermented soy-based frozen desserts, soy-based products, such as soy yogurt, soy curd, soy product fermented with Kefir, rice-based yogurt, oat, barley, and malt-based products. Besides, even sorghum has already been found as an option to produce yogurt.

Marinho, da Silva, Mazzocato, Tulini, and Favaro-Trindade (2019) used jussara (*Euterpe edulis*) pulp to develop a probiotic sorbet and evaluated the formulations' stability during the storage period. For this purpose, methanol acidified with HCl (85:15 v/v) was used to extract polyphenols from jussara pulp. TPC ranged from 133 to 125 mg GAE/100 g of sorbet during the shelf-life test. Total anthocyanin content ranged from 81 (initial time) to 76 mg/100 g after 120 days of storage. Moreover, the authors reported that the *probiotics Lactobacillus acidophilus* and *Lactobacillus paracasei* remained viable during the storage time, reaching near 9 log colony-forming unit (CFU)/g. It was possible to combine the beneficial health effects of polyphenols from jussara with the benefits of probiotics. In summary, the study confirmed that the jussara pulp could be applied in the manufacture of a non-dairy sorbet.

New techniques have been applied to improve and optimize processes or products in the food sector. In this aspect, a study recently published a development process using nanotechnology to produce rice bran oil and soy protein nanoemulsion to obtain a non-dairy yogurt with functional properties (Sengupta, Bhattacharyya, Goswami, & Bhowal, 2019). The main objective of the approach' was to evaluate the applicability and the nanoemulsion effects on the physical and microbiological properties, antioxidant activity, oxidative stability, and textural properties of the soy yogurts. Data showed that the yogurt formulated with nanoemulsion (SPNY) provided higher radical scavenging ability (DPPH assay; 11% versus 34% of inhibition) and ferric reducing antioxidant property (FRAP assay; 0.96 versus 2.20 mg/kg) compared to the control yogurt (SYC). In addition, SPNY had higher TPC (1 versus 5.3 g/kg) than SYC during the storage period of 45 days. Finally, fortified soy yogurt had significantly higher oxidative stability. Thus, rice bran oil and soy protein nanoemulsion could be potential sources for the development of potentially functional soy-based yogurts.

The sensory appeal, nutritional, techno-functional qualities, and health-promotion properties focus on new non-dairy polyphenol-enriched foods. In this sense, Cáceres, Peñas, Martínez-Villaluenga, García-Mora, and Frías (2019) developed a multifunctional yogurt-like product from germinated brown rice. Results showed that the germination and fermentation processes influenced the bioactivities and proximal composition. Fermented formulations obtained from raw brown rice after 48 h of germination were sensory accepted and exhibited a higher TPC (9.81 versus 12.41 mg GAE/100 g), angiotensin I-converting enzyme (25.17% versus 44.04%) in comparison to non-fermented counterparts. On the other hand, after 96 h of germination, fermented products demonstrated a slight and non-significant decrease in the TPC (15.20 versus 16.19 mg GAE/100 g) and oxygen radical absorbance capacity (ORAC) values (46.9 versus 49.48 μg TE/100 g) values compared to non-fermented formulations. Germinated brown rice showed an increase in the angiotensin I-converting enzyme inhibitory activity (ACE) (61.47% versus 24.62%) compared to soaked brown rice samples. Finally, although there is a difference between types of products, germinated brown rice demonstrated overall satisfactory bioactivity results, with a higher TPC, antioxidant, and antihypertensive activities than raw and soaked brown rice, proving to be an attractive and viable substrate for non-dairy yogurt-like foods/beverages.

Kemsawasd and Chaikham (2020) evaluated the effects of frozen storage' on the probiotics viability and antioxidant capacity of riceberry and sesame-riceberry milk ice creams. As a tentative to bring synbiotic effects, initially, dried riceberry (*Oryza sativa* L.) and black sesame seeds (*Sesamum orientale* L.) added with riceberry (ratio 1:5, w/w) were combined with boiling water (1:5, w/v) and extracted employing a soybean milk extractor. In the development stage, ice cream formulations were supplemented with probiotics (*Lactobacillus casei* 01 and *Lactobacillus acidophilus* LA5) and inulin. Although combinations of prebiotics have been considered useful in enhancing probiotic ice cream quality, the reduction in pH followed by an increase in total acidity over prolonged storage decreased color and flavor scores of the ice cream. Additionally, findings revealed a decline in antioxidant levels during storage for all combinations. For example, considering samples with mixed riceberry and sesame-riceberry, probiotics (*Lactobacillus* species), and inulin, the antioxidant activity measured by the DPPH assay decreased from 55% to 48% of inhibition. The same trend was observed for the FRAP assay (44 to 10 mmol/L FeSO4/100 g, approximately).

2.4 Application of polyphenol-rich extracts in beverages

Flowers have emerged as potential raw materials for extracting different bioactive compounds, mainly phenolic acids and flavonoids, which are recognized antimicrobial and antioxidant agents. Following this trend, Escher et al. (2018) used water acidified with acetic acid (0.545 mol/L, pH 2.5) to extract anthocyanins from cornflower (*Centaurea cyanus*) petals using different extraction times and temperatures. The main objective of the approach was to obtain the optimal extraction conditions to develop a model beverage rich in anthocyanins. The optimal conditions (extraction at 68.3 °C/51.2 min) were used to obtain a model beverage containing bioactive compounds (anthocyanins and other flavonoids and phenolic acids). Authors concluded the freeze-dried extract inhibited the lipid oxidation of egg yolk homogenate, presented antioxidant capacity and inhibited the ACE activity, protected human erythrocytes against hemolysis and did not present toxic effects in HepG2, Caco-2 and A549 cell lines. The optimal extract also presented reversibility of anthocyanins when pH was changed from 2.79 to 7.01, with a slight decrease in the DPPH radical.

Not only flower petals have been used as raw materials to develop model beverages rich in phenolic compounds but also culinary herbs and edible vegetables. Migliorini et al. (2019) varied the extraction time (5–34 min) and temperature (21–79 °C) to obtain a phenolic-rich extract source of anthocyanins from red chicory (*Chicorium intybus*) leaves. Water acidified with acetic acid (pH 2.5) was used in the extractions. Anthocyanins ranged from 43.7 to 73 mg/100 g, while total flavonoids varied from 48 to 93 mg/100 g, and all extracts presented free-radical scavenging activity toward DPPH radical. Red chicory extract was exposed to different temperatures (60 to 100 °C) and the half-life of anthocyanins was higher at 60 °C (74.2 h) than at 70 °C (17.2 h) or 100 °C (2.2 h). The thermal stability of phenolic compounds is an important technological factor when beverages are manufactured. As thermal processes partially degrade these compounds, especially anthocyanins, compounds confer a red/violet color to liquids. When the red chicory was exposed to different pH values (2.5–5.5) using a reverse titration, anthocyanins were shown to be poorly reversible (38.6%), which may impact in absorption and bioaccessibility of these compounds in humans.

Although agro-industrial by-products for further utilization as ingredients by food companies are not regulated worldwide, experimental research has shown massive potential for some materials. For example, Leonarski et al. (2021) produced a kombucha-like beverage adding acerola (*Malpighia emarginata*) by-product, which contains a significant amount of ascorbic acid and phenolic compounds. The acerola by-product was used at different concentrations (1–5% w/v) and added to the formulation basis (green tea added with glucose and fructose). Beverages were assessed for their ascorbic acid content, TPC, and antioxidant activity (DPPH). The TPC ranged from 200 to 1400 mg gallic acid equivalent per liter, and the DPPH ranged from 2 to 6 mmol Trolox equivalent per liter, showing a dose-dependency. Although no sensory evaluation of the beverages was performed using potential consumers as panelists, the authors concluded that the acerola by-product proved to be a promising raw material for producing kombucha-like beverages.

Corn, also known as maize, is a grain plant cultivated for food. Its production reached 1.2 billion metric tons in which the USA, Brazil, and China are the largest producers, totaling 50% of the worldwide amount. During maize processing, various by-products are produced, but little is known on how to take advantage of the by-products for the development of foods and ingredients. Trying to bring a new perspective in the actual scenario, Buitimea-Cantúa, Antunes-Ricardo, Villela-Castrejón, and Gutiérrez-Uribe (2019) used white dent maize (*Zea mays* L.) to develop beverages enriched with nejayote solids (a watery by-product from the maize processing that contains phenolic compounds, organic acids, and calcium). Different concentrations (3%, 6%, or 9%, w/w) of nejayote were used to produce the beverages, and the physicochemical characterization, analysis of phenolic acids, and cellular-based antioxidant activity in Caco-2 cells (immortalized *cell* line of human colorectal adenocarcinoma *cells*) and the anti-inflammatory activity was tested using a murine macrophage cell line (RAW 264.7). Higher nejayote concentrations increased the concentration of calcium, ferulic, *p*-coumaric, dehydro-diferulic, and dehydro-triferulic acids. Similarly, higher nejayote concentrations were associated with lower intracellular production of reactive oxygen species, ROS (thus higher cellular antioxidant activity) and higher inhibition of nitrix oxide on RAW 264.7 cells (thus, higher anti-inflammatory activity). In fact, compared to the control beverage, the anti-inflammatory potential of the beverages added with the phenolic-rich by-product was increased in 16.7, 27.3, and 55.8-times with the addition of 3%, 6%, and 9% of nejayote solids, respectively.

In the same trend, considering that brewers' spent grains (BSG) are generated as waste side streams during the beer production process, where global yield is estimated at up to 40×10^6 tons per year. Tan, Mok, and Chen (2020) developed a beverage using submerged fermentation with *Bacillus subtilis* WX-17. The fermentation of BSG was able to release leucine, valine, threonine, phenylamine, methionine, lysine, and tyrosine and increase the TPC (447 versus 126 μg GAE/mL) and antioxidant activity (DPPH assay;

7.31 versus 3.51 μg Trolox equivalent/mL) compared to the unfermented BSG sample. Complementary sensory analysis should be conducted to assess the product's acceptability by potential consumers.

Using a similar approach, Escher, Wen, Zhang, Rosso, and Granato (2020) produced an aqueous extract from the petals of butterfly pea flower (*Clitoria ternatea*), which has been proven to be a toxicological safe material that presented many bioactivities, such as antioxidant, antimicrobial, antidiabetic and antihypertensive potential (Escher, Wen, et al., 2020). The aqueous extract, a source of special anthocyanins called ternatins, showed stability in relation to pH (e.g., 2.2–10.2), providing different colors at different pH (from blue to green). In addition, when fructooligosaccharides were added, the aqueous extract exhibited thermal stability (60–100 °C) and photostability. Considering the complementary work of food chemistry, biochemistry and toxicology, authors concluded that ternatin anthocyanins extracted from butterfly pea petals have the potential to be used as natural colorants in beverage models rich in phenolic compounds.

Rodríguez et al. (2020) used *Prosopis alba* pods, popularly known as alfarrobo blanco, to develop two traditional non-dairy beverages—aloja (fermented) and anapa (unfermented)—rich in polyphenols, especially *C*-glycosidic flavonoids and *O*-flavonol glycosides. Authors also assessed the toxicity of the beverages using the brine shrimp assay, and a consumer test was undertaken to evaluate the products' acceptability. Data showed that aloja had a TPC of 150–180 mg/L while anapa had only 21 mg/L, showing that the fermentation process increased the beverage's bioactive compounds considerably. Vicenin II and vitexin, two apigenin *C*-glycoside compounds, together with quercetin and cinnamic acid were detected in the beverages, and the antioxidant activity measured by the ABTS and oxidation of β-carotene methods showed that anapa was more effective than aloja, which shows that, in this specific case, TPC was not correlated with higher antioxidant activity. The toxicity evaluation also showed that the lethal dose (LD_{50}) was between 2500 and 8000 mg/kg, which is far lower than the international health organizations.

Other fermented beverages made from sugarcane (*Saccharum officinarum*) molasses were developed by Samaniego-Sánchez, Marín-García, and Quesada-Granados (2020). Authors analyzed their TPC, flavonoids, tannins, and antioxidant activity using DPPH, FRAP, and ABTS assays and verified that the sugarcane fermented beverages were not a good source of TPC (20–140 mg GAE/L), flavonoids (2–57 mg catechin equivalent (CTE)/L), and tannins (1–16 mg CTE/L). The antioxidant activity measured by the three different methods presented low values (2–80 mmol Trolox equivalent per liter). These values are much lower than those usually found in other alcoholic beverages, such as beer and wine.

Arya and Shakya (2021) developed a multigrain functional beverage using minor millets such as barnyard, foxtail, and kodo millet added with fructooligosaccharides galactooligosaccharides as sources of dietary fibers. For this purpose, a design of experiment was used to optimize the formulation based on the pH, total phenolic content, antioxidant activity (DPPH), and overall acceptability of the beverages individually as a function of the three grains. Results showed that TPC of all the 16 multigrain beverages ranged between 40.03 and 59.27 mg ferulic acid equivalent/100 mL, and the inhibition of DPPH was very modest (20–30%). The multigrain formulation's optimized ratio was 7.00:10.00:8.00 for barnyard, foxtail, and kodo millet, respectively, and this sample presented TPC of 47.7 mg/100 mL and inhibited 25.7% of DPPH. The optimal multigrain sample was shown to have prebiotic activity, low glycemic index, and suitable sensory properties.

Another possible strategy to produce beverages rich in phenolic compounds is by adding high-purity extracts or isolated compounds directly to the beverage of interest. This approach's feasibility may be questioned because of financial reasons. Maybe in the near future, upscaling phenolic compounds' production may be incentivized to decrease the market value of such bioactives. Considering the technological innovation and the scientific gap in the beverage sector, Xu, Gao, and Granato (2021) used a simplex centroid design to optimize a mixture of epigallocatechin gallate (EGCG), epigallocatechin (EGC), and epicatechin gallate (ECG) based on the chemical and cell-based antioxidant activity. For this purpose, the authors measured the Cu^{2+} chelating ability, cupric ion reducing antioxidant capacity (CUPRAC), ferric ion antioxidant power (FRAP), Folin-Ciocalteu reducing capacity (FCRC), and the free-radical scavenging activity toward DPPH. Cell-based antioxidant activity was measured using OVCAR-3 (human ovarian carcinoma), HEK293 (human embryonic kidney cells), and HFL1 (human fetal lung fibroblast) cell lines by inducing the formation of ROS via stimulation with H_2O_2. Results showed that the optimal polyphenol mixture should be composed of 67.4% ECG and 32.6% EGCG. This combination of phenolic compounds was added in a catechin-free beverage model composed of chrysanthemum tea (CS), and there was a sharp increase in total phenolic content (2.15-fold), DPPH (3.17-fold), FRAP (3.39-fold), and FCRC (2.04-fold). Interestingly, the total phenolic content and antioxidant activity of CS was unaffected by slow pasteurization conducted at 63 °C/30 min. CS added with the optimal combination of phenolic compounds had a more robust perception of astringency compared to the control CS. Thus, when developing beverages fortified with phenolic compounds, not only the antioxidant activity should be considered but also the sensory properties of the final product.

Although some genetic efforts have been made to improve the sensory attributes of soy and derived products (e.g., water-soluble extract and tofu), some consumers tend to reject soy-based foods. Voss et al. (2021) took advantage of an agro-industrial side-stream, okara, to manufacture a non-dairy potentially functional beverage rich in isoflavones, which are flavonoids commonly found in soy. Fermented beverages were produced using *Lactobacillus rhamnosus* and *Bifidobacterium animalis* spp. *lactis* and okara previously submitted to enzymatic hydrolysis (proteases). Beverages contained a TPC of 20–30 mg gallic acid equivalent per 100 mL, in which daidzein and genistein were the major isoflavones found. The antioxidant activity using the ABTS assay of the beverages was between 5 and 10 mg Trolox equivalent per 100 mL. TPC and the antioxidant activity were roughly stable over refrigerated storage (28 days).

Additionally, okara fermented beverages presented inhibition of angiotensin I-converting enzyme after a simulated digestion process, which implies potential functional properties. Although interesting results were obtained from the functionality standpoint, the beverages' sensory analysis showed the consumers did not find the beverages appealing as hedonic scores ranged from "dislike moderately" to "neither like nor dislike". This is a clear example that the prototypes' sensory analysis should always be assessed when developing more nutritious products. In addition, bioactivity and any alleged functional effects should always be allied to suitable sensory acceptance.

3. Concluding remarks and trends

Non-dairy products brought global attention as sources of nutrients and bioactive compounds for individuals who look for vegetarian, lactose-free foods and beverages. Because of continuous advertising and combining marketing projects, notable consumer awareness has been created for many products. The association between the consumption of non-dairy foods and phenolic compounds with potential health benefits can also be considered a decision-making factor for such a trend. Besides the importance of studying the health benefits that non-dairy products can provide, factors like stability in the manufacturing process and storage, sensory attributes, consumer attraction, and price also play a crucial role in the product's success at commercial levels. In this chapter, the most essential technological aspects for developing non-dairy foods added with polyphenol-rich extracts were comprehensively described and explained. Most importantly, for new non-dairy foods containing natural polyphenol rich extracts, not only acceptance, shelf life, and technological issues should be considered but also the toxicological safety and the effects on the extracts on human's cells. In addition, any health claim and efficacy should be supported by comprehensive in vivo data, such as clinical trials following the recommendation of international health agencies.

References

Ahmadi, A., Gandomi, H., Derakhshandeh, A., Misaghi, A., & Noori, N. (2021). Phytochemical composition and in vitro safety evaluation of *Ziziphora clinopodioides* lam. ethanolic extract: Cytotoxicity, genotoxicity and mutagenicity assessment. *Journal of Ethnopharmacology*, *266*, 113428.

Al-Ghanim, K. A., Mahboob, S., Vijayaraghavan, P., Al-Misned, F. A., Kim, Y. O., & Kim, H. (2020). Sub-lethal effect of synthetic pyrethroid pesticide on metabolic enzymes and protein profile of non-target zebra fish, Danio rerio. *Saudi Journal of Biological Sciences*, *27*(1), 441–447.

Arya, S. S., & Shakya, N. K. (2021). High fiber, low glycaemic index (GI) prebiotic multigrain functional beverage from barnyard, foxtail and kodo millet. *LWT-Food Science and Technology*, *135*, 109991. https://doi.org/10.1016/j.lwt.2020.109991. In this issue.

Battacchi, D., Verkerk, R., Pellegrini, N., Fogliano, V., & Steenbekkers, B. (2020). The state of the art of food ingredients' naturalness evaluation: A review of proposed approaches and their relation with consumer trends. *Trends in Food Science & Technology*, *106*, 434–444.

Bellucci, E. R. B., Munekata, P. E. S., Pateiro, M., Lorenzo, J. M., & Barretto, A. C. S. (2021). Red pitaya extract as natural antioxidant in pork patties with total replacement of animal fat. *Meat Science*, *171*, 108284.

Brezoiu, A., Matei, C., Deaconu, M., Stanciuc, A., Trifan, A., Gaspar-Pintiliescu, A., et al. (2019). Polyphenols extract from grape pomace. characterization and valorisation through encapsulation into mesoporous silica-type matrices. *Food and Chemical Toxicology*, *133*, 110787.

Buitimea-Cantúa, N. E., Antunes-Ricardo, M., Villela-Castrejón, J., & Gutiérrez-Uribe, J. A. (2019). Changes in cellular antioxidant and anti-inflammatory activity after 12 months storage of roasted maize-based beverages supplemented with nejayote solids. *Journal of Cereal Science*, *89*, 102807.

Cáceres, P. J., Peñas, E., Martínez-Villaluenga, C., García-Mora, P., & Frías, J. (2019). Development of a multifunctional yogurt-like product from germinated brown rice. *LWT- Food Science and Technology*, *99*, 306–312.

Candela, F. M., Giordano, W. F., Quiroga, P. L., Escobar, F. M., Mañas, F., Roma, D. A., et al. (2020). Evaluation of cellular safety and the chemical composition of the peanut (Arachis hypogaea L.) ethanolic extracts. *Heliyon*, *6*(10), e05119.

Chung, C., Rojanasasithara, T., Mutilangi, W., & McClements, D. J. (2016). Stabilization of natural colors and nutraceuticals: Inhibition of anthocyanin degradation in model beverages using polyphenols. *Food Chemistry*, *212*, 596–603.

Costa, N.d. A., de Almeida Paula, D., Brêda, J. D., Rufino Vieira, É. N., Furtado Martins, E. M., & Ramos, A. M. (2020). A symbiotic dessert composed of yam (*Dioscorea* sp.) and ubá mango pulp (*Mangifera indica* L.). *LWT- Food Science and Technology*, *133*, 110074.

Croft, K. D., Yamashita, Y., O'Donoghue, H., Shirasaya, D., Ward, N. C., & Ashida, H. (2018). Screening plant derived dietary phenolic compounds for bioactivity related to cardiovascular disease. *Fitoterapia*, *126*, 22–28.

Cropotova, J., Popel, S., Draganova, E., Parshakova, L., & Colesnicenco, A. (2017). The use of seedless grape varieties for developing dairy-free creams. *Annals of the University Dunarea De Jos of Galati Fascicle VI: Food Technology*, *41*, 127–140.

Damerau, A., Kakko, T., Tian, Y., Tuomasjukka, S., Sandell, M., Hopia, A., et al. (2020). Effect of supercritical CO2 plant extract and berry press cakes on stability and consumer acceptance of frozen Baltic herring (*Clupea harengus membras*) mince. *Food Chemistry*, *332*, 127385.

de Araújo, F. F., de Paulo Farias, D., Neri-Numa, I. A., & Pastore, G. M. (2021). Polyphenols and their applications: An approach in food chemistry and innovation potential. *Food Chemistry*, *338*, 127535.

Dey, G. (2018). Non-dairy probiotic foods: Innovations and market trends. In S. K. Panda, & P. H. Shetty (Eds.), *Innovations in technologies for fermented food and beverage industries* (pp. 159–173). Springer International Publishing.

Escher, G. B., Marques, M. B., do Carmo, M. A. V., Azevedo, L., Furtado, M. M., Sant'Ana, A. S., et al. (2020). *Clitoria ternatea* L. petal bioactive compounds display antioxidant, antihemolytic and antihypertensive effects, inhibit α-amylase and α-glucosidase activities and reduce human LDL cholesterol and DNA induced oxidation. *Food Research International*, *128*, 108763a.

Escher, G. B., Santos, J. S., Rosso, N. D., Marques, M. B., Azevedo, L., do Carmo, M. A. V., et al. (2018). Chemical study, antioxidant, antihypertensive, and cytotoxic/cytoprotective activities of *Centaurea cyanus* L. petals aqueous extract. *Food and Chemical Toxicology*, *118*, 439–453.

Escher, G. B., Wen, M., Zhang, L., Rosso, N. D., & Granato, D. (2020). Phenolic composition by UHPLC-Q-TOF-MS/MS and stability of anthocyanins from *Clitoria ternatea* L. (butterfly pea) blue petals. *Food Chemistry*, *331*, 127341b.

Ferreira, S. C., Fernández, A. M., Bilbao, M. D.d. C., & Fernández, A. M. (2019). New functional ingredients from agroindustrial by-products for the development of healthy foods. In P. Ferranti, E. M. Berry, & J. R. Anderson (Eds.), *Encyclopedia of food security and sustainability* (pp. 351–359). Oxford: Elsevier.

Fidelis, M., Santos, J. S., Escher, G. B., Vieira do Carmo, M., Azevedo, L., Cristina da Silva, M., et al. (2018). *In vitro* antioxidant and antihypertensive compounds from camu-camu (*Myrciaria dubia* McVaugh, Myrtaceae) seed coat: A multivariate structure-activity study. *Food and Chemical Toxicology*, *120*, 479–490.

Foffe, H. A. K., Ndomou Houketchang, S. C., Djikeng, F. T., Teboukeu, G. B., Tsopmo, A., & Womeni, H. M. (2020). Effect of *Syzigium aromaticum* and *Allium sativum* spice extract powders on the lipid quality of groundnuts (*Arachis hypogaea*) pudding during steam cooking. *Heliyon*, *6*, 05166.

Forte, C., Branciari, R., Pacetti, D., Miraglia, D., Ranucci, D., Acuti, G., et al. (2018). Dietary oregano (*Origanum vulgare* L.) aqueous extract improves oxidative stability and consumer acceptance of meat enriched with CLA and n-3 PUFA in broilers. *Poultry Science*, *97*, 1774–1785.

Freire, A. L., Ramos, C. L., Souza, P. N. C., Cardoso, M. G. B., & Schwan, R. F. (2017). Non-dairy beverage produced by controlled fermentation with potential probiotic starter cultures of lactic acid bacteria and yeast. *International Journal of Food Microbiology, 248*, 39–46.

Granato, D., Barba, F. J., Bursac Kovacevic, D., Lorenzo, J. M., Cruz, A. G., & Putnik, P. (2020). Functional foods: Product development, technological trends, efficacy testing, and safety. *Annual Review of Food Science and Technology, 11*, 93–118.

Hasselgren, C., Ahlberg, E., Akahori, Y., Amberg, A., Anger, L. T., Atienzar, F., et al. (2019). Genetic toxicology in silico protocol. *Regulatory Toxicology and Pharmacology, 107*, 104403.

Hyacienth, B. M.d. S., Picanço, K. R. T., Sánchez-Ortiz, B. L., Silva, L. B., Pereira, A. C. M., Góes, L. D. M., et al. (2020). Hydroethanolic extract from *Endopleura uchi* (Huber) cuatrecasas and its marker bergenin: Toxicological and pharmacokinetic studies *in silico* and *in vivo* on zebrafish. *Toxicology Reports*, 7, 217–232.

Ivana, B., & Marija, L. (2019). 8—Role of phenols in energy and functional beverages. In A. M. Grumezescu, & A. M. Holban (Eds.), *Sports and energy drinks* (pp. 229–268). Woodhead Publishing.

Kemsawasd, V., & Chaikham, P. (2020). Effects of frozen storage on viability of probiotics and antioxidant capacities of synbiotic riceberry and sesame-riceberry milk ice creams. *Current Research in Nutrition and Food Science, 8*(1), 107–121.

Kiptisia, R. T., Maiyoh, G. K., Macharia, B. N., & Tuei, V. C. (2020). Safety evaluation and antihyperglycaemic effect of root extract of *Maerua decumbens* (Brongn.) DeWolf in Wistar rats. *Scientific African, 8*, e00379.

Lamothe, S., Guérette, C., Dion, F., Sabik, H., & Britten, M. (2019). Antioxidant activity of milk and polyphenol-rich beverages during simulated gastrointestinal digestion of linseed oil emulsions. *Food Research International, 122*, 149–156.

Leonarski, E., Cesca, K., Zanella, E., Stambuk, B. U., de Oliveira, D., & Poletto, P. (2021). Production of kombucha-like beverage and bacterial cellulose by acerola by-product as raw material. *LWT- Food Science and Technology, 135*, 110075.

Linan, J., Arroyo, P., & Carrete, L. (2019). Conceptualizing healthy food: How consumer's values influence the perceived healthiness of a food product. *Journal of Food and Nutrition Research*, 7(9), 679–687.

Locatelli, M., Yerlikaya, S., Baloglu, M. C., Zengin, G., Altunoglu, Y. C., Cacciagrano, F., et al. (2018). Investigations into the therapeutic potential of *Asphodeline liburnica* roots: *In vitro* and *in silico* biochemical and toxicological perspectives. *Food and Chemical Toxicology, 120*, 172–182.

Lopusiewicz, L., Drozlowska, E., Tarnowiecka-Kuca, A., Bartkowiak, A., Mazurkiewicz-Zapalowicz, K., & Salachna, P. (2020). Biotransformation of flaxseed oil cake into bioactive camembert-analogue using lactic acid bacteria, *Penicillium camemberti* and *Geotrichum candidum*. *Microorganisms, 8*(9), 1266.

Lu, K., Ou, G., Chang, H., Chen, W., Liu, S., & Sheen, L. (2020). Safety evaluation of water extract of gastrodia elata blume: Genotoxicity and 28-day oral toxicity studies. *Regulatory Toxicology and Pharmacology, 114*, 104657.

Luo, X., Wu, J., Li, Z., Jin, W., Zhang, F., Sun, H., et al. (2020). Safety evaluation of *Eucommia ulmoides* extract. *Regulatory Toxicology and Pharmacology, 118*, 104811.

Macho-Gonzalez, A., Lopez-Oliva, M. E., Merino, J. J., Garcia-Fernandez, R. A., Garcimartin, A., Redondo-Castillejo, R., et al. (2020). Carob fruit extract-enriched meat improves pancreatic beta-cell dysfunction, hepatic insulin signaling and lipogenesis in late-stage type 2 diabetes mellitus model. *The Journal of Nutritional Biochemistry, 84*, 108461.

Manohar, S., Rehman, V., & Sivakumaran, B. (2021). Role of unfamiliarity and information on consumers' willingness to try new healthy foods. *Food Quality and Preference, 87*, 104037.

Marinho, J. F. U., da Silva, M. P., Mazzocato, M. C., Tulini, F. L., & Favaro-Trindade, C. S. (2019). Probiotic and synbiotic sorbets produced with Jussara (*Euterpe edulis*) pulp: Evaluation throughout the storage period and effect of the matrix on probiotics exposed to simulated gastrointestinal fluids. *Probiotics and Antimicrobial Proteins*, *11*(1), 264–272.

Martinović, N., Polak, T., Ulrih, N. P., & Abramovič, H. (2020). Mustard seed: Phenolic composition and effects on lipid oxidation in oil, oil-in-water emulsion and oleogel. *Industrial Crops and Products*, *156*, 112851.

Migliorini, A. A., Piroski, C. S., Daniel, T. G., Cruz, T. M., Escher, G. B., Vieira do Carmo, M. A., et al. (2019). Red chicory (*Cichorium intybus*) extract rich in anthocyanins: Chemical stability, antioxidant activity, and antiproliferative activity in vitro. *Journal of Food Science*, *84*(5), 990–1001.

Mugaranja, K. P., & Kulal, A. (2020). Alpha glucosidase inhibition activity of phenolic fraction from *Simarouba glauca*: An in-vitro, in-silico and kinetic study. *Heliyon*, *6*, e04392.

Myatt, G. J., Ahlberg, E., Akahori, Y., Allen, D., Amberg, A., Anger, L. T., et al. (2018). In silico toxicology protocols. *Regulatory Toxicology and Pharmacology*, *96*, 1–17.

Papillo, V. A., Locatelli, M., Travaglia, F., Bordiga, M., Garino, C., Coïsson, J. D., et al. (2019). Cocoa hulls polyphenols stabilized by microencapsulation as functional ingredient for bakery applications. *Food Research International*, *115*, 511–518.

Peterson, M., Minton, E. A., Liu, R. L., & Bartholomew, D. E. (2021). Sustainable marketing and consumer support for sustainable businesses. *Sustainable Production and Consumption*, *27*, 157–168.

Rocha-Guzmán, N. E., González-Laredo, R. F., Vázquez-Cabral, B. D., Moreno-Jiménez, M. R., Gallegos-Infante, J. A., Gamboa-Gómez, C. I., et al. (2019). Oak leaves as a new potential source for functional beverages: Their antioxidant capacity and monomer flavonoid composition. In A. M. Grumezescu, & A. M. Holban (Eds.), *Functional and medicinal beverages* (pp. 381–411). Academic Press.

Rodino, S., Butu, M., Grumezescu, A. M., & Holban, A. M. (2019). Herbal extracts—New trends in functional and medicinal beverages. In *Functional and medicinal beverages* (pp. 73–108). Academic Press.

Rodríguez, I. F., Cattaneo, F., Zech, X. V., Svavh, E., Pérez, M. J., Zampini, I. C., et al. (2020). Aloja and añapa, two traditional beverages obtained from *Prosopis alba* pods: Nutritional and functional characterization. *Food Bioscience*, *35*, 100546.

Saadiya, A. R., Jan, T., & Sharma, P. (2018). Non-dairy probiotics—An emerging trend in health care products. *International Journal of Current Microbiology and Applied Sciences*, *7*, 131–145.

Saganuwan, S. (2017). Toxicity studies of drugs and chemicals in animals: An overview. *Bulgarian Journal of Veterinary Medicine*, *20*, 291–318.

Samaniego-Sánchez, C., Marín-García, G., & Quesada-Granados, J. J. (2020). A new fermented beverage from sugarcane (*Saccharum officinarum* L.) molasses: Analysis of physicochemical properties and antioxidant capacity, and comparison with other industrial alcohol products. *LWT- Food Science and Technology*, *128*, 109505.

Santana, I., Castelo-Branco, V. N., Guimarães, B. M., Silva, L.d. O., Peixoto, V. O. D. S., Cabral, L. M. C., et al. (2019). Hass avocado (*Persea americana* Mill.) oil enriched in phenolic compounds and tocopherols by expeller-pressing the unpeeled microwave dried fruit. *Food Chemistry*, *286*, 354–361.

Santos, J. S., Escher, G. B., do Carmo, M. C., Azevedo, L., Boscacci Marques, M., Daguer, H., et al. (2020). A new analytical concept based on chemistry and toxicology for herbal extracts analysis: From phenolic composition to bioactivity. *Food Research International*, *132*, 109090.

Schechtel, S. L., de Matos, V. C. R., Santos, J. S., Cruz, T. M., Marques, M. B., Wen, M., et al. (2019). Flaxleaf fleabane leaves (*Conyza bonariensis*), A new functional nonconventional edible plant? *Journal of Food Science*, *84*(12), 3473–3482.

Sengupta, S., Bhattacharyya, D. K., Goswami, R., & Bhowal, J. (2019). Emulsions stabilized by soy protein nanoparticles as potential functional non-dairy yogurts. *Journal of the Science of Food and Agriculture*, *99*(13), 5808–5818.

Silva, M., Castellanos, L., & Ottens, M. (2018). Capture and purification of polyphenols using functionalized hydrophobic resins. *Industrial & Engineering Chemistry Research*, *57*(15), 5359–5369.

Smith, S. (2019). *Global non-dairy milk market to reach values of $38 billion during the period 2018–2024- market research by Arizton*. PR Newswire.

Soto-Covasich, J., Reyes-Farias, M., Torres, R. F., Vasquez, K., Duarte, L., Quezada, J., et al. (2020). A polyphenol-rich Calafate (*Berberis microphylla*) extract rescues glucose tolerance in mice fed with cafeteria diet. *Journal of Functional Foods*, *67*, 103856.

Talaei, M., Pan, A., Yuan, J., & Koh, W. (2018). Dairy intake and risk of type 2 diabetes. *Clinical Nutrition*, *37*(2), 712–718.

Tan, Y. X., Mok, W. K., & Chen, W. N. (2020). Potential novel nutritional beverage using submerged fermentation with *Bacillus subtilis* WX-17 on brewers' spent grains. *Heliyon*, *6*(6), e04155.

Tesfaye, W., Suarez-Lepe, J. A., Loira, I., Palomero, F., & Morata, A. (2019). Dairy and nondairy-based beverages as a vehicle for probiotics, prebiotics, and symbiotics: Alternatives to health versus disease binomial approach through food. In A. M. Grumezescu, & A. M. Holban (Eds.), *Milk-based beverages* (pp. 473–520). Woodhead Publishing.

Voss, G. B., Monteiro, M. J. P., Jauregi, P., Valente, L. M. P., & Pintado, M. E. (2021). Functional characterisation and sensory evaluation of a novel synbiotic okara beverage. *Food Chemistry*, *340*, 127793.

Waidyanatha, S., Pierfelice, J., Cristy, T., Mutlu, E., Burback, B., Rider, C. V., et al. (2020). A strategy for test article selection and phytochemical characterization of *Echinacea purpurea* extract for safety testing. *Food and Chemical Toxicology*, *137*, 111125.

Wang, W. C., Worsley, A., & Cunningham, E. G. (2008). Social ideological influences on reported food consumption and BMI. *The International Journal of Behavioral Nutrition and Physical Activity*, *5*, 20.

Wang, D., Zhang, M., Wang, T., Liu, T., Guo, Y., & Granato, D. (2020). Green tea polyphenols mitigate the plant lectins-induced liver inflammation and immunological reaction in C57BL/6 mice via NLRP3 and Nrf2 signaling pathways. *Food and Chemical Toxicology*, *144*, 111576.

Xu, Y., Gao, Y., & Granato, D. (2021). Effects of epigallocatechin gallate, epigallocatechin and epicatechin gallate on the chemical and cell-based antioxidant activity, sensory properties, and cytotoxicity of a catechin-free model beverage. *Food Chemistry*, *339*, 128060.

Zanoni, F., Primiterra, M., Angeli, N., & Zoccatelli, G. (2020). Microencapsulation by spray-drying of polyphenols extracted from red chicory and red cabbage: Effects on stability and color properties. *Food Chemistry*, *307*, 125535.

Zdunić, G., Aradski, A. A., Gođevac, D., Živković, J., Laušević, S. D., Milošević, D. K., et al. (2020). *In vitro* hypoglycemic, antioxidant and antineurodegenerative activity of chokeberry (*Aronia melanocarpa*) leaves. *Industrial Crops and Products*, *148*, 112328.

Ziani, B. E. C., Carocho, M., Abreu, R. M. V., Bachari, K., Alves, M. J., Calhelha, R. C., et al. (2020). Phenolic profiling, biological activities and *in silico* studies of *Acacia tortilis* (forssk.) hayne ssp. *raddiana* extracts. *Food Bioscience*, *36*, 100616.

Zokti, J. A., Baharin, B. S., Abdulkarim, S. M., & Abas, F. (2016). Microencapsulation of green tea extracts and its effects on the physico-chemical and functional properties of mango drinks. *International Journal of Basic & Applied Sciences*, *16*, 16–32.

CHAPTER FOUR

Underutilized Northern plant sources and technological aspects for recovering their polyphenols

Pirjo Mattila[a,*], Nora Pap[b], Eila Järvenpää[b], Minna Kahala[b], and Sari Mäkinen[b]

[a]Natural Resources Institute Finland (Luke), Turku, Finland
[b]Natural Resources Institute Finland (Luke), Jokioinen, Finland
*Corresponding author: e-mail address: mattilapirjoh@gmail.com

Contents

Abstract

Consumers worldwide are increasingly interested in the authenticity and naturalness of products. At the same time, the food, agricultural and forest industries generate large quantities of sidestreams that are not effectively utilized. However, these raw materials are rich and inexpensive sources of bioactive compounds such as polyphenols. The exploitation of these raw materials increases income for producers and processors, while reducing transportation and waste management costs. Many Northern sidestreams and other underutilized raw materials are good sources of polyphenols.

Advances in Food and Nutrition Research, Volume 98
ISSN 1043-4526
https://doi.org/10.1016/bs.afnr.2021.02.007

These include berry, apple, vegetable, softwood, and rapeseed sidestreams, as well as underutilized algae species. Berry sidestreams are especially good sources of various phenolic compounds.

This chapter presents the properties of these raw materials, providing an overview of the techniques for refining these materials into functional polyphenol-rich ingredients. The focus is on economically and environmentally sound technologies suitable for the pre-treatment of the raw materials, the modification and recovery of the polyphenols, as well as the formulation and stabilization of the ingredients. For example, sprouting, fermentation, and enzyme technologies, as well as various traditional and novel extraction methods are discussed. Regarding the extraction technologies, this chapter focuses on safe and green technologies that do not use organic solvents. In addition, formulation and stabilization that aim to protect isolated polyphenols during storage and extend shelflife are reviewed. The formulated polyphenol-rich ingredients produced from underutilized renewable resources could be used as sustainable, active ingredients—for example, in food and nutraceutical industries.

1. Introduction

Natural ingredients in daily life are more acceptable for most people than their synthetic counterparts. According to Asioli et al. (2017), consumers' major motives are health issues, while increased consumer awareness of environmental concerns enhances the market for the growth of bioactive ingredients. Plant extracts are currently widely used in the food and beverages industry as colorants, flavorants, and reinforcing products' nutritional content. The global functional food market size is growing. It was estimated at USD 161 billion in 2018, and a compound annual growth rate of 7.9% for 2019–2025 was predicted. The global organic personal care ingredients market size was estimated at USD 7.7 billion in 2018, exhibiting a growth rate of 5.4% for 2019–2025 (Market Research Report, 2019). The European Commission and the Environmental Protection Agency have framed supportive regulations to augment the production and consumption of naturally derived ingredients in light of growing concern about reducing greenhouse gas emissions (Market Research Report, 2019).

The valorization of waste, residuals, and other underutilized sources from the agricultural, forest, and food industries constitutes a beneficial solution for societal health and contributes to a sustainable industry system from an environmental perspective. The utilization of such waste to produce value-added ingredients for potential applications in the foods, cosmetic, nutraceutical, and pharmaceutical industries also benefits the whole chain financially. In this chapter, the term "sidestream" is used for all residuals, waste, by-products, and co-products.

Losses and waste occur throughout the supply chain, from production to consumer. Currently, most of these sidestreams are utilized as biofuel, feed, or organic fertilizer. However, many of this waste contains a high content of several bioactive compounds such as polyphenols. Agricultural and industrial sidestreams of fruit, berries, vegetables, and cereals are especially good sources of polyphenols. There are also other underutilized polyphenol-rich sources like wood sidestreams, rapeseed press cake, and algae.

Recently, the availability of green and eco-efficient extraction and enrichment technologies has been increasingly publicized, leading to more effective opportunities to utilize under- or unutilized sources for producing value-added products. These technologies include enzyme-assisted, microwave-assisted, ultrasound-assisted extraction, subcritical water extraction, and supercritical fluid extraction, followed by concentration by membrane technologies, for example. Using these novel and green extraction methods, the valuable components can be recovered in an economically and environmentally sustainable way.

This chapter introduces the under- and unutilized Northern sources of polyphenols, their extraction and enrichment methods for recovery and for use as ingredients in foods, nutraceuticals, and cosmetics, for example. The valorization of these raw materials constitutes a beneficial business solution for societal health and contributes to sustainable production system from an environmental and economic perspective.

2. Underutilized Northern sources of polyphenols for supplement and ingredient production

2.1 Berry sidestreams

Berries are widely used in the North. They are gathered in forests and bogs, as well as being cultivated on an industrial scale. The following wild berry species have had at least some commercial use: bilberry (*Vaccinium myrtillus*); lingonberry (*Vaccinium vitis-idaea*); cloudberry (*Rubus chamaemorus*); crowberry (*Empetrum nigrum* spp. *hermaphroditum*); cranberry (*Vaccinium oxycoccos*); Arctic bramble (*Rubus arcticus*); raspberry (*Rubus idaeus*); sea buckthorn (*Hippophaë rhamnoides*); bogberry (*Vaccinium uliginosum*) (https://www.arktisetaromit.fi/en/berries/wild+berries/). Common examples of Northern cultivated berries are strawberries (*Fragaria × ananassa*), raspberries (*Rubus idaeus*), currants (*Ribes* spp), and blueberries (*Vaccinium cyanococcus × V. angustifolium*).

Berries are an exceptionally rich source of polyphenols (Fig. 1). They are also a very versatile source, because they contain a wide variety of different phenolic compounds. Recent clinical studies have focused globally on the

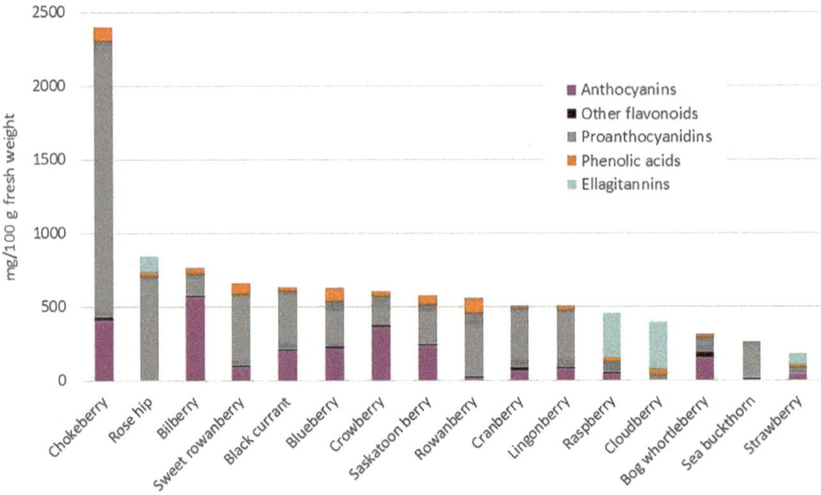

Fig. 1 The content of polyphenols in berries (Hellström, Törrönen, & Mattila, 2009; Koponen, Happonen, Mattila, & Törrönen, 2007; Mattila, Astola, & Kumpulainen, 2000; Mattila, Hellström, & Törrönen, 2006).

ability of berries to prevent cardiovascular diseases, type 2 diabetes, beneficial effects on brain health, and the prevention of memory disorders (Edirisinghe & Burton-Freeman, 2016; Guo, Yang, Tan, Jiang, & Li, 2016; Pribis & Shukitt-Hale, 2014; Rodriguez-Mateos, Heiss, Borges, & Crozier, 2013; Subash et al., 2014). In the background of these and many other chronic diseases is a low-grade chronic inflammation, and including berries in the diet may be a simple way of controlling this (Joseph, Edirisinghe, & Burton-Freeman, 2014). In addition to the prevention of these chronic diseases, berry polyphenols can have strong antimicrobial effects and growth inhibition potential against pathogens such as *E. coli* and *S. aureus*, while the colonization of probiotics, such as *Lactobacillus* and *Bifidobacterium*, is unaffected (Puljula, Walton, Woodward, & Karonen, 2020).

The main underutilized berry sidestreams are press cake residues from the juice industry and berry leaves as an agro-industrial sidestream. Fractionated bioactive compounds of these materials can be valorized for use as natural antioxidants, antimicrobials, health-promoting extracts, prebiotics, aroma compounds, skin enhancing compounds, and biostimulants, for example (Fierascu, Fierascu, Avramescu, & Sieniawska, 2019; Struck, Plaza, Turner, & Rohm, 2016).

In the manufacture of juices, 20%–30% of the mass of the original berries ends with the press cake being mainly berry skins, seeds, and stems. The press cake is still underutilized, often ending up as waste, despite the fact that is a rich source of phenolic compounds and has a potential use as a raw material for polyphenol-rich ingredients (Struck et al., 2016). The press cake contains as much or even more polyphenols as whole berries (Klavins, Kviesis, Nakurte, & Klavins, 2018; Struck et al., 2016). Anthocyanins are especially valued for their health effects (He & Giusti, 2010; Klavins et al., 2018; Ziberna et al., 2010). The price of anthocyanin isolates and hydrolysates ranges approximately between €30 and €300 per kg, depending on the raw material and anthocyanin content. In total, the global anthocyanins market was valued at US$300 million in 2015, and the market is projected to reach US$387.4 million by 2021 (Transparency Market Research, 2016). Anthocyanin-rich berry press cakes include berries with black, blue, or red peel like chokeberry, bilberry, blueberry, and blackcurrant (Fig. 1). It is estimated that the EU's juice concentrate industry uses around 80,000 t of blackcurrant annually, which means that the amount of blackcurrant press cake is as much as 16,000–24,000 t per year in the EU area (https://ec.europa.eu/agriculture/publi/reports/fruitveg/softfruit/workdoc_en.pdf). In addition to peel, the seeds of the berries are valuable, because they are rich in oils containing essential fatty acids with a favorable n6/n3 ratio. Seeds also contain antioxidants such as tocopherols, carotenoids, and polyphenols. Pressing residues of berry seeds after seed oil extraction and seed meals have been shown to exhibit high levels of antioxidant capacity and total phenolics (Ayoub, Costa de Camargo, & Shahidi, 2016; Helbig, Böhm, Wagner, Schubert, & Jahreis, 2008).

Berry leaves are rich sources of tannins, flavonoids, and phenolic acids and other phytochemicals, usually even more than the fruit (Ferlemi & Lamari, 2016; Tian et al., 2017). In contrast with berry fruit, the leaves are seldom used, despite the fact that they have traditionally been used in the (self-)treatment of several diseases such as the common cold, inflammation, diabetes, and ocular inflammation (Ferlemi & Lamari, 2016).

Leaves are the main sidestream of berry harvesting and cultivation. For example, in Finland, the berry raw material harvested from the forest contains ca. 5%–10% of discarded leaves. This means that about 0.2–1 million kg of both lingonberry and bilberry leaves go to waste annually, despite containing valuable bioactive compounds which could be utilized especially in cosmetics products as preservatives and skincare agents. If the leaves are harvested efficiently instead of berries their annual harvest would be more than 10-fold. The long cane cultivation of berries is becoming increasingly popular in Europe, because it has been shown to possess very high cropping potential. The long cane berry is a special plant type that has been induced to flower and is cold stored to produce a harvest in the planting year. After harvesting the berries, massive amounts of leaves containing high amounts of polyphenols are discarded.

2.2 Apple sidestreams

Oranges, pineapples, bananas, watermelons, and mangos are the dominant fruit in tropical and subtropical areas, while grapes and apples are the most important fruit in the temperature zone (Schieber, 2017). In the Northern hemisphere, the only economically important fruit is apple.

The combined phytochemical and nutrient profiles in apples and apple products suggest their potential to be powerful in preventing several chronic conditions in humans such as cardiovascular disease, Alzheimer's disease, asthma, diabetes, and cancer. Apple products may also be associated with improved outcomes related to cognitive decline in normal aging, weight management, bone health, pulmonary function, and gastrointestinal protection (Hyson, 2011). However, according to Ravn-Haren et al. (2013), ingestion of whole apples, apple pomace, turbid apple juice, and clear apple juice has no effect on serum HDL-cholesterol, TAG, weight, waist-to-hip ratio, blood pressure, inflammation (hs-CRP), the composition of the gut microbiota, or markers of glucose metabolism (insulin, IGF1 and IGFBP3). However, all other products except clear apple juice decreased total cholesterol and LDL values in the serum, and pomace most predominantly. Clear apple juice had the opposite effect, increasing serum total and LDL values. Ravn-Haren et al. (2013) concluded that the fiber component was necessary for the cholesterol-lowering effect of apples in healthy humans.

Apple polyphenols are predominantly localized in pomace. It is only to a minor extent extracted into the juice during processing. Fig. 2 shows that the peel of four examined varieties (Red Delicious, Granny Smith, Lobo, and Valkea kuulas) contained more polyphenols than the whole apples, and the polyphenol content of commercial apple juice was low. Among polyphenols, apple pomace is a rich source of catechins, hydroxycinnamates, phloretin glycosides, quercetin glycosides, and procyanidins (Djilas, Čanadanović-Brunet, & Ćetković, 2009). Phloridzin is an interesting polyphenol compound in apple. It can be used as a treatment for type 2 diabetes, as a weight-loss agent for obesity, and in the acute management of hyperglycemia (Ehrenkranz, Lewis, Ronald Kahn, & Roth, 2005).

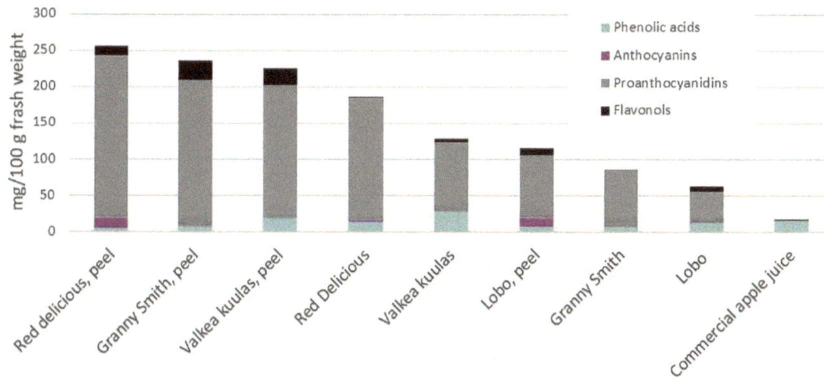

Fig. 2 Content of polyphenols in the whole apples and peel of various apple varieties. *Compiled from Hellström, J., Törrönen, R., & Mattila, P. (2009). Proanthocyanidins in common food products of plant origin.* Journal of Agricultural and Food Chemistry, 57, *7899–7906; Koponen, J., Happonen, A., Mattila, P., & Törrönen, R. (2007). Contents of anthocyanins and ellagitannins in foods consumed in Finland.* Journal of Agricultural and Food Chemistry, 55, *1612–1619; Mattila, P., & Hellström, J., unpublished data.*

The utilization of apple pomace needs careful action from processors, because the apple seeds contain toxic cyanogenic glucosides. In addition, enzyme-catalyzed browning reactions may adversely affect the color, flavor, and nutritional quality of pomace.

2.3 Vegetable sidestreams

Vegetables are generally not as versatile a source of polyphenols as berries and fruit. The main phenolic group found in vegetables is phenolic acids (Hellström et al., 2009; Koponen et al., 2007; Mattila et al., 2000; Mattila & Hellström, 2007). However, red, blue, and purple vegetables like red cabbage, red potato varieties, red onion, red radish, red salad varieties, and eggplant also contain anthocyanins (Koponen et al., 2007; Fig. 3), and some vegetables (like onion) have significant content of other flavonoids (Mattila et al., 2000). In addition to having beneficial health effects, vegetables rich in colored anthocyanins diversify the human diet and make food products more attractive to the consumer. Sidestreams of colored vegetables can also be valuable sources of natural coloring agents (Di Gioia et al., 2020). Proanthocyanins are rarely found in vegetables, and the determined content is generally low. However, rhubarb, for example, is a moderately good source of proanthocyanidins (Hellström et al., 2009).

Large quantities of plant-based sidestreams are generated by the food industry, where this underutilised plant material is considered and treated

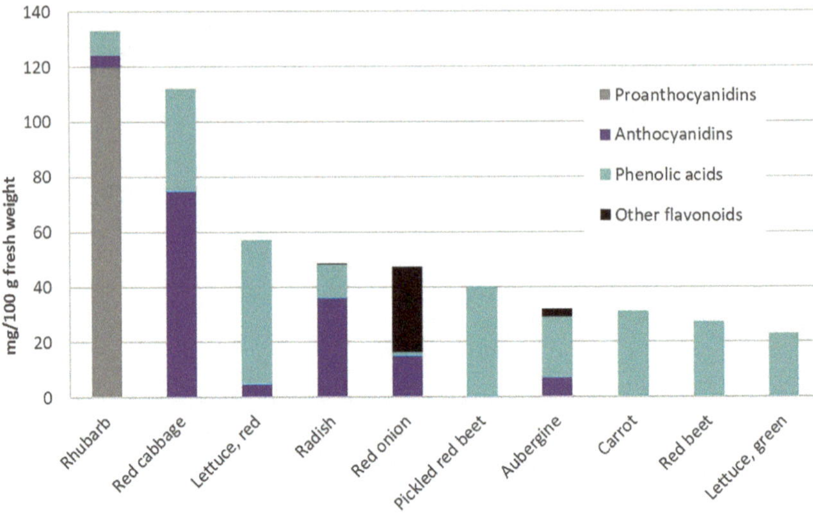

Fig. 3 Content of phenolic compounds in some vegetables. *Compiled from Koponen, J., Happonen, A., Mattila, P., & Törrönen, R. (2007). Contents of anthocyanins and ellagitannins in foods consumed in Finland.* Journal of Agricultural and Food Chemistry, 55, *1612–1619; Mattila, P. & Hellström, J. (2007). Phenolic acids in potatoes, vegetables, and some of their products.* Journal of Food Composition and Analysis, 20, *152–160; Hellström, J., Törrönen, R., & Mattila, P. (2009). Proanthocyanidins in common food products of plant origin.* Journal of Agricultural and Food Chemistry, 57, *7899–7906; Mattila, P., Astola, J., & Kumpulainen, J. (2000). Determination of flavonoids in plant material by HPLC with diode array and electro array detections.* Journal of Agricultural and Food Chemistry, 48, *5834–5841.*

as biodegradable waste. There is therefore a strong need to optimize the vegetable waste treatment system (Varzakas, Zakynthinos, & Verpoort, 2016). In addition, vegetable waste like leaves, twigs, and stems are generated at the farm during the harvesting and post-harvesting stages.

2.3.1 Potato

Potato is the world's fourth-largest food crop, after rice, wheat, and maize (Mahgouba, Eisab, & Youssefc, 2015). Peel is the major sidestream of potato processing (Schieber, 2017). It was estimated that the total world potato production was 368 million tons in 2018; of this, 115 million tons were produced in the European Union (FAOSTAT, 2018). Fresh peel weight is approximately 21.1%–23.9% of the total weight of potatoes (Im et al., 2008). Processed potatoes are increasingly consumed, and industrial processing of potatoes creates huge amounts of peel as a sidestream, and this generates environmental problems while being a good source of phenolic

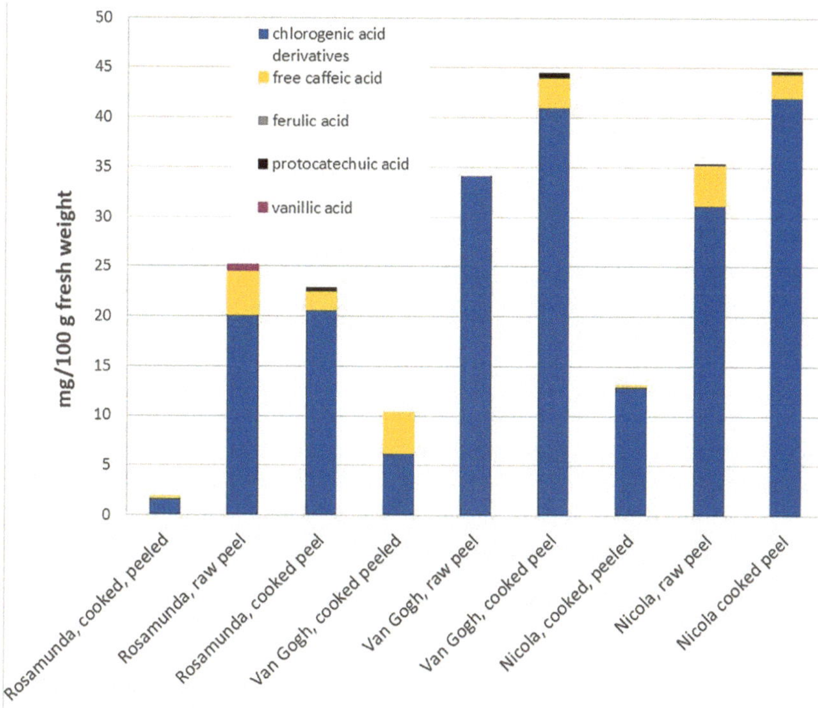

Fig. 4 Phenolic acid content of peeled potatoes and their peel as aglycones. Chlorogenic acid derivatives are expressed as caffeic acid (Mattila & Hellström, 2007).

compounds (Fig. 4). Phenolic acids are the most abundant phenolic compounds in potato, but flavonols, flavanols, and anthocyanins (in red/purple varieties) have also been found (Deußer, Guignard, Hoffmann, & Evers, 2012; Koponen et al., 2007; Oertel et al., 2017).

The dominant phenolic acid in potato is chlorogenic acid, because it constitutes up to 90% or even more of the total phenolic content of tubers. Chlorogenic and caffeic acids are also major phenolic compounds in potato peel, but other phenolic acids such as gallic, ferulic, *p*-hydroxybenzoic, *p*-coumaric, protocatechuic, vanillic, and *trans*-o-hydroxycinnamic acids are also found (Fig. 4; Akyol, Riciputi, Capanoglu, Fiorenza Caboni, & Verardo, 2016; Amado, Franco, Sánchez, Zapata, & Vázquez, 2014; Mohdaly, Hassanien, Mahmoud, Sarhan, & Smetanska, 2013). The phenolic acids are concentrated in peel (Fig. 4; Im et al., 2008). The total phenolic acid levels (in g/100 g fresh weight) of the peel of five potato varieties grown in Korea ranged from 6.5 to 42.1, and of the flesh (pulp) from 0.5 to 16.5, with peel/pulp ratios ranging from 2.6 to 21.1. Red and purple potatoes contained the highest amounts of phenolic compounds (Im et al., 2008). Anthocyanins are reported to be in acylated forms in purple, blue, and red fleshed potatoes. Petunidin and malvidin are the major anthocyanidin compound in purple and blue ones, and pelargonidin in red ones (Oertel et al., 2017; Yin et al., 2016).

2.3.2 Carrot

Carrot is widely cultivated throughout the world. It was estimated that the total world production of carrot and turnip was 40 million tons in 2018. Of this, 8.3 million tons were produced in the European Union (FAOSTAT, 2018). Orange carrots are the most popular, but there is a wide range of other shades of carrot flesh like red, white, yellow, black, purple, or even multicolor, which attract interest due to their attractive appearance, nutritional value and potential health effects. The beneficial health effects of purple and black carrot especially have been reported, and they are widely used as blue colorants and functional ingredients in the food industry. The purple and black color is caused by acylated derivatives of anthocyanins (Petropoulos et al., 2019).

Sidestreams of carrot include press cake from carrot juice processing, and the peel and other mash from the processing of shredded and other fresh-cut carrots, for example. Sidestreams of orange carrots, carotenoids, and dietary fiber have been found to be the components with the most potential to be further utilized (Ben-Othman, Jõudu, & Bhat, 2020; Sagar, Pareek, Sharma, Yahia, & Lobo, 2018; Schieber, 2017). Carrot also contains significant amounts of phenolic acids, especially caffeoyl-quinic acids (Hellström, Granato, & Mattila, 2020). Carrot phenolics are mainly situated in the upper layers of the carrot root. Zhang and Hamauzu (2004) reported that carrot peel only accounted for 11.0% of the carrot fresh weight, but it provided 54.1% of the total phenolics. Wounding further increases the biosynthesis of phenolics in fresh-cut carrots, for example. This is a result of the stress induced increase in PAL activity, which is well demonstrated in various studies (e.g., Hellström et al., 2020; Howard & Griffin, 1993; Reyes, Villarreal, & Cisneros-Zevallos, 2007).

2.3.3 Onion

Onion is one of the most cultivated vegetables worldwide, and its cultivation is increasing. Generally, there are three color varieties of onion (red, yellow/brown, and white). In the European Union, 50,000 t of onion sidestream are generated annually. These streams mainly consist of onion skins, roots, the two outer fleshy scales of the onion, the top and bottom of bulbs, and undersized, malformed, diseased, or damaged bulbs (Ren, Nian, & Perussello, 2020; Sharma, Mahato, Shivraj, & Rok Lee, 2016).

The dominant flavonoid group in onions is flavonols, and red/purple varieties also contain anthocyanins. Depending on varieties, the major flavonols in onions are quercetin aglycone, quercetin-4′-O-monoglucoside, and quercetin-3,4′-O-diglucoside. The total flavonoid content expressed as quercetin equivalents was found to vary from ca. 284 to 2275 mg/kg fresh weight in six onion varieties. The amounts were highest in red skinned varieties, followed by yellow-skinned and white-skinned (Sharma, Mahato, & Rok Leef, 2018). The content of flavonoids in onion skins are significantly higher (up to 10 g/kg) than in the inner parts (Ren et al., 2020; Sharma et al., 2016). According to Albishi, John, Al-Khalifa, and Shahid (2013) the content of phenolics extracted from onion skins was approximately six times higher than that of their flesh counterparts. In addition to flavonols, purple and red onions contain several anthocyanins in the form of glycosides. The dominant anthocyanin is cyanidin 3 glucoside (Sharma et al., 2016). The content of total anthocyanins in dried red onion skin was ca. 10 mg/100 g (Albishi et al., 2013). Onions may have many potential health benefits, including anticancer, antioxidant, and antiproliferative benefits, for example (Sharma et al., 2016).

2.4 Cereal sidestreams

Cereals are one of the most important food sources for human consumption, with a production of more than 2 billion tons/year. Approximately 30% of this amount is wasted for various reasons. Globally, wheat, maize, rice, barley, sorghum, millet, oat, and rye are the most important cereals. In the Northern hemisphere, the most important cereals are wheat, rye, oat, and barley. The main sidestreams of cereal processing are germ and bran, which are separated from the grain's starchy endosperm during milling. However, they are rich sources of many unique health-promoting bioactive components (Călinoiu & Vodnar, 2018; Luithui, Nisha, & Meera, 2019; Verni, Rizzello, & Coda, 2019). Both animal and human studies have indicated that whole grains (and bran and germ fractions) help the treatment of metabolic diseases by favoring bacteria and metabolites involved in the control host immunity, appetite, lipogenesis, gluconeogenesis, gut barrier, and oxidant stress (Gong et al., 2018).

Cereals contain phenolic acids, especially hydroxycinnamic acids, and lignans. These phytochemicals may be important from a nutritional health perspective. Most phenolics and other phytochemicals are bound to dietary fiber polysaccharides and are predominantly concentrated in the bran fraction. Ferulic acid is the major phenolic acid in most cereals. It is abundant especially in the aleurone, pericarp, and embryo cell walls. Other hydroxycinnamic acids found in cereals include caffeic acid, *p*-coumaric acid, and sinapic acid, as well as hydroxybenzoic acids. The hydroxycinnamic acid derivatives are mainly present in the bound form, linked to the cell wall structural components. For example, wheat and rye bran are good sources of ferulic acids (Luithui et al., 2019; Knudsen, Nørskov, Bolvig, Hedemann, & Lærke, 2017; Mattila, Pihlava, & Hellström, 2005).

The main lignans in cereals are secoisolaricinol, 7-hydroxymatairesinol, matairesinol, lariciresinol, pinoresinol, medioresinol, and syringaresinol. The content is higher in rye than in wheat and oat. The bran contains substantially higher concentrations of lignans than whole grain and refined flour (Knudsen et al., 2017). Cereals also contain alkylresorcinols and species-specific phenolic compounds like avenanthramides in oat, for example (Mattila et al., 2005).

2.5 Other underutilized potential sources of polyphenols

2.5.1 Rapeseed press cake

Rapeseed meal is generated as a sidestream of rapeseed oil production. It is commonly used as feed due to its high nutritional value (Downey & Bell, 1990). Rapeseed meal contains approximately 40% of proteins with balanced amino acid composition (review by Naczk, Amarowicz, Sullivan, & Shahidi, 1998). It is also rich in fiber and contains minerals such as calcium, magnesium, zinc, and copper, a number of vitamins, and other bioactive compounds. For example, tocopherols, B vitamins, and choline increase the nutritional value of rapeseed meal. Rapeseed, and especially rapeseed meal, is rich in phenolic compounds. Indeed, rapeseed has been shown to be the richest source of phenolic compounds among all oilseed plants (Naczk et al., 1998). Rapeseed phenolics include esterified phenolic acids, free phenolic acids, and tannins (Krygier, Sosulski, & Hogge, 1982). The total content of phenolic acids in rapeseed meal varies between 6400 and 18,400 Pg/g, depending on the variety and oil recovery method.

Esterified phenolic acids. Phenolic acids in rapeseed are present mainly in esterified form. Phenolic esters constitute more than 90% of the total phenolics in rapeseed flour. The main phenolic ester in rapeseed is sinapine, the choline ester of sinapic acid, which comprises approximately 80% of the total phenolic compounds in rapeseed (Kozlowska, Naczk, Shahidi, & Zadernowski, 1990).

Besides choline, sinapic acid can be esterified with other phenolic acids, sugars, or kaempferols (Kozlowska et al., 1990). Other phenolic acids from rapeseed such as *p*-hydroxybenzoic acid, ferulic acid, and syringic acid can also form ester bonds (Kozlowska, Zadernowski, & Sosulski, 1983). Cultivation and growing conditions affect the levels of esterified phenolic acids, but the composition is determined by genetic control. Rapeseed contains ca. 0.39%–1.06% of sinapine, depending on the species, growing conditions, and degree of maturation (Kozlowska et al., 1990; review by Naczk et al., 1998).

Sinapine plays an important biological role in rapeseed. In young plants, sinapic acid and choline are stored as sinapine, and during seed maturation, some of the sinapic acid and choline are released. Choline is an important compound for the methylation cycle, while sinapic acid is a precursor molecule of lignin and 15 flavonoids (Kozlowska et al., 1990).

Free phenolic acids constitute 6.5%–9.0% of the total phenolics in rapeseed, and up to 15% in rapeseed meal. However, some varieties are exceptional in containing only trace amounts or even no detectable quantity of free phenolic acids. The most prevalent phenolic compounds in rapeseed are sinapic acid derivatives. The typical amount of sinapic acid derivatives in rapeseed meal varies between 6390 and 18,370 Pg/g, depending on the variety of oilseed plant and the oil processing method (Kozlowska et al., 1990). Sinapic acid constitutes 70.2%–85.4% of free phenolic acids in defatted rapeseed meals. In addition to sinapic acid, which belongs to the hydroxycinnamic acid group, rapeseed meal contains ferulic acid, *o*-coumaric, *p*-coumaric, caffeic, *p*-hydroxybenzoic, vanillic, gentisic, protocatechic, syringic, and chlorogenic acids (Kozlowska, Rotkiewicz, & Zadernovski, 1983; Kozlowska, Zadernowski, et al., 1983; Shahidi & Naczk, 1992). Cinnamic and salicylic acids have also been found in rapeseed (Kozlowska, Zadernowski, et al., 1983).

Tannins are polyphenolic compounds that are widely distributed in plants. The amount of tannin in rapeseed meal varies from 0.2% to 3% of defatted rapeseed meal, depending on the variety and degree of maturation (Naczk, Amarowicz, Pink, & Shahidi, 2000). Most of the tannins in rapeseed are condensed tannins, formed by the polymerization of flavan-3-ols or flavan-3,4-diols (Naczk et al., 2000).

2.5.2 Algae

Microalgae and seaweeds (macroalgae) are a valuable source of bioactive compounds with a wide potential application in food and cosmetics as colorants, natural preservatives, stabilizers and gelling agents. Algae also provide a variety of other functional ingredients with a potential for medical, industrial, and biotechnological applications (Nova et al., 2020). The functional compounds of algae include polysaccharides, vitamins, minerals, pigments, enzymes, proteins and peptides, polyunsaturated fatty acids (PUFA), phenolics, and other secondary metabolites. These compounds have shown a potential for tackling health disorders such as type 2 diabetes, cardiovascular, and neurological diseases (Nova et al., 2020).

Polyphenolic compounds of seaweed include phlorotannins, bromophenols, flavonoids, phenolic terpenoids, and mycosporine-like amino acids. Of all the seaweed phenolic metabolites, most attention has been focused on tannins because of their interesting bioactive properties (Cotas et al., 2020). Phlorotannins are located at intracellular vesicles called physodes in the periphery of the cell (Audibert, Fauchon, Blanc, Hauchard, & Ar Gall, 2010). Only brown seaweeds contain phlorotannins, because phlorotannins are oligomers of phloroglucinol, which is restricted to brown seaweeds (Audibert et al., 2010). In green and red seaweed, the largest proportion of phenolic compounds are bromophenols, flavonoids, phenolics acids, and phenolic terpenoids, and in red seaweed, also mycosporine-like amino acids (Cotas et al., 2020). These molecules play an important role in seaweeds' defense mechanisms against herbivores and UV radiation (Audibert et al., 2010). Studies of seaweed phenolic acids are scarce. In green seaweeds, coumarins and vanilic acid derivatives have been identified, while in brown seaweeds, rosmarinic acid and quinic acid derivatives have been characterized. Benzoic acid, *p*-hydroxybenzoic acid, salicylic acid, gentisic acid, protocatechuic acid, vanillic acid, gallic acid, and syringic acid have also been detected in the genus *Gracilaria* (Cotas et al., 2020).

Information about the phenolic compounds in microalgae is rather limited, although studies concerning the characterization and identification of phenolic compounds in marine microalgae have recently been published (e.g., Goiris et al., 2012, 2014; Haoujar et al., 2019). Haoujar and colleagues studied the phenolic compounds of *Phaedactylium tricornitum* (Bacillariophyceae), *Nannochloropsis gaditana* (Eustigmatophyceae), *Nannochloris sp* (Trebouxiophyceae), and *Tetraselmis suecica* (Chlorodendrophycea). *P. tricornutum* extract showed the highest phenolic content among the samples (39.94 mg GAE/g), followed by *Nannochloris* sp and *T. suecica*, with 33.23 and 28.03 mg GAE/g, respectively. The phenolic content of *N. gaditana* was the lowest, at 22.94 mg GAE/g. A lower phenolic content level has been reported for *P. tricornutum* (16.80 mg GAE/g) and *T. suecica* (25.5 mg GAE/g) (López, Rico, Rivero, & Suaréz de Tanguil, 2011), and higher levels (32.0 mg GAE/g) for *N. gaditana* (Maadane et al., 2015).

The polyphenolic content and composition of microalgae are known to vary substantially, depending on microalgae growth conditions such as nutrient availability, temperature, and light intensity (Goiris et al., 2012). Polyphenolic content can also increase with exposure to UV-light (Jahnke, 1999), which opens an interesting possibility of enriching these bioactive compounds in microalgae biomass.

Flavonoids have been shown to represent the most abundant class of phenolic compounds in certain microalgae strains (Haoujar et al., 2019). Haoujar and colleagues determined four classes of phenolic compound in the four species of microalgae (*P. tricornutum*, *T. suecica*, *Nannochloris* sp., and *N. gaditana*). For example, dimethoxyflavone, quercetin, catechin and kaempferol, protocatechuic acid, caffeic acid, caffeoyl glucoside, and *p*-coumaroyl tyrosine were identified (Haoujar et al., 2019). Previously, single phenolic compounds and hydroxycinnamic acids has been found in industrially cultivated species *Phaeodactylum tricornutum* and *Tetraselmis suecica* species (Goiris et al., 2014).

2.5.3 Softwood sidestreams

Scots pine (*Pinus sylvestris*) and Norway spruce (*Picea abies*), the most abundant (conifer) trees in the Nordic countries, are sources of various wood-based products, from solid lumber to cellulose. Industrial activities produce large volumes of various sidestreams (Verkasalo et al., 2019) that could be utilized more efficiently in a modern circular bioeconomy. Conifer tree sidestreams, branches, bark, and sawdust have been used to produce techno-chemical products, bioenergy, and gardening products. They contain various polyphenolic compounds that have been extensively studied in different European countries. Most structural groups are present in the different parts or organs of softwood trees, including stilbenes, different flavonoid structures (flavanones, flavanols, dihydroflavonols, flavan-3-ols), anthocyanidins, benzoic and cinnamic acid derivatives, and lignans. Studies published before 2019 have been reviewed by Metsämuuronen and Sirén (2019), including a description of the biosynthesis pathways of polyphenols in conifer species, extraction and fractionation, and bioactivity characterization.

The commercial utilization of *Pinus* sp. extracts as nutraceutical with various claims was already established several years ago, including Mediterranean and Nordic species (Latva-Mäenpää, 2017). Recenty, especially chemical compounds found in *Picea* sp. and spruce bark have been of interest for their bioactivities in many European countries.

In the utilization of softwood sidestreams, various isolation process approaches have been used. The review by Metsämuuronen and Sirén (2019) includes data until 2019. Many research groups target more environmentally friendly approaches, especially with reduced organic solvent use (Co et al., 2012; Raitanen et al., 2020). Recent solvent-based reports use enhanced extraction methodologies, including pressurized or cavitational systems, and/or microwave or sonication to facilitate improved mixing with less organic solvent use (Albanese, Bonetti, D'Acqui, Meneguzzo, & Zabini, 2019; Chmelová, Škulcová, Legerská, Horník, & Ondrejovič, 2020; Spinelli et al., 2019; Strižincová et al., 2019). In addition to spruce bark, other wood production sidestreams have been studied with the aim of identifying suitability for edible applications. For food or nutraceutical applications, the quality of raw materials and (thermo-chemical) processes used during the whole value chain determine the quality and suitability of extracts as new ingredients (Albanese et al., 2019; Raitanen et al., 2020).

3. Modification and enrichment of phenolic compounds

3.1 Effect of sprouting on phenolic content

Studies have shown that germination and sprout formation can induce a significant change in the nutritional and bioactive content of plants. Special attention has been paid to the total phenolic content with antioxidant potential. Germination was reported to increase the phenolic content of legumes (Cevallos-Casals & Cisneros-Zevallos, 2010; Gharachorloo, Tarzi, & Baharinia, 2013; Mekky et al., 2020), cereals such as oats (Tang et al., 2020; Tian et al., 2010), and wheat (Cevallos-Casals & Cisneros-Zevallos, 2010), as well as vegetables like broccoli (Olszewska et al., 2020), kale, onion, and radish (Cevallos-Casals & Cisneros-Zevallos, 2010). The germination of fava beans increases the quantities of flavonoids, phenolic acids, and saponins, thus enhancing the antioxidant capacity (Mekky et al., 2020).

The concentration and availability of nutrients and bioactive compounds can be influenced by adjusting the steeping and germinating temperatures. In oat, the content of total phenolics has been found to be stable or to slightly decrease in the early phase of steeping and germination, and to increase when germination proceeds further (Tian et al., 2010; Xu et al., 2009). The optimal steeping and germination conditions for increasing the nutritional value of oat have been shown to be 18 °C for 12 h and 20 °C for 48 h, respectively. In these conditions, the total phenolic and flavonoid content increased significantly, being 13.5 and 7.2 mg/g, respectively, in sprouted oat (Tang et al., 2020). Steeping and germination were reported to enhance the antioxidant activity of oat, correlating significantly with the content of phenolic compounds (Tang et al., 2020; Xu et al., 2009).

In pulses, germination can decrease some anti-nutritional factors and increase beneficial nutrients, including phenolic compounds. Compared with oat, relatively long germination times have been applied for pulses. For example, soaking and consequent germination at 20 °C for 5 days increased the phenolic content and antioxidant activity of white bean (*Phaseolus vulgaris L*), common vetch (*Vicia sativa*), lentil (*Lens culinaris*), and chickpea (*Cicer arietinum*) seeds. The greatest increase in phenolic content from dormant seed to sprouted seed was in chickpea and common vetch. However, in all samples, the phenolic content increased significantly (Gharachorloo et al., 2013).

3.2 Effect of fermentation on phenolic content

The use of micro-organisms in processing is an alternative for solvent extraction to obtain bioactive compounds from natural sources via metabolism or the microbes' exogenous enzymes. Several studies have been carried out on the production and extraction of phenolic compounds from agro-industrial sidestreams by biotransformation (Madeira Junior, Teixeira, & Macedo, 2015; Martins et al., 2011). Phenolic compounds are often present in plant material as insoluble, and they are bound with sugars, fatty acids, or amino acids, which makes extraction of these compounds more difficult (Acosta-Estrada, Gutiérrez-Uribe, & Serna-Saldívar, 2014; Dey & Kuhad, 2014; Shen et al., 2018). Microbial processing of plant-based raw materials by yeast, fungi, and bacteria is considered a promising economically and environmentally friendly technique for releasing phenolic compounds from natural resources and increasing the yield of their extraction (Huynh, Van Camp, Smagghe, & Raes, 2014; Madeira Junior et al., 2015). Fermented extracts can serve as potential sources of natural antioxidants for the food and pharmaceutical industries (Dey et al., 2016). Information about the microbial conversion of phenolic compounds and the effect of this process on products' health benefits has increased significantly in recent years. Microbial hydrolysis has been shown to increase the concentration of free phenolics and the solubility of phenolic compounds and enhance bioactivity, while converting phenolic compounds into different metabolites, thus producing new bioactive compounds by a controlled submerged or solid-state fermentation process (Dey, Chakraborty, Jain, Sharma, & Kuhad, 2016; Huynh et al., 2014; Madeira Junior et al., 2015; Martin & Matar, 2005). It is also known that phenolic compounds may inhibit microbial processes.

Micro-organisms can produce hydrolyzing enzymes such as cellulases, β-glucosidase, xylanase, pectinases, β-xylosidase, β-galactosidase, α-amylases, and esterase. They can therefore release bound phenolics in soluble form (Dey & Kuhad, 2014; Septembre-Malaterre, Remize, & Poucheret, 2018). It is also suggested some other unknown biochemical pathways should be involved in the fermentation process, leading to increases in the total phenolic content (TPC) (Dey & Kuhad, 2014). The release and modification of phenolic compounds has also been connected with changes in antioxidant properties (Septembre-Malaterre et al., 2018; Yang et al., 2018). Lactic acid bacteria (LAB) are a group of food-grade micro-organisms widely used in food fermentation. Some LAB are known to have more resistance to phenolic compounds compared to other bacterial groups. Lactic acid fermentation has been shown to result in metabolites that are potentially more biologically active than the parent compounds and to enhance the level of health-promoting compounds (Filannino, Bai, Di Cagno, Gobbetti, & Gänzle, 2015; Rodríguez et al., 2009). LAB species are known to metabolize phenolic acids through decarboxylation and/or reduction activities (Filannino et al., 2015; Rodríguez et al., 2009). They can produce glycosyl hydrolases, which can convert glycosides to the corresponding aglycones, and esterases, which degrade methyl gallate, tannins, or phenolic acid esters (Filannino et al., 2015; Landete, Curiel, Rodríguez, de las Rivas, & Muñoz, 2014; Septembre-Malaterre et al., 2018). Deglycosylation of aryl glycosides has been shown to increase the antioxidant activity and bioavailability of glycosylated phenolic compounds (Landete et al., 2014). Fungi and yeasts also produce different types of enzymes during fermentation, including glycoside hydrolase, cellulases, xylanases, and esterases (Kadiri, 2017; Moore et al., 2007). For example, in cereal fermentation, these enzymes enable the softening of the kernel structure, break down cereal cell walls, and hydrolyze the β-glucosidic bonds of several phenolic compounds that occur mainly as conjugates (Kadiri, 2017).

Fermentation has a potential for the modification and extraction of the phenolics from different kinds of raw material such as berries and fruit (Cheng et al., 2020; Puupponen-Pimiä et al., 2016), legumes (Gan, Shah, Wang, Lui, & Corke, 2016), and cereals (Dey et al., 2016; Huynh et al., 2014). The number of studies focusing on the fermentation of fruit and berry sidestreams by fungi and bacteria has increased recently (Cheng et al., 2020). The press cake is a sidestream of fruit and berry processes, and can be targeted at the production of value-added bioactive ingredients. Total phenolics, antioxidative activity, and radical scavenging activity have been shown to increase in blueberry press cake or juice by fermentation with single or mixed strains (Cheng et al., 2020; Martin & Matar, 2005; Yan et al., 2019). In cherries, protocatechuic acid decreased by 70% in *L. plantarum*. Meanwhile, catechol was accumulated (Filannino et al., 2015). The degradation of ellagitannins with simultaneous ellagic acid accumulation has been evidenced in fermented cloudberries (Puupponen-Pimiä et al., 2016). Solid-state fermentation of cranberry press cake using a food-grade fungus, *Lentinus edodes*, resulted in high enrichment in ellagic acid content (Vattem & Shetty, 2003). An increase in total phenolics and antioxidant activity during the fermentation of berries and fruit is widely demonstrated, whereas a reduction in anthocyanin content is evidenced in some studies (Mousavi et al., 2013; Wei et al., 2018).

Fermented legume powders rich in antioxidant polyphenols have been proposed for use in developing novel functional foods (Gan et al., 2016). In fermented cowpea, increasing the bioavailability of legume polyphenols and compounds such as catechin was shown to increase with natural fermentation (Gan et al., 2016), while the content of total extractable phenolic compounds and anthocyanins, with antioxidant activities, increased in black beans in a process with filamentous fungi (Lee, Hung, & Chou, 2008). Fermentation by natural process and inoculation with LAB improved the antioxidant properties of red cabbage in a study by Hunaefi, Akumo, and Smetanska (2013). At the same time, the level of total anthocyanins and phenolic acids decreased. Higher levels of some phenolics and a release of the bound phenolics of the sweet potato were detected after fermentation with *L. acidophilus*, and compounds such as p-coumaric acid and ferulic acid were produced (Shen et al., 2018). In broccoli, *L. reuteri* was shown to hydrolyze chlorogenic acid, leading to the synthesis of caffeic acid and quinic acid. Metabolites like dihydrocaffeic acid are known to have high human bioavailability and higher antioxidant capacity than their precursor (Filannino et al., 2015).

Fermentation can be used as a method to increase the bioactive potential of cereals. During the fermentation of cereals, enzymes, such as amylases, xylanases, and proteases derived from microbes, but also from grains, contribute to the modification of the raw material composition. Fermentation affected the bioactive constituent, antioxidative activity, and total phenolic content of selected cereals (buckwheat, wheat, rye, and barley) in the study of Đorđević, Šiler-Marinković, and Dimitrijević-Branković (2010). An increase in TPC was evidenced in all raw materials with the *S. cerevisiae* yeast and *L. rhamnosus* bacterium. Filamentous fungi, belonging to the genus *Aspergillus*, changed the phenolic acid profile and increased the content of caffeic acid and ferulic acid in oat extract remarkably compared to native oat, while LAB strains resulted in a significant increase of ferulic acid and *p*-coumaric acid in grain barley (Huynh et al., 2014). An improvement in the bio-accessibility of free phenolic acids, such as ferulic acid, *p*-coumaric acid, and sinapic acid, was demonstrated with the combined effect of yeast fermentation and hydrolytic enzyme treatment in wheat bran (Kadiri, 2017). *S. cerevisiae* increased the content of soluble free syringic, *p*-coumaric, and ferulic acid up to 333%, 48%, and 51%, respectively, in wheat bran (Moore et al., 2007).

These examples show that fermentation can be applied as an affordable mild processing technique to increase the content of free phenolics and recover biologically active compounds from different types of plant-based material. However, both decreases and increases in phenolic compounds have been shown, depending on raw material, starter strains, time, and temperature, indicating the importance of the selection of the starter strains and fermentation parameters (Filannino et al., 2015; Huynh et al., 2014; Madeira Junior et al., 2015; Septembre-Malaterre et al., 2018).

3.3 Effect of enzyme technologies on phenolic content

The use of food-grade enzymes is considered a safe and efficient method for unfolding the embedded compounds in the cell wall matrix of the raw material or sidestream. This is usually considered a pre-processing step and is followed by a secondary extraction method. The aim of the enzymatic treatment is usually to dismantle of the structure of cellulose, hemicellulose, lignin, and pectin in the case of wood barks, press cakes, or leaves in sidestream utilization. Cellulases, hemicellulases, xylanases, β-glucosidases, endo-polygalacturonases, pectinesterases (Ptichkina, Markina, & Rumyantseva, 2008), and sometimes laccases (Osma, Toca-Herrera, & Rodrıguez-Couto, 2010) are used in these processes. However, unfortunately even today the economic issues in larger-scale hinder the widespread use of laccases.

The enzymes used in sidestream processing are in fact often including cellulases, polygalacturonases, pectin lyases, and rhamnogalacturonan lyases activities at the same time, and can be successfully used to extract pectin from gold kiwifruit, for example (Yuliarti, Goh, Matia-Merino, Mawson, & Brennan, 2015). The press cakes from berry processing are rich in polyphenols such as anthocyanins, flavonols, and phenolic acids, and pectinase enzymes are used in their exploitation (Pap et al., 2010). The same authors performed a secondary pectinase treatment of the juice with the aim of increasing the filterability of the blackcurrant juice, however, they observed that also the anthocyanin and flavonol content of the juices increased in the treatments. Complex enzyme preparations, including cellulase, polygalacturonase, pectin lyase and rhamnogalacturonan lyase complex (Cellulast 1.5 L), xylanase, and beta-glucosidase complex were found effective in the processing of the apple pomace. The results indicated that although other factors such as the solid-liquid ratio determine the efficiency of enzymatic hydrolysis, the Cellulast 1.5 L enzyme recovered the greatest amount of pectin from apple pomace, corresponding to 18.95% (Wikiera, Mika, Starzyńska-Janiszewska, & Stodolak, 2016).

The main advantages of these enzymatic processes are mild conditions, and although being multistage but simple and reduced number of stages then in conventional processing (Gligor et al., 2019). However, challenges may arise from the consumer perspective, because enzymes are inactivated at the end of the processes yet remain in the product (Gligor et al., 2019). They are therefore counted as additives.

4. Extraction of phenolic compounds

4.1 Stabilization and pre-processing of raw materials before extraction

All operations before the extraction of phenolic compounds, including raw material selection, collection, packaging, storage, and transportation, have a significant impact on the quality of raw materials and their products. Approach and models of hazard/quality assurance and analysis of perishable foods, e.g., fresh-cut vegetables and fruits (Barrett, Beaulieu, & Shewfelt, 2010; Mariquele, Hagelaar, & Lakemond, 2012), could provide basic tools for the development of quality assurance protocols for side streams. Appropriate storage, packaging, and transfer systems for raw materials and sidestreams are needed if the extraction is not close to the site of the material's origin. Controlling the temperature during transportation and storage is as important as in the food chain, where the chemical, enzymatic, and microbial deterioration of the material is kept to a minimum. The stabilization of the raw and sidestream materials for extraction and ingredient production seeks to stop spoilage and polyphenol degradation. Stabilization technologies, including freezing, drying, conservation, and fermentation processes typical of the food industry, can be used if the materials must be stored before extraction.

Drying methods are well suited to extraction schemes using non-aqueous or non-mixing extraction vessels. Low temperatures are preferred. Maillard reactions may occur in sidestreams at 100–150 °C or higher, and as a result, new bioactive compounds may also be formed (Plaza, Abrahamsson, & Turner, 2013). Polyphenol oxidase (PPO) inhibitors may also be a choice for inhibiting enzymatic browning and improving the stability of the polyphenolic content in slurries and other types of sidestream with high a_w and small particle size, regardless of their utilization pathway.

When considering targeted ways to stabilize the polyphenols in sidestreams, actions should also inhibit polyphenol oxidase activity. Rapid transfer chains and efficient unit processes are most important when other alternatives are not often feasible for PPO inhibition (Tinello & Lante, 2018). Controlling enzymatic browning is essential for the external quality of cut and sliced ready-to-eat fruit and vegetable products, but also for the stability of polyphenol compounds. Reductive additives, such as ascorbic acid and sulfur compounds, have been used for dried or drinkable fruit and vegetable products. The inhibition of PPO is one key outcome here, and is achieved by controlling the temperature and $p(O_2)$ during storage. Inhibitions have also been shown in juices using UV light or high-pressure processing. PPO chemical inhibitors in foods can be categorized as antioxidants, reducing agents, chelating agents, acidulants, and/or mixed-type inhibitors (Moon, Kwon, Lee, & Kim, 2020; Tinello & Lante, 2018). Even certain volatile compounds have been found to inhibit polyphenol oxidase (Lee et al., 2020).

A recent paper proposes sourcing PPO inhibiting chemicals from edible vegetable/fruit sources and their by-products (Moon et al., 2020). In a practical case, the inhibiting polyphenols were sourced from other plant species from where they would be applied, i.e., strawberry sidestream polyphenols for reducing apple PPO activity (Villamil-Galindo, Van de Velde, & Piagentini, 2020).

4.2 Extraction technologies

Inevitably, huge numbers of sidestreams are formed in every fruit, vegetable, animal, or marine industrial processing. These sidestreams can be valorized using economically sound extraction methods to recover different bioactive compounds that have great importance in nutrition or as pharmaceuticals. Such compounds can be carotenoids, proteins, vitamins, minerals, polysaccharides, carbohydrates, pectins, anthocyanins, and other phenolic compounds to mention some that are of increasing interest to food and pharmaceutical companies (Kowalska, Czajkowska, Cichowska, & Lenart, 2017).

Several extraction technologies have been developed to extract polyphenols from different sidestreams. Traditionally, liquid extraction methods with mixtures of ethanol/alcohols and water with acids or alkalis have been and are currently used. These methodologies are valid and in use on both analytical and preparative/process scales. The extraction efficiency of liquid extraction methods can be improved using the enhanced mixing systems described later in this chapter. The focus of this chapter is safe and green technologies that do not use organic solvents. The techno-economic feasibility, that is, the availability from the analytical lab scale to at least the pilot scale in an economically and ecologically sound manner, is also discussed in general for these novel extraction techniques.

Li, Chen, Zhang, and Fu (2017) describe the hampering of the wide industrial use of novel extraction technologies due to persistently high equipment-, installation-, and operation costs. However, there is still a great interest in developing these processes further for different raw materials to make them competitive alternative processing methods on the industrial scale as well. Recently, May and Guenther (2020) utilized material flow cost accounting in the berry-processing chain to compare the process performance from the perspective of CO_2 emissions and the economic outcome. The results showed that CO_2 emissions were substantially lower, and profitability remained the same, when the berry juice processing by-product was refined and sold as a pharmaceutical. This tool is a promising possibility for estimating the economic potential and environmental effects of using the by-products of any sidestream of plant, animal, or marine origin to produce high-value ingredients.

Barba, Zhu, Koubaa, Sant'Ana, and Orlien (2016) collate alternative extraction methods to valorize polyphenols from the sidestreams of wine production. The authors introduce an extensive and detailed list of extraction technologies, including accelerated solvent extraction (ASE), microwave assisted extraction (MAE), pulsed electric field extraction (PEF), supercritical fluid extraction (SFE), and hydroponic solvent extraction. The operating conditions used for grape sidestreams might be applicable, with adjustments, to other similar materials, e.g., sidestreams of berry and fruit juice production, and in some cases, also slurries and wet vegetable-processing sidestreams, from the vegetable juice, canning, and freezing processes. Panja (2018) also describes the principles of operation of emerging or green extraction techniques, e.g., enzyme-assisted extraction (EAE), pressurized liquid extraction (PLE), ultrasound assisted extraction (UAE), MAE, PEF, and SFE, and gives examples of the recovery of polyphenols from vegetable sidestreams. Recent examples of using SFE, MAE, and UAE for the extraction of phenolics are given later in this chapter.

Below, a more detailed example is given of subcritical water extraction (SWE), which uses only water as a solvent. Pure water is also used in hot water extraction (HWE) and pressurized hot water extraction (PHWE). SWE is based on water, which maintains the liquid state in temperatures between the boiling point of 100 °C and critical point of 374 °C under adequate pressure. HWE and PHWE also concern water in temperatures below or above subcritical conditions. Pressurized extraction processes and other high-pressure processes also reduce enzymatic activity, including polyphenol oxidase (Liu, Sandahl, Sjöberg, & Turner, 2014), an option for better recovery.

Although the most frequently used technologies at an industrial level today are still the conventional extraction methods (solvent extraction and maceration) due to their economic feasibility and relatively high process yields, there is a demand for change in this industry and novel/hybrid technologies are becoming increasingly prevalent in the valorization of the sidestreams. Few industrial applications of new techniques in bioactive compound recovery are yet available, but MAE has already been adopted. In describing the demand for novel technologies, the following subchapter will provide information about the different results of process development that uses more efficient novel approaches.

4.2.1 Subcritical water extraction (SWE)

SWE is an eco-efficient extraction method for bioactive compounds. Subcritical water is defined as water that maintains its liquid state under adequate pressure at a temperature between the boiling point of 100 °C and the critical point of 374 °C. Subcritical water can extract both polar and non-polar analytes, depending on the temperature and pressure applied. SWE can result in high-quality products at a lower production cost, higher efficiency, and lower environmental effects compared to conventional extraction methods (Shitu, Izhar, & Tahir, 2015).

SWE has been applied in recovering phenolic compounds—for example, from berry leaves (Kumar, Dutta, Prasad, & Misra, 2011; Mäkinen, Hellström, Mäki, Korpinen, & Mattila, 2020), fruit peel, shells, seeds, and food matrices (Shitu et al., 2015). SW extracts of sea buckthorn and bilberry leaves have shown slightly lower antioxidant capacity compared with ethanolic extracts, but significantly higher than that of conventional water extracts (Kumar et al., 2011; Mäkinen et al., 2020). Sea buckthorn leaf and bilberry leaf SW extracts have also been shown to prevent lipid oxidation in conceptual meat products with high efficacy (Mäkinen et al., 2020).

Extraction temperatures of 100–150 °C with 1–15 min extraction times have been found efficient for recovering phenolic compounds from plant materials (Kumar et al., 2011; Mäkinen et al., 2020; Shitu et al., 2015). The optimal SWE conditions for recovering phenolic compounds depend on the material. For example, 180 °C with an extraction time of 30 and 90 min has been used for recovering phenolic compounds from potato peel (Singh & Saldaña, 2011) and mango peel (Tunchaiyaphum, Eshtiaghi, & Yoswathana, 2013), respectively. In general, the extractability of compounds increases with higher temperatures and prolonged extraction times. However, high temperatures may also induce the degradation of heat-sensitive molecules.

4.2.2 Supercritical fluid extraction (SFE)

SFE uses compressed carbon dioxide as the solvent, typically at temperatures of 30–70 °C and pressures of 150–450 bar. In these conditions, the extraction of volatile compounds (aromas, fragrances) and oil-soluble constituents (e.g., fatty acids and their derivatives, oils, carotenoids, sterols) is easily attainable when the water content of the raw material is relatively low; drying preprocesses may therefore be needed (Herrero, Mendiola, Cifuentes, & Ibáñez, 2010; Tyśkiewicz, Konkol, & Rój, 2018). When the material source for polyphenols contains large amounts of undesirable aromas, oil-soluble pigments, or oils, an SFE process can be used as a preprocess or a technique in the cascade processing of sidestreams. The main asset of SFE is that after extraction, no solvent residues remain in the extract or the exhausted material unless the raw material contains organic solvents, or solvent has been intentionally added to the process. The addition of ethanol or other polar solvent improves the solvating power and yield of polyphenols. Typically, 15%–20% *v*/v of ethanol is added to supercritical CO_2 for the extraction of polyphenols, yielding an ethanolic extract. Liquid solvents end up with the targeted extract, and solvents can be removed using air flow-, vacuum- or supercritical fluid-based evaporation processes (e.g., Quintana, Villanueva-Bermejo, Reglero, García-Risco, & Fornari, 2019). Tyśkiewicz et al. (2018) reviews and reports a large number of experiments. Recent examples of different approaches include the cascade extraction of cranberry press residue (Tamkutė, Liepuoniūtė, Pukalskienė, & Venskutonis, 2020) and germinated lupine (*Lupinus luteus* L.) seeds (Buszewski et al., 2019), both collecting and analyzing lipophilic compounds and phenolic compound fractions; and the extraction and fractionation of lipophilic fat derivatives and phlorotannins from red and brown macroalgae (De Corato, Salimbeni, De Pretis, Avella, & Patruno, 2017).

4.2.3 Microwave-assisted extraction (MAE)

The efficiency of MAE is manifested by the difference of heating mechanism compared with conventional extraction. In conventional extraction, the vessel is heated for a finite time to reach the desired temperature. In contrast, in heating with microwaves, volumetric heating takes place and the liquid or suspension is heated directly. MAE has been found to provide very good results in the improvement of extraction yield and the richness of the extract with respect to conventional solid–liquid extraction, while shortening the extraction time and reducing the amount of solvent use. The yield improvement is thought to be related to the enhancement of internal mass transfer, because it has been shown for several raw materials that microwave radiation disrupts their solid structure, facilitating the leakage of active compounds.

In MAE, the recovery of compounds takes place using a frequency in a range between 300 MHz and 300 GHz (Azmir et al., 2013). The extraction process is manifested in three different stages: first, the solute separates from the matrix under temperature and pressure; then the solvent diffuses across the sample matrix; and finally the solute is released from the matrix to the solvent (Alupului, Călinescu, & Lavric, 2012). Several factors influence the efficiency of the MAE performance, such as the solvent and the nature of the raw material, microwave power, extraction temperature and extraction time/cycle, and stirring (Chana, Yusoff, Ngoha, & Kung, 2011).

Ferreira et al. (2020) used a 1 *w*/*v* % citric acid solution as an anthocyanin stabilization agent, microwave powers of 525 W and 700 W, and a 5-min treatment time for the extraction of anthocyanins from blueberry bagasse. Anthocyanin recovery was low, and the reason was probably due to the degradation of the anthocyanins in the selected conditions. Based on these results, processing with MAE would not be a viable option due to the excessive need for organic solvent compared to the amount of recovered compound. Citric acid was also used as an anthocyanin stabilization agent in sweet potato processing by microwaves by Liu, Yang, Zhou, Wen, and Dong (2019). The authors also state that the gelatinization process of the starch present in the raw material was avoided in MAE.

In general, MAE is usually considered a fast and economic process, with the only disadvantage being the formation of temperature spots and overheating in the matrix, thus manifesting the possible thermal degradation of the target compounds.

4.2.4 Ultrasound-assisted extraction (UAE)

The advantage of UAE is that the rupture of the cells of the raw material happens due to acoustic cavitation (Saini & Keum, 2018). This technique applies high-frequency sound waves that are above 20 kHz for cell eruption, followed by the extraction of the target compounds. In most studies published thus far, the extraction processes were carried out between 20 and 40 kHz (Kumar, Srivastav, & Sharanagat, 2021). The mechanism in the process is that cavitation bubbles are generated, grow, and finally collapse, creating a hotspot that causes a series of biochemical processes to occur. The solvent in these extractions is often, e.g., acidified water, ethanol, methanol, acetone, or water. In general, UAE is a low-cost technology with high yields, avoiding the thermal degradation of the sensitive compounds. The usual application of UAE is in the extraction of pectin (from peel, press cakes, and sunflower heads), the extraction of dietary fiber (stems, seeds, and peel), polysaccharides (peel, seeds, and stems), carotenoids (tomato pomace and citrus peel), and polyphenols (pomace and peel).

Kumar et al. (2021) have published a review article on the extraction of bioactive compounds from different fruit and vegetables. They found that in most studies the optimization by response surface methodology is implemented to optimize the process for high recoveries, considering frequency, cycles, time, temperature, solvent to solid ratio, and the type of solvent. UAE was found feasible in the recovery of many different compounds, and the most important factors for the success of the processing were analyzed. Ochoa, Durango-Zuleta, and Osorio-Tobón (2020) carried out an economic feasibility study of the extraction of anthocyanins. On the other hand, Bouchez, Vauchel, D'Alessandro, and Dimitrov (2020) also took into account the environmental performance of the UAE process by performing life cycle assessment (LCA) and carrying out multi-objective optimization, using beetroot seeds for modeling. Kumar et al. (2020) investigated the extraction of flavonols and total phenols accordingly.

UAE may also be a potential application in the valorization of different leaves. Saifullah, McCullum, McCluskey, and Vuong (2020) compared conventional solvent extraction with UAE in terms of total phenolics, proanthocyanidins, flavonoids, and antioxidant activity. Meanwhile, de

Sousa et al. (2020) used ultrasound power to recover different phenolic compounds (catechin, gallic acid, quercetin, vanillic acid, ellagic acid, trans-cinnamic acid, *p*-coumaric acid, syringic acid, ferulic acid, and chlorogenic acid) from the leaves of Croton heliotropiifolius Kunth. In both studies, good results were achieved in the UAE process, which indicates that there is potential and a need for process development for this purpose.

In both studies, good results were achieved in the UAE process, which indicates that there is potential and a need for process development for this purpose.

4.2.5 Hybrid extraction technologies

Hybrid extraction techniques describe processes in which novel extraction techniques (SWE, MAE, UAE, SFE, etc.) serve as the pre-extraction or secondary extraction step to maximize the yield of the desired compounds. The technologies that are usually combined with pre-treatment methods, such are the enzymatic hydrolyses or conventional extractions.

Sometimes the sidestream contains multiple valuable compounds, and the goal is to recover most of them as separate fractions and with good techno-functional condition. For example, Ninčević Grassino et al. (2020) used ultrasound power along with high hydrostatic pressure extraction (HHPE) for tomato peel. HHPE was used to recover the pectin, and although recovery yield was not enhanced too much, the time of extraction was shortened significantly when the results were compared to the conventional extraction. After the recovery of the pectin, the peel was exploited in UAE for the recovery of phenolics, in which process a 70% ethanol solution was used, and the UAE duration was 5 min. To gain maximum profit from the processing, the residue after the two extractions were still subjected to Soxhlet extraction of the fatty acids. The tomato peel extraction residue was still a good source of lauric, palmitic, and stearic acids.

According to Dassoff and Li (2019), ultrasound-assisted supercritical CO_2 extraction may improve yields, reduce pressure and/or temperature requirements, and/or reduce processing time. However, several factors in the operating conditions should be considered and optimized. Romero-Díez et al. (2019) reported on the efficiency of anthocyanin recovery when a conventional ethanol/water solvent extraction was carried out using MAE or sonication as a pre-treatment method. The authors concluded that MAE as a pre-treatment was successful for increasing the mass transfer of anthocyanins and reduced the processing time, while ultrasound was only capable of reducing the processing time, with no positive benefit on the anthocyanin yield using the reported operation parameters. Meanwhile,

Sillero, Prado, and Labidi (2020) coupled UAE with MAE in utilizing *Larix decidua* bark for the recovery of phenolic compounds. An interaction was observed between the extraction time, and the amplitude of treatment was observed in the UAE step. The recovery was almost twice that of the optimized conventional extraction process.

4.2.6 Recent advances in process developments

As a general tendency in the last couple of years of development in the field of used solvents, the application of new solvents has been increasing. The most promising novel solvents for the extraction of different health-promoting compounds are ionic liquids, deep eutectic solvents, aqueous two-phase solvents, and non-ionic surfactants. As Ekezie, Sun, and Cheng (2017) describe, their application in MAE is very diverse, because it is suitable for protein, phenolics, polysaccharides, lipids, essential oils, and pectin recovery from different raw material sources. Ionic liquids, deep eutetic solvents, and aqueous two-phase solvent were also used in UAE (Bosiljkov et al., 2017; Đorđević & Antov, 2017; Murador, Braga, Martins, Mercadante, & de Rosso, 2019). Except for the ionic liquids, the price of the solvents mentioned earlier are considered unexpensive. Some of these solvents require special attention due to toxicity issues, as in the case of ionic liquids. A general need for improvement lies in the selection of new, but safe and economic solvents.

In UAE, most process studies are optimized using the tool of response surface methodology (Kumar et al., 2021), which also helps to establish the interaction between the different process factors. Some studies also include environmental issues, such as carbon footprint, in the optimization. A general need for improvement lies in the selection of new, safe, and economic solvents in the UAE as well.

More recently, hybrid processing concept has been evolving. Here, one novel processing method is combined with another. In these cases, the first technology serves the preparation of the sample. It can be enzymatic treatment, ultrasound extraction, etc. These hybrid processes are a promising option for process efficiency enhancement and require more study and development in future.

5. Concentration of phenolic compounds

Concentration processes of different phenolic compounds pose challenges in compromising between the efficiency of the concentration and the preservation of the compounds that are relevant from the nutritional perspective. Although multi-stage vacuum evaporation is still the most widely used fruit juice concentration method, for example, (Wenten, Khoiruddin, Reynard, Lugito, & Julian, 2021), more novel technologies in which the concentration is carried out at a low temperature are slowly eliminating the application of high temperatures in concentration practices. These technologies rely on the application of an appropriate membrane filtration process in nanofiltration or a reverse osmosis range. Both processes can be carried out at substantially lower temperatures than evaporation, but the burden of the technology is usually the low final concentrations achieved compared to evaporation. In evaporation, a berry juice containing polyphenols can be concentrated up to 65°Brix; but in reverse osmosis it will not exceed 20–25°Brix (Pap et al., 2010). The reason for this phenomenon is usually the increased osmotic pressure, which will eventually decrease the effective process pressure. Other reasons connected with membrane technology that the decrease of efficiency are relatively common due to concentration polarization and fouling, and therefore requiring the selection of appropriate membranes and operating conditions to reduce these drawbacks to a minimum level. Wenten et al. (2021) reviewed some of the most promising membrane technologies that have a good agreement in terms of the concentration factor but still operate at room temperature. The authors collected information from the area in which osmotic/membrane distillation (OD/MD) and forward osmosis (FO) were studied. In these processes, filtration does not occur due to hydraulic pressure, but to the concentration difference in feed and draw solutions which are circulated through the membrane until the desired concentration is reached. In the study of Nayak and Rastogi (2010), the anthocyanin extract was concentrated from the initial 2°Brix to 52°Brix, corresponding to an anthocyanin concentration increase from 49.6 mg/L to 2.7 g/L, using a cellulose acetate membrane (Osmotek Inc.) and 6 M NaCl as a draw solution. The concentration was performed within 18 h, and there was no transfer in osmotic agent. Julian, Yaohanny, Devina, Purwadi, and Wenten (2019), on the other hand, suggests a submerged direct contact membrane distillation (SDCMD) technique as an advanced method for concentration in the food industry which is successful in concentrating apple juice up to 35°Brix and preserving its nutrients.

Besides the novel technologies for concentration, there are also trials to intensify the already well adopted evaporation process. Sabanci and Icier (2017) coupled ohmic heating with evaporation to concentrate cherry juice from 19.2°Brix to 65°Brix, while the total processing time was reduced by half. On the other hand, Norouzi, Fadavi, and Darvishi (2021) studied the effect of ohmic heating and concentration on the quality of cherry juice, stating that ohmic heating has no negative effect on the quality attributes of the cherry juice. Color attributes especially were not altered in terms of the L*a*b* coordinates. This may indicate that the anthocyanins were sufficiently preserved during this processing. Good results were also achieved in the processing of pomegranate juice with combining ohmic heating with evaporation. The pomegranate juice was concentrated to 35°Brix in less than 1h, with a voltage gradient of 12.5V/cm (Icier, Yildiz, Sabanci, Cevik, & Cokgezme, 2017).

Such an improvement in the total concentration time of juices or other extracts can contribute to better preservation of the quality of the end product, without compromising the achievable concentration factor.

Bhattacharjee, Saxena, and Dutta (2017) reviewed the different membrane processes used for concentration purposes. Besides FD and MD they also gave an insight into nanofiltration and reverse osmosis for concentration and pervaporation for aroma recovery. Additionally, the review also discusses the process engineering issues regarding these processes for improving efficiency, e.g., reducing fouling and concentration polarization.

Given the potential benefits of changing thermal concentration methods to membrane processing, there is a need to study if these processes are economically sound. In their study, Álvarez et al. (2000) calculated that the capital investment compared with the manufacturing cost, which were approximately equal in thermal concentration and integrated membrane technology (including clarification and pre-concentrations by membranes) for concentrating apple juice. Although membrane-based processing faces some engineering issues in reducing concentration polarization and fouling by selecting the appropriate membrane or defining optimal hydrodynamic conditions, there have been many developments in recent decades, both in the area of the development of new membranes and in recognizing the need for process optimization.

6. Formulation and stabilization of extracts

Formulation and stabilization processes are a group of different unit operations of the food and pharmaceutical industries, in which the aim is to protect isolated bioactivities during storage and extend their shelf-life. Protection may be targeted at maintaining the specific activity, pigment stability, and/or shelf-life of nutrients and/or controlled release of the bioactivity. Process phases often include the improvement of the applicability of the extracts for the targeted product type. Concentrated liquids (syrups) and dried powders are the most typical forms, but emulsified bioactive ingredients also have their uses, for example, in emulsified foods and cosmetics. The emulsification of polyphenol extracts with oil phases has also been tested for improved bioavailability or storage stability.

Microcapsules, microparticles, and microemulsions are different forms for protecting bioactivity, but coating materials specific for each form affect the usability of the polyphenolic ingredient. Coating materials and the process form are selected according to the use of the material, e.g., for food, cosmetic, or pharmaceutical ingredients. Particle sizes of micro-capsules and particles depend on the technology used for mixing (and drying), but is typically a few μm to mm in size. Popović et al. (2019) present the most suitable methodologies for the encapsulation of polyphenolics for the food industry.

The recent review by Macías-Cortés et al. (2020) describes the basic differences between different encapsulation techniques: mixing and drying with polymeric supporting material; encapsulation using waxes, alginates, or pectin; and liposome/phospholipid particles in liquid or dried form. Microencapsulation improves the storage stability of the extracts. It modifies the applicability of the extracts for different product models and controls the release of the bioactivity. Microcapsules for food and cosmetics models are made by mixing extracts with alginate, pectin, or synthetic polymers in water or water-soluble mixtures. Using the gel extrusion process, microcapsules are formed as droplets in a (water-based) hardening solution (e.g., $CaCl_2$), collected, and dried.

Microemulsion forms modify the applicability and storage stability of the polyphenolic extracts via the formation of water-in-oil emulsions from water-soluble extracts and oil-in-water emulsions from all other extracts. The process involves mixing the extract with water or oil, then oil or water, and emulsifiers and additives—compatible with the target end use—are mixed to form an emulsion using a high-energy (high-speed, ultrasound, high-pressure) mixer with micron-scale emulsion droplets, including the bioactive extracts (Kaimainen, Marze, Järvenpää, Anton, & Huopalahti, 2015). Microemulsions are applied in wet or dry (powder) forms after spray-drying.

Bioavailability is improved by reducing particle size. Nanotechnological solutions for polyphenols were reviewed by Aguirre and Borneo (2019). Briefly, the authors present the use of food-derived polymers, polysaccharides, and proteins in the formulation of polyphenols from sub-micron to nano-sized delivery systems, with improved applicability for edible systems and bioavailability. However, most examples of polyphenolics protected within nanoemulsions, nanocapsules, and nanoparticles have been targeted at pharmaceutical uses. The carriers and coatings therefore often contain polymers and chemicals used in pharmaceutical products (Barras et al., 2009). The review by Rezaei, Fathi, and Jafari (2019) has comprehensive tables of scientific research examples where bioactive compounds, including pigments and phenolic compounds, were protected using different nano-formulation techniques. Their examples contain several types of coating materials, which could control the bioavailability, and, at the same time have their specific techno-functional properties in different product applications, based on their surface properties.

Polyphenol extracts individually, or mixed with polymers such as maltodextrins or other edible polymers, are the most usual form for food use. Powder forms are easy to handle, mix, and add to other powders or granular ingredients. Spray-dry processes for polyphenol extracts from berries and fruit with *co*-drying polymers have been studied for a long time, and commercial products exist. These dried mixtures can be defined as encapsulated ingredients, but depending on the polymers, processing method, and conditions used (e.g., spray-drying, freeze drying), the particles may have a porous structure. A porous structure offers poor oxygenation or moisture protection to the polyphenols, and careful packaging is needed. Recently, vegetable polyphenols have received more interest, probably due to the circular economy and the need to improve material efficiency in the food system. The process itself is easy, inexpensive, and versatile in its selection of the matrix material and process conditions. In addition to berry and fruit extracts, a recent paper shows the utilization of anthocyanin and other polyphenols containing vegetable sidestreams. Zanoni, Primiterra, Angeli, and Zoccatelli (2020) used octenyl succinic anhydride (OSA) modified waxy starch, which forms more resistant particles, or capsules, than normal starches.

The dried press cakes from juice processing contain polyphenols, polysaccharides, and fibers that can be used as a food ingredient. A low O_2 content and low temperatures are preferred, e.g., freeze drying. A recent report on a bench-scale cyclone dryer shows that more rapid drying at a slightly elevated temperature may sometimes preserve polyphenols better if the mixing of the particulate material is very effective. A dried press cake of blackcurrant or blueberry contained a total phenolic content of 5–8 g GAE/100 g DM (Grimm, Nyström, Mossing, Östman, & Geladi, 2020) with different types of drying device. Another example for the utilization of berry polyphenolics is an extruded snack process, which simultaneously reduces microbiological risks and dries the material for improved stability. The example contains 20 g GAE/100 g DM with a wet blueberry press cake, and 6 g GAE/100 g DM as dried powder. The extruded snack product, containing 10%–20% *w*/w of dried berry powder, had a total phenolic content of 0.4–0.8 g GAE/100 g DM (Höglund et al., 2018).

7. Conclusions and perspectives

There is increasing global demand for sustainable and renewable raw materials to replace the utilization of fossil sources. An important trend in the food, feed, nutraceutical, and cosmetics industries today is the use of natural ingredients to produce biological activities and reduce the need for synthetic chemical additives. It also seems that increased consumer awareness of the environmental concerns and health issues enhances the market for bioactive ingredients such as polyphenols.

Underutilized and polyphenol-rich raw materials in the Northern hemisphere include berry, apple, vegetable, softwood, and rapeseed sidestreams, as well as underutilized algae species. These raw materials can be applied to the production of natural ingredients and additives with many bioactivities. Environmentally friendly pre-treatment, enrichment, extraction, concentration, and stabilization techniques should be developed and applied for ingredient production. This requires systematic research and development work in close collaboration between business and academics. In addition, completely new business and value chains with cross-sectoral interconnections need to be established, and new concepts need to be designed to unlock the economic, technological, logistic, and regulatory bottlenecks in natural ingredient production.

References

Đorđević, T., & Antov, M. (2017). Ultrasound assisted extraction in aqueous two-phase system for the integrated extraction and separation of antioxidants from wheat chaff. *Separation and Purification Technology*, *182*, 52–58.

Đorđević, T. M., Šiler-Marinković, S. S., & Dimitrijević-Branković, S. I. (2010). Effect of fermentation on antioxidant properties of some cereals and pseudo cereals. *Food Chemistry*, *119*(3), 957–963.

Acosta-Estrada, B. A., Gutiérrez-Uribe, J. A., & Serna-Saldívar, S. O. (2014). Bound phenolics in foods, a review. *Food Chemistry*, *152*, 46–55.

Aguirre, A., & Borneo, R. (2019). Improving bioavailability of polyphenols using nanodelivery systems based on food polymers. In R. R. Watson (Ed.), *Polyphenols in plants. Isolation, purification and extract preparation* (2nd ed., pp. 59–65). London: Academic Press.

Akyol, H., Riciputi, Y., Capanoglu, E., Fiorenza Caboni, M., & Verardo, V. (2016). Phenolic compounds in the potato and its byproducts: An overview. *International Journal of Molecular Sciences*, *17*, 835.

Albanese, L., Bonetti, A., D'Acqui, L. P., Meneguzzo, F., & Zabini, F. (2019). Affordable production of antioxidant aqueous solutions by hydrodynamic cavitation processing of silver fir (*Abies alba* Mill.) needles. *Food*, *8*(2), 65.

Albishi, T., John, J. A., Al-Khalifa, A. S., & Shahid, F. (2013). Antioxidative phenolic constituents of skins of onion varieties and their activities. *Journal of Functional Foods*, *5*, 1191–1203.

Alupului, A., Călinescu, I., & Lavric, V. (2012). Microwave extraction of active principles from medicinal plants. *UPB Scientific Bulletin, Series B: Chemistry and Materials Science*, *74*(2), 130–142.

Álvarez, S., Riera, F. A., Álvarez, R., Coca, J., Cuperus, F. P., Bouwer, S. T., et al. (2000). A new integrated membrane process for producing clarified apple juice and apple juice aroma concentrate. *Journal of Food Engineering*, *46*, 109–125.

Amado, I. R., Franco, D., Sánchez, M., Zapata, C., & Vázquez, J. A. (2014). Optimisation of antioxidant extraction from Solanum tuberosum potato peel waste by surface response methodology. *Food Chemistry*, *165*, 290–299.

Asioli, D., Aschemann-Witzel, J., Caputo, V., Vecchio, R., Annunziata, A., Næs, T., et al. (2017). Making sense of the "clean label" trends: A review of consumer food choice behavior and discussion of industry implications. *Food Research International*, *99*, 58–71.

Audibert, L., Fauchon, M., Blanc, N., Hauchard, D., & Ar Gall, E. (2010). Phenolic compounds in the brown seaweed Ascophyllum nodosum: Distribution and radical-scavenging activities. *Phytochemical Analysis*, *21*, 399–405.

Ayoub, M., Costa de Camargo, A., & Shahidi, F. (2016). Antioxidants and bioactivities of free, esterified and insoluble-bound phenolics from berry seed meals. *Food Chemistry*, *197*(*Pt. A*), 221–232.

Azmir, J., Zaidul, I. S. M., Rahman, M. M., Sharif, K. M., Mohamed, A., Sahena, F., et al. (2013). Techniques for extraction of bioactive compounds from plant materials: A review. *Journal of Food Engineering*, *117*(4), 426–436.

Barba, F. J., Zhu, Z., Koubaa, M., Sant'Ana, A. S., & Orlien, V. (2016). Green alternative methods for the extraction of antioxidant bioactive compounds from winery wastes and by-products: A review. *Trends in Food Science & Technology*, *49*, 96–109.

Barras, A., Mezzetti, A., Richard, A., Lazzaroni, S., Roux, S., Melnyk, P., et al. (2009). Formulation and characterization of polyphenol-loaded lipid nanocapsules. *International Journal of Pharmaceutics*, *379*(2), 270–277.

Barrett, D. M., Beaulieu, J. C., & Shewfelt, R. (2010). Color, flavor, texture, and nutritional quality of fresh-cut fruits and vegetables: desirable levels, instrumental and sensory measurement, and the effects of processing. *Critical Reviews in Food Science and Nutrition*, *50*(5), 369–389.

Ben-Othman, S., Jõudu, I., & Bhat, R. (2020). Bioactives from agri-food wastes: Present insights and future challenges. *Molecules*, *25*, 510.

Bhattacharjee, C., Saxena, V. K., & Dutta, S. (2017). Fruit juice processing using membrane technology: A review. *Innovative Food Science & Emerging Technologies*, *43*, 136–153.

Bosiljkov, T., Dujmić, F., Bubalo, M. C., Hribar, J., Vidrih, R., Brnčić, M., et al. (2017). Natural deep eutectic solvents and ultrasound-assisted extraction: Green approaches for extraction of wine lees anthocyanins. *Food and Bioproducts Processing*, *102*, 195–203.

Bouchez, A., Vauchel, P., D'Alessandro, L. G., & Dimitrov, K. (2020). Multi-objective optimization tool for ultrasound-assisted extraction including environmental impacts. *Chemical Engineering Research and Design*, *164*, 324–337.

Buszewski, B., Rafińska, K., Cvetanović, A., Walczak, J., Krakowska, A., Rudnicka, J., et al. (2019). Phytochemical analysis and biological activity of Lupinus luteus seeds extracts obtained by supercritical fluid extraction. *Phytochemistry Letters*, *30*, 338–348.

Călinoiu, L. F., & Vodnar, D. C. (2018). Whole grains and phenolic acids: A review on bioactivity, functionality, health benefits and bioavailability. *Nutrients*, *10*, 1615.

Cevallos-Casals, B. A., & Cisneros-Zevallos, L. (2010). Impact of germination on phenolic content and antioxidant activity of 13 edible seed species. *Food Chemistry*, *119*, 1485–1490.

Chana, C.-H., Yusoff, R., Ngoha, G.-C., & Kung, F. W.-L. (2011). Microwave-assisted extractions of active ingredients from plants. *Journal of Chromatography A*, *1218*(37), 6213–6225.

Cheng, Y., Wu, T., Chu, X., Tang, S., Cao, W., Liang, F., et al. (2020). Fermented blueberry pomace with antioxidant properties improves fecal microbiota community structure and short chain fatty acids production in an in vitro mode. *LWT*, *125*, 109260.

Chmelová, D., Škulcová, D., Legerská, B., Horník, M., & Ondrejovič, M. (2020). Ultrasonic-assisted extraction of polyphenols and antioxidants from *Picea abies* bark. *Journal of Biotechnology*, *314–315*, 25–33.

Co, M., Fagerlund, A., Engman, L., Sunnerheim, K., Sjöberg, P. J., & Turner, C. (2012). Extraction of antioxidants from spruce (*Picea abies*) bark using eco-friendly solvents. *Phytochemical Analysis*, *23*(1), 1–11.

Cotas, J., Leandro, A., Monteiro, P., Pacheco, D., Figueirinha, A., Gonçalves, A. M. M., et al. (2020). Seaweed phenolics: From extraction to applications. *Marine Drugs*, *18*, 384.

Dassoff, E., & Li, Y. O. (2019). Mechanisms and effects of ultrasound-assisted supercritical CO_2 extraction. *Trends in Food Science & Technology*, *86*, 492–501.

De Corato, U., Salimbeni, R., De Pretis, A., Avella, N., & Patruno, G. (2017). Antifungal activity of crude extracts from brown and red seaweeds by a supercritical carbon dioxide technique against fruit postharvest fungal diseases. *Postharvest Biology and Technology*, *131*, 16–30.

de Sousa, C., dos Anjos, G. L., Nóbrega, R. S. A., da S. Magaton, A., de Miranda, F. M., & de S. Dias, F. (2020). Greener ultrasound-assisted extraction of bioactive phenolic compounds in Croton heliotropiifolius Kunth leaves. *Microchemical Journal*, *159*, 105525.

Deußer, H., Guignard, C., Hoffmann, L., & Evers, D. (2012). Polyphenol and glycoalkaloid contents in potato cultivars grown in Luxembourg. *Food Chemistry*, *135*, 2814–2824.

Dey, T. B., & Kuhad, R. C. (2014). Enhanced production and extraction of phenolic compounds from wheat by solid-state fermentation with *Rhizopus oryzae* RCK2012. *Biotechnology Reports*, *4*, 120–127.

Dey, T. B., Chakraborty, S., Jain, K. K., Sharma, A., & Kuhad, R. C. (2016). Antioxidant phenolics and their microbial production by submerged and solid-state fermentation process: A review. *Trends in Food Science & Technology*, *53*, 60–74.

Di Gioia, F., Tzortzakis, N., Rouphael, Y., Kyriacou, M. C., Sampaio, S. L., Ferreira, I. C. F. R., et al. (2020). Grown to be blue—antioxidant properties and health effects of colored vegetables. Part II: Leafy, fruit, and other vegetables. *Antioxidants*, *9*, 97.

Djilas, S., Čanadanović-Brunet, J., & Ćetković, G. (2009). By-products of fruits processing as a source of phytochemicals. *Chemical Industry and Chemical Engineering*, *15*, 191–202.

Downey, R. K., & Bell, J. M. (1990). New developments in canola research. In F. Shahidi (Ed.), *Canola and rapeseed: Production, chemistry, nutrition and processing technology* (pp. 37–45). New York, NY: Springer.

Edirisinghe, I., & Burton-Freeman, B. (2016). Anti-diabetic actions of berry polyphenols—Review on proposed mechanisms of action. *Journal of Berry Research*, *6*, 237–250.

Ehrenkranz, J. R. L., Lewis, N. G., Ronald Kahn, C., & Roth, J. (2005). Phlorizin: A review. *Diabetes/Metabolism Research and Reviews*, *21*, 31–38.

Ekezie, F.-G. C., Sun, D.-W., & Cheng, J.-H. (2017). Acceleration of microwave-assisted extraction processes of food components by integrating technologies and applying emerging solvents: A review of latest developments. *Trends in Food Science & Technology*, *67*, 160–172.

FAOSTAT. (2018). *Potatoes production in the world*. Available online http://www.fao.org/faostat/en/#data/QC.

Ferlemi, A.-V., & Lamari, F. N. (2016). Berry leaves: An alternative source of bioactive natural products of nutritional and medicinal value. *Antioxidants*, *5*, 17.

Ferreira, L. F., Minuzzi, N. M., Rodrigues, R. F., Pauletto, R., Rodrigues, E., Emanuelli, T., et al. (2020). Citric acid water-based solution for blueberry bagasse anthocyanins recovery: Optimization and comparisons with microwave-assisted extraction (MAE). *LWT- Food Science and Technology*, *133*, 110064.

Fierascu, R. C., Fierascu, I., Avramescu, S. M., & Sieniawska, E. (2019). Recovery of natural antioxidants from agro-industrial side streams through advanced extraction techniques. *Molecules*, *24*(23), 4212.

Filannino, P., Bai, Y., Di Cagno, R., Gobbetti, M., & Gänzle, M. G. (2015). Metabolism of phenolic compounds by *Lactobacillus* spp. during fermentation of cherry juice and broccoli puree. *Food Microbiology*, *46*, 272–279.

Gan, R. Y., Shah, N. P., Wang, M. F., Lui, W. Y., & Corke, H. (2016). Fermentation alters antioxidant capacity and polyphenol distribution in selected edible legumes. *International Journal of Food Science & Technology*, *51*(4), 875–884.

Gharachorloo, M., Tarzi, B. G., & Baharinia, M. (2013). The effect of germination on phenolic compounds and antioxidant activity of pulses. *Journal of the American Oil Chemists' Society*, *90*(3), 407–411.

Gligor, O., Mocan, A., Moldovan, C., Locatelli, M., Crişan, G., & Ferreira, I. C. F. R. (2019). Enzyme-assisted extractions of polyphenols—A comprehensive review. *Trends in Food Science & Technology*, *88*, 302–315.

Goiris, K., Muylaert, K., Fraeye, I., Foubert, I., De Brabanter, J., & De Cooman, L. (2012). Antioxidant potential of microalgae in relation to their phenolic and carotenoid content. *Journal of Applied Phycology*, *24*, 1477–1486.

Goiris, K., Muylaert, K., Voorspoels, S., Noten, B., De Paepe, D. E., Baart, G. J., et al. (2014). Detection of flavonoids in microalgae from different evolutionary lineages. *Journal of Phycology*, *50*, 483–492.

Gong, L., Cao, W., Chi, H., Wang, J., Zhang, H., Liu, J., et al. (2018). Review whole cereal grains and potential health effects: Involvement of the gut microbiota. *Food Research International*, *103*, 84–102.

Grimm, A., Nyström, J., Mossing, T., Östman, U. B., & Geladi, P. (2020). Novel drying treatment to stabilize bilberry, blackcurrant, and cloudberry press cakes: Dryer performance and product quality characteristics. *LWT- Food Science and Technology*, *128*, 109478.

Guo, X., Yang, B., Tan, J., Jiang, J., & Li, D. (2016). Associations of dietary intakes of anthocyanins and berry fruits with risk of type 2 diabetes mellitus: A systematic review and meta-analysis of prospective cohort studies. *European Journal of Clinical Nutrition*, *70*, 1360–1367.

Haoujar, I., Cacciola, F., Abrini, J., Mangraviti, D., Giuffrida, D., Oulad El Majdoub, Y., et al. (2019). The contribution of carotenoids, phenolic compounds, and flavonoids to the antioxidative properties of marine microalgae isolated from Mediterranean Morocco. *Molecules*, *24*, 4037.

He, J., & Giusti, M. M. (2010). Anthocyanins: Natural colorants with health-promoting 380 properties. *Annual Review of Food Science and Technology*, *1*, 163–187.

Helbig, D., Böhm, V., Wagner, A., Schubert, R., & Jahreis, G. (2008). Berry seed press residues and their valuable ingredients with special regard to black currant seed press residues. *Food Chemistry*, *111*, 1043–1049.

Hellström, J., Törrönen, R., & Mattila, P. (2009). Proanthocyanidins in common food products of plant origin. *Journal of Agricultural and Food Chemistry*, *57*, 7899–7906.
Hellström, J., Granato, D., & Mattila, P. (2020). Accumulation of phenolic acids during storage over differently handled fresh carrots. *Food*, *9*, 1515.
Herrero, M., Mendiola, J. A., Cifuentes, A., & Ibáñez, E. (2010). Supercritical fluid extraction: Recent advances and applications. *Journal of Chromatography A*, *1217*(16), 2495–2511.
Höglund, E., Eliasson, L., Oliveira, G., Almli, V. L., Sozer, N., & Alminger, M. (2018). Effect of drying and extrusion processing on physical and nutritional characteristics of bilberry press cake extrudates. *LWT- Food Science and Technology*, *92*, 422–428.
Howard, L. R., & Griffin, L. E. (1993). Lignin formation and surface discolouration of minimally processed carrot sticks. *Journal of Food Science*, *58*, 1065–1067.
Hunaefi, D., Akumo, D. N., & Smetanska, I. (2013). Effect of fermentation on antioxidant properties of red cabbages. *Food Biotechnology*, *27*(1), 66–85.
Huynh, N. T., Van Camp, J., Smagghe, G., & Raes, K. (2014). Improved release and metabolism of flavonoids by steered fermentation processes: A review. *International Journal of Molecular Sciences*, *15*(11), 19369–19388.
Hyson, D. A. (2011). A comprehensive review of apples and apple components and their relationship to human health. *Advances in Nutrition*, *2*, 408–420.
Icier, F., Yildiz, H., Sabanci, S., Cevik, M., & Cokgezme, O. F. (2017). Ohmic heating assisted vacuum evaporation of pomegranate juice: Electrical conductivity changes. *Innovative Food Science & Emerging Technologies*, *39*, 241–246.
Im, H. W., Suh, B. S., Lee, S. U., Kozukue, N., Ohnisi-Kameyama, M., Levin, C. E., et al. (2008). Analysis of phenolic compounds by high-performance liquid chromatography and liquid chromatography/mass spectrometry in potato plant flowers, leaves, stems, and tubers and in home-processed potatoes. *Journal of Agricultural and Food Chemistry*, *56*(9), 3341–3349.
Jahnke, L. S. (1999). Massive carotenoid accumulation in Dunaliella bardawil induced by ultraviolet-A radiation. *Journal of Photochemistry and Photobiology B - Biology*, *48*(1), 68–74. https://doi.org/10.1016/S1011-1344(99)00012-3.
Joseph, S. V., Edirisinghe, I., & Burton-Freeman, B. M. (2014). Berries: Anti-inflammatory effects in humans. *Journal of Agricultural and Food Chemistry*, *62*, 3886–3903.
Julian, H., Yaohanny, F., Devina, A., Purwadi, R., & Wenten, I. G. (2019). Apple juice concentration using submerged direct contact membrane distillation (SDCMD). *Journal of Food Engineering*, *272*, 109807.
Kadiri, O. (2017). A review on the status of the phenolic compounds and antioxidant capacity of the flour: Effects of cereal processing. *International Journal of Food Properties*, *20*(sup1), S798–S809.
Kaimainen, M., Marze, S., Järvenpää, E., Anton, M., & Huopalahti, R. (2015). Encapsulation of betalain into w/o/w double emulsion and release during in vitro intestinal lipid digestion. *LWT- Food Science and Technology*, *60*(2), 899–904.
Klavins, L., Kviesis, J., Nakurte, I., & Klavins, M. (2018). Berry press residues as a valuable source of polyphenolics: Extraction optimisation and analysis. *LWT- Food Science and Technology*, *93*, 583–591.
Knudsen, K. E. B., Nørskov, N. P., Bolvig, A. K., Hedemann, M. S., & Lærke, H. N. (2017). Dietary fibers and associated phytochemicals in cereals. *Molecular Nutrition & Food Research*, *61*, 1600518.
Koponen, J., Happonen, A., Mattila, P., & Törrönen, R. (2007). Contents of anthocyanins and ellagitannins in foods consumed in Finland. *Journal of Agricultural and Food Chemistry*, *55*, 1612–1619.
Kowalska, H., Czajkowska, K., Cichowska, J., & Lenart, A. (2017). What's new in biopotential of fruit and vegetable by-products applied in the food processing industry. *Trends in Food Science & Technology*, *67*, 150–159.

Kozlowska, H., Zadernowski, R., & Sosulski, F. W. (1983). Phenolic acids in oilseed flours. *Die Nahrung, 27*, 449–453.

Kozlowska, H., Rotkiewicz, D. A., & Zadernovski, R. (1983). Phenolic acids in rapeseed and mustard. *Journal of the American Oil Chemists' Society, 60*, 1119–1123.

Kozlowska, H., Naczk, M., Shahidi, F., & Zadernowski, R. (1990). Phenolic acids and tannins in rapeseed and canola. In F. Shahidi (Ed.), *Canola and rapeseed: Production, chemistry, nutrition and processing technology* (pp. 193–210). New York, NY: Springer.

Krygier, K., Sosulski, F., & Hogge, L. (1982). Free, esterified, and insoluble-bound phenolic acids. 1. Extraction and purification procedure. *Journal of Agricultural and Food Chemistry, 30*, 330–334.

Kumar, M. S. Y., Dutta, R., Prasad, D., & Misra, K. (2011). Subcritical water extraction of antioxidant compounds from Sea buckthorn (*Hippophae rhamnoides*) leaves for the comparative evaluation of antioxidant activity. *Food Chemistry, 127*, 1309–1316.

Kumar, Y., Singhal, S., Tarafdar, A., Pharande, A., Ganesan, M., & Badgujar, P. C. (2020). Ultrasound assisted extraction of selected edible macroalgae: Effect on antioxidant activity and quantitative assessment of polyphenols by liquid chromatography with tandem mass spectrometry (LC-MS/MS). *Algal Research, 52*, 102114.

Kumar, K., Srivastav, S., & Sharanagat, V. S. (2021). Ultrasound assisted extraction (UAE) of bioactive compounds from fruit and vegetable processing by-products: A review. *Ultrasonics Sonochemistry, 70*, 105325.

Landete, J. M., Curiel, J. A., Rodríguez, H., de las Rivas, B., & Muñoz, R. (2014). Aryl glycosidases from *Lactobacillus plantarum* increase antioxidant activity of phenolic compounds. *Journal of Functional Foods, 7*, 322–329.

Latva-Mäenpää, H. (2017). *Bioactive and protective polyphenolics from roots and stumps of conifer trees (Norway spruce and Scots pine) [Academic Dissertation]* (p. 74). *Chemistry Faculty of Science University of Helsinki Finland.*

Lee, I. H., Hung, Y. H., & Chou, C. C. (2008). Solid-state fermentation with fungi to enhance the antioxidative activity, total phenolic and anthocyanin contents of black bean. *International Journal of Food Microbiology, 121*(2), 150–156.

Lee, J. H., Kasote, D. M., Jayaprakasha, G. K., Avila, C. A., Crosby, K. M., & Patil, B. S. (2020). Effect of production system and inhibitory potential of aroma volatiles on polyphenol oxidase and peroxidase activity in tomatoes. *Journal of the Science of Food and Agriculture, 101*, 307–314.

Li, F., Chen, G., Zhang, B., & Fu, X. (2017). Current applications and new opportunities for the thermal and non-thermal processing technologies to generate berry product or extracts with high nutraceutical contents. *Food Research International, 100*, 19–30.

Liu, J., Sandahl, M., Sjöberg, P. J., & Turner, C. (2014). Pressurised hot water extraction in continuous flow mode for thermolabile compounds: Extraction of polyphenols in red onions. *Analytical and Bioanalytical Chemistry, 406*(2), 441–445.

Liu, W., Yang, C., Zhou, C., Wen, Z., & Dong, X. (2019). An improved microwave-assisted extraction of anthocyanins from purple sweet potato in favor of subsequent comprehensive utilization of pomace. *Food and Bioproducts Processing, 115*, 1–9.

López, A., Rico, M., Rivero, A., & Suaréz de Tanguil, M. (2011). The effects of solvents on the phenolic contents and antioxidant activity of *Stypocaulon scoparium* algae extracts. *Food Chemistry, 125*, 1104–1109.

Luithui, T., Nisha, R. B., & Meera, M. S. (2019). Cereal by-products as an important functional ingredient: Effect of processing. *Journal of Food Science and Technology, 56*, 1–11.

Maadane, A., Merghoub, N., Ainane, T., El Arroussi, H., Benhima, R., Amzazi, S., et al. (2015). Antioxidant activity of some Moroccan marine microalgae: Pufa profiles, carotenoids and phenolic content. *Journal of Biotechnology, 215*, 13–19. https://doi.org/10.1016/j.jbiotec.2015.06.400.

Macías-Cortés, E., Gallegos-Infante, J. A., Rocha-Guzmán, N. E., Moreno-Jiménez, M. R., Medina-Torres, L., & González-Laredo, R. F. (2020). Microencapsulation of phenolic compounds: Technologies and novel polymers. *Revista Mexicana de Ingenieria Quimica, 19*(2), 491–521.

Madeira Junior, J. V., Teixeira, C. B., & Macedo, G. A. (2015). Biotransformation and bioconversion of phenolic compounds obtainment: An overview. *Critical Reviews in Biotechnology, 35*(1), 75–81.

Mahgouba, H. A. M., Eisab, G. S. A., & Youssefc, M. A. H. (2015). Molecular, biochemical and anatomical analysis of some potato (Solanum tuberosum L.) cultivars growing in Egypt. *Journal, Genetic Engineering & Biotechnology, 13*, 39–49.

Mäkinen, S., Hellström, J., Mäki, M., Korpinen, R., & Mattila, P. H. (2020). Bilberry and sea buckthorn leaves and their subcritical water extracts prevent lipid oxidation in meat products. *Food, 9*(3), 265.

Mariquele, B. E., Hagelaar, G., & Lakemond, C. (2012). *Critical Quality Points in fresh-cut vegetables supply chain: Understanding critical factors that influence shelf life of fresh-cut vegetables at retailers from a techno-managerial approach.* Thesis work report. Pdf-format published. https://doi.org/10.13140/2.1.3220.0000.

Market Research Report. (2019). *Market research report.* San Francisco, CA, USA: Grand View Research, Inc.

Martin, L. J., & Matar, C. (2005). Increase of antioxidant capacity of the lowbush blueberry (Vaccinium angustifolium) during fermentation by a novel bacterium from the fruit microflora. *Journal of the Science of Food and Agriculture, 85*(9), 1477–1484.

Martins, S., Mussatto, S. I., Martínez-Avila, G., Montañez-Saenz, J., Aguilar, C. N., & Teixeira, J. A. (2011). Bioactive phenolic compounds: Production and extraction by solid-state fermentation. A review. *Biotechnology Advances, 29*(3), 365–373.

Mattila, P., & Hellström, J. (2007). Phenolic acids in potatoes, vegetables, and some of their products. *Journal of Food Composition and Analysis, 20*, 152–160.

Mattila, P., Astola, J., & Kumpulainen, J. (2000). Determination of flavonoids in plant material by HPLC with diode array and electro array detections. *Journal of Agricultural and Food Chemistry, 48*, 5834–5841.

Mattila, P., Pihlava, J.-M., & Hellström, J. (2005). Contents of phenolic acids, alkyl- and alkenylresorcinols, and avenanthramides in Commercial Grain Products. *Journal of Agricultural and Food Chemistry, 53*, 8290–8295.

Mattila, P., Hellström, J., & Törrönen, R. (2006). Phenolic acids in berries, fruits, and beverages. *Journal of Agricultural and Food Chemistry, 52*, 4477–4486.

May, N., & Guenther, E. (2020). Shared benefit by Material Flow Cost Accounting in the food supply chain–The case of berry pomace as upcycled by-product of a black currant juice production. *Journal of Cleaner Production, 245*, 118946.

Mekky, R. H., Thabet, M. M., Rodríguez-Pérez, C., Elnaggar, D. M. Y., Mahrous, E. A., Segura-Carretero, A., et al. (2020). Comparative metabolite profiling and antioxidant potentials of seeds and sprouts of three Egyptian cultivars of Vicia faba L. *Food Research International, 136*, 109537.

Metsämuuronen, S., & Sirén, H. (2019). Bioactive phenolic compounds, metabolism and properties: A review on valuable chemical compounds in Scots pine and Norway spruce. *Phytochemistry Reviews, 18*(3), 623–664.

Mohdaly, A. A. A., Hassanien, M. F. R., Mahmoud, A., Sarhan, M. A., & Smetanska, I. (2013). Phenolic extracted from potato, sugar beet, and sesame processing by-products. *International Journal of Food Properties, 16*, 1148–1168.

Moon, K. M., Kwon, E. B., Lee, B., & Kim, C. Y. (2020). Recent trends in controlling the enzymatic browning of fruit and vegetable products. *Molecules, 25*(12), 2754.

Moore, J., Cheng, Z., Hao, J., Guo, G., Liu, J. G., Lin, C., et al. (2007). Effects of solid-state yeast treatment on the antioxidant properties and protein and fiber compositions of common hard wheat bran. *Journal of Agricultural and Food Chemistry, 55*(25), 10173–10182.

Mousavi, Z. E., Mousavi, S. M., Razavi, S. H., Hadinejad, M., Emam-Djomeh, Z., & Mirzapour, M. (2013). Effect of fermentation of pomegranate juice by Lactobacillus plantarum and Lactobacillus acidophilus on the antioxidant activity and metabolism of sugars, organic acids and phenolic compounds. *Food Biotechnology*, *27*(1), 1–13.

Murador, D. C., Braga, A. R. C., Martins, P. L. G., Mercadante, A. Z., & de Rosso, V. (2019). Ionic liquid associated with ultrasonic-assisted extraction: A new approach to obtain carotenoids from orange peel. *Food Research International*, *126*, 108653.

Naczk, M., Amarowicz, R., Sullivan, A., & Shahidi, F. (1998). Current research developments on polyphenolics of rapeseed/canola: A review. *Food Chemistry*, *62*, 489–502.

Naczk, M., Amarowicz, R., Pink, D., & Shahidi, F. (2000). Insoluble condensed tannins of Canola/Rapeseed. *Journal of Agricultural and Food Chemistry*, *48*, 1758–1762.

Nayak, C. A., & Rastogi, N. K. (2010). Comparison of osmotic membrane distillation and forward osmosis membrane processes for concentration of anthocyanin. *Desalination and Water Treatment*, *16*, 134–145.

Ninčević Grassino, A., Ostojić, J., Miletić, V., Djaković, S., Bosiljkov, T., Zorić, Z., et al. (2020). Application of high hydrostatic pressure and ultrasound-assisted extractions as a novel approach for pectin and polyphenols recovery from tomato peel waste. *Innovative Food Science & Emerging Technologies*, *64*, 102424. ISSN 1466-8564.

Norouzi, S., Fadavi, A., & Darvishi, H. (2021). The ohmic and conventional heating methods in concentration of sour cherry juice: Quality and engineering factors. *Journal of Food Engineering*, *291*, 110242.

Nova, P., Pimenta-Martins, A., Silva, J. L., Silva, A. M., Gomes, A. M., & Freitas, A. C. (2020). Health benefits and bioavailability of marine resources components that contribute to health—What's new? *Critical Reviews in Food Science and Nutrition*, *60*(21), 3680–3692. https://doi.org/10.1080/10408398.2019.1704681.

Ochoa, S., Durango-Zuleta, M. M., & Osorio-Tobón, J. F. (2020). Techno-economic evaluation of the extraction of anthocyanins from purple yam (*Dioscorea alata*) using ultrasound-assisted extraction and conventional extraction processes. *Food and Bioproducts Processing*, *122*, 111–123.

Oertel, A., Matros, A., Hartmann, A., Arapitsas, P., Dehmer, K. J., Martens, S., et al. (2017). Metabolite profiling of red and blue potatoes revealed cultivar and tissue specific patterns for anthocyanins and other polyphenols. *Planta*, *246*, 281–297.

Olszewska, M. A., Granica, S., Kolodziejczyk-Czepas, J., Magiera, A., Czerwińska, M. E., Nowak, P., et al. (2020). Variability of sinapic acid derivatives during germination and their contribution to antioxidant and anti-inflammatory effects of broccoli sprouts on human plasma and human peripheral blood mononuclear cells. *Food & Function*, *11*(8), 7231–7244.

Osma, J. F., Toca-Herrera, J. F., & Rodrıguez-Couto, S. (2010). Use of laccases in the food industry. *Enzyme Research*, *2010*, 918761. 8 pp.

Panja, P. (2018). Green extraction methods of food polyphenols from vegetable materials. *Current Opinion in Food Science*, *23*, 173–182.

Pap, N., Pongrácz, E., Jaakkola, M., Tolonen, T., Virtanen, V., Turkki, A., et al. (2010). The effect of pre-treatment on the anthocyanin and flavonol content of black currant juice (*Ribes nigrum* L.) in concentration by reverse osmosis. *Journal of Food Engineering*, *98*(4), 429–436.

Petropoulos, S. A., Sampaio, S. L., Di Gioia, F., Tzortzakis, N., Rouphael, Y., Kyriacou, M. C., et al. (2019). Grown to be blue—Antioxidant properties and health effects of colored vegetables. Part I: Root vegetables. *Antioxidants*, *8*, 617.

Plaza, M., Abrahamsson, V., & Turner, C. (2013). Extraction and neoformation of antioxidant compounds by pressurized hot water extraction from apple byproducts. *Journal of Agricultural and Food Chemistry*, *61*(23), 5500–5510.

Popović, D. A., Milinčić, D. D., Pešić, M. B., Kalušević, A. M., Tešić, Ž. L., & Nedović, V. A. (2019). Encapsulation technologies for polyphenol-loaded microparticles in food industry. In F. Chemat, & E. Vorobiev (Eds.), *Green food processing techniques* (pp. 335–367). London: Academic Press.

Pribis, P., & Shukitt-Hale, B. (2014). Cognition: The new frontier for nuts and berries. *American Journal of Clinical Nutrition*, *100*(suppl), 347S–352S.

Ptichkina, N. M., Markina, O. A., & Rumyantseva, G. N. (2008). Pectin extraction from pumpkin with the aid of microbial enzymes. *Food Hydrocolloids*, *22*(1), 192–195.

Puljula, E., Walton, G., Woodward, M. J., & Karonen, M. (2020). Antimicrobial activities of ellagitannins against *Clostridiales perfringens*, *Escherichia coli*, *Lactobacillus plantarum* and *Staphylococcus aureus*. *Molecules*, *25*, 3714.

Puupponen-Pimiä, R., Nohynek, L., Juvonen, R., Kössö, T., Truchado, P., Westerlund-Wikström, et al. (2016). Fermentation and dry fractionation increase bioactivity of cloudberry (*Rubus chamaemorus*). *Food Chemistry*, *197*, 950–958.

Quintana, S. E., Villanueva-Bermejo, D., Reglero, G., García-Risco, M. R., & Fornari, T. (2019). Supercritical antisolvent particle precipitation and fractionation of rosemary (*Rosmarinus officinalis* L.) extracts. *Journal of CO_2 Utilization*, *34*, 479–489.

Raitanen, J. E., Järvenpää, E., Korpinen, R., Mäkinen, S., Hellström, J., Kilpeläinen, P., et al. (2020). Tannins of conifer bark as nordic piquancy—Sustainable preservative and aroma? *Molecules*, *25*(3), 567.

Ravn-Haren, G., Dragsted, L. O., Buch-Andersen, T., Jensen, E. N., Jensen, R. I., Nemeth-Balogh, M., et al. (2013). Intake of whole apples or clear apple juice has contrasting effects on plasma lipids in healthy volunteers. *European Journal of Nutrition*, *52*(8), 1875–1889.

Ren, F., Nian, Y., & Perussello, C. A. (2020). Effect of storage, food processing and novel extraction technologies on onions flavonoid content: A review. *Food Research International*, *132*, 108953.

Reyes, L. F., Villarreal, J. E., & Cisneros-Zevallos, L. (2007). The increase in antioxidant capacity after wounding depends on the type of fruit or vegetable tissue. *Food Chemistry*, *101*, 1254–1262.

Rezaei, A., Fathi, M., & Jafari, S. M. (2019). Nanoencapsulation of hydrophobic and low-soluble food bioactive compounds within different nanocarriers. *Food Hydrocolloids*, *88*, 146–162.

Rodríguez, H., Curiel, J. A., Landete, J. M., de Las Rivas, B., de Felipe, F. L., Gómez-Cordovés, C., et al. (2009). Food phenolics and lactic acid bacteria. *International Journal of Food Microbiology*, *132*(2–3), 79–90.

Rodriguez-Mateos, A., Heiss, C., Borges, G., & Crozier, A. (2013). Berry (poly)phenols and cardiovascular health. *Journal of Agricultural and Food Chemistry*, *62*(18), 3842–3851.

Romero-Díez, R., Matos, M., Rodrigues, L., Bronze, M. R., Rodríguez-Rojo, S., Cocero, M. J., et al. (2019). Microwave and ultrasound pre-treatments to enhance anthocyanins extraction from different wine lees. *Food Chemistry*, *272*, 258–266.

Sabanci, S., & Icier, F. (2017). Applicability of ohmic heating assisted vacuum evaporation for concentration of sour cherry juice. *Journal of Food Engineering*, *212*, 262–270.

Sagar, N. A., Pareek, S., Sharma, S., Yahia, E. M., & Lobo, M. G. (2018). Fruit and vegetable waste: Bioactive compounds, their extraction, and possible utilization. *Comprehensive Reviews in Food Science and Food Safety*, *17*, 512–531.

Saifullah, M., McCullum, R., McCluskey, A., & Vuong, Q. (2020). Comparison of conventional extraction technique with ultrasound assisted extraction on recovery of phenolic compounds from lemon scented tea tree (Leptospermum petersonii) leaves. *Heliyon*, *6*(4), e03666.

Saini, R. K., & Keum, Y.-S. (2018). Carotenoid extraction methods: A review of recent developments. *Food Chemistry*, *240*, 90–103.

Schieber, A. (2017). Side streams of plant food processing as a source of valuable compounds: Selected samples. *Annual Review of Food Science and Technology*, *8*, 97–112.

Septembre-Malaterre, A., Remize, F., & Poucheret, P. (2018). Fruits and vegetables, as a source of nutritional compounds and phytochemicals: Changes in bioactive compounds during lactic fermentation. *Food Research International, 104*, 86–99.

Shahidi, F., & Naczk, M. (1992). An overview of the phenolics of canola and rapeseed: Chemical, sensory and nutritional significance. *Journal of the American Oil Chemists' Society, 69*, 917–924.

Sharma, K., Mahato, N., Shivraj, N., & Rok Lee, Y. (2016). Economical and environment-friendly approaches for usage of onion (Allium cepa L.) wastes. *Food & Function, 7*, 3354.

Sharma, K., Mahato, N., & Rok Leef, Y. (2018). Systematic study on active compounds as antibacterial and antibiofilm agent in aging onions. *Food and Drug Analysis, 26*, 518–528.

Shen, Y., Sun, H., Zeng, H., Prinyawiwatukul, W., Xu, W., & Xu, Z. (2018). Increases of phenolics, fatty acids, phytosterols, and anticancer activity of sweet potato after fermented by lactobacillus acidophilus. *Journal of Agricultural and Food Chemistry, 66*, 2735–2741.

Shitu, A., Izhar, S., & Tahir, T. M. (2015). Sub-critical water as a green solvent for production of valuable materials from agricultural waste biomass: A review of recent work. *Global Journal of Environmental Science and Management, 1*, 255–264.

Sillero, L., Prado, R., & Labidi, J. (2020). Simultaneous microwave-ultrasound assisted extraction of bioactive compounds from bark. *Chemical Engineering and Processing Process Intensification, 156*, 108100.

Singh, P. P., & Saldaña, M. D. A. (2011). Subcritical water extraction of phenolic compounds from potato peel. *Food Research International, 44*, 2452–2458.

Spinelli, S., Costa, C., Conte, A., La Porta, N., Padalino, L., & Del Nobile, M. A. (2019). Bioactive compounds from Norway spruce bark: Comparison among sustainable extraction techniques for potential food applications. *Food, 8*(11), 524.

Strižincová, P., Ház, A., Burčová, Z., Feranc, J., Kreps, F., Šurina, I., et al. (2019). Spruce bark—A source of polyphenolic compounds: Optimizing the operating conditions of supercritical carbon dioxide extraction. *Molecules, 24*(22), 4049.

Struck, S., Plaza, M., Turner, C., & Rohm, H. (2016). Berry pomace—A review of processing and chemical analysis of its polyphenols. *International Journal of Food Science and Technology, 51*, 1305–1318.

Subash, S., Essa, M. M., Al-Adawi, S., Memon, M. A., Manivasagam, T., & Akbar, M. (2014). Neuroprotective effects of berry fruits on neurodegenerative diseases. *Neural Regeneration Research, 9*, 1557–1566.

Tamkutė, L., Liepuoniūtė, R., Pukalskienė, M., & Venskutonis, P. R. (2020). Recovery of valuable lipophilic and polyphenolic fractions from cranberry pomace by consecutive supercritical CO_2 and pressurized liquid extraction. *The Journal of Supercritical Fluids, 159*, 104755.

Tang, S., Mao, G., Yuan, Y., Weng, Y., Zhu, R., Cai, C., et al. (2020). Optimization of oat seed steeping and germination temperatures to maximize nutrient content and antioxidant activity. *Journal of Food Processing and Preservation, 44*(9), 1–11. https://doi.org/10.1111/jfpp.14683.

Tian, B., Xie, B., Shi, J., Wu, J., Cai, Y., Xu, T., et al. (2010). Physicochemical changes of oat seeds during germination. *Food Chemistry, 119*(3), 1195–1200.

Tian, Y., Liimatainen, J., Alanne, A.-L., Lindstedt, A., Liu, P., Sinkkonen, J., et al. (2017). Phenolic compounds extracted by acidic aqueous ethanol from berries and leaves of different berry plants. *Food Chemistry, 222*, 266–281. https://doi.org/10.1016/j.foodchem.2016.09.145.

Tinello, F., & Lante, A. (2018). Recent advances in controlling polyphenol oxidase activity of fruit and vegetable products. *Innovative Food Science & Emerging Technologies, 50*, 73–83.

Transparency Market Research. (2016). Anthocyanins market (product type—Cyanidin, malvidin, delphinidin, and peonidin; application—Food beverage, pharmaceuticals

products, and personal care)—Global industry analysis. *Size, Share, Growth, Trends, and Forecast*, 2015–2021.
Tunchaiyaphum, S., Eshtiaghi, M. N., & Yoswathana, N. (2013). Extraction of bioactive compounds from mango peels using green technology. *International Journal of Chemical Engineering and Applications*, *4*, 194–198.
Tyśkiewicz, K., Konkol, M., & Rój, E. (2018). The application of supercritical fluid extraction in phenolic compounds isolation from natural plant materials. *Molecules*, *23*(10), 2625.
Varzakas, T. H., Zakynthinos, G., & Verpoort, F. (2016). Plant food residues as a source of nutraceuticals and functional foods. *Food*, *5*(4), 88.
Vattem, D. A., & Shetty, K. (2003). Ellagic acid production and phenolic antioxidant activity in cranberry pomace (*Vaccinium macrocarpon*) mediated by *Lentinus edodes* using a solid-state system. *Process Biochemistry*, *39*(3), 367–379.
Verkasalo, E., Leppälä, J., Muhonen, T., Korpinen, R., Kurppa, S., & Möttönen, V. (2019). Novel industrial ecosystems and value chains to utilize side-streams of wood products industries—Finnish perspective. In L. Gurau, M. Campean, & M. Ispas (Eds.), *Proceedings of the international conference wood science and engineering in the third millenum–icwse 2019* (12th ed., pp. 277–294). Brasov: Editura Universitatii Transilvania din Brasov.
Verni, M., Rizzello, C. G., & Coda, R. (2019). Fermentation biotechnology applied to cereal industry by-products: Nutritional and functional insights. *Frontiers in Nutrition*, *6*, 42.
Villamil-Galindo, E., Van de Velde, F., & Piagentini, A. M. (2020). Extracts from strawberry by-products rich in phenolic compounds reduce the activity of apple polyphenol oxidase. *LWT*, *133*, 110097.
Wei, M., Wang, S., Gu, P., Ouyang, X., Liu, S., Li, Y., et al. (2018). Comparison of physicochemical indexes, amino acids, phenolic compounds and volatile compounds in bog bilberry juice fermented by *Lactobacillus plantarum* under different pH conditions. *Journal of Food Science and Technology*, *55*(6), 2240–2250.
Wenten, I. G., Khoiruddin, K., Reynard, R., Lugito, G., & Julian, H. (2021). Advancement of forward osmosis (FO) membrane for fruit juice concentration. *Journal of Food Engineering*, *290*, 110216.
Wikiera, A., Mika, M., Starzyńska-Janiszewska, A., & Stodolak, B. (2016). Endo-xylanase and endo-cellulase-assisted extraction of pectin from apple pomace. *Carbohydrate Polymers*, *142*, 199–205.
Xu, J. G., Tian, C. R., Hu, Q. P., Luo, J. Y., Wang, X. D., & Tian, X. D. (2009). Dynamic changes in phenolic compounds and antioxidant activity in oats (Avena nuda L.) during steeping and germination. *Journal of Agricultural and Food Chemistry*, *57*, 10392–10398.
Yan, Y., Zhang, F., Chai, Z., Liu, M., Battino, M., & Meng, X. (2019). Mixed fermentation of blueberry pomace with *L. rhamnosus* GG and *L. plantarum*-1: Enhance the active ingredient, antioxidant activity and health-promoting benefits. *Food and Chemical Toxicology*, *131*, 110541.
Yang, X., Zhou, J., Fan, L., Qin, Z., Chen, Q., & Zhao, L. (2018). Antioxidant properties of a vegetable–fruit beverage fermented with two *Lactobacillus plantarum* strains. *Food Science and Biotechnology*, *27*(6), 1719–1726.
Yin, L., Chen, T., Li, Y., Fu, S., Li, L., Xu, M., et al. (2016). Comparative study on total anthocyanin content, composition of anthocyanidin, total phenolic content and antioxidant activity of pigmented potato peel and flesh. *Food Science and Technology Research*, *22*, 219–226.
Yuliarti, O., Goh, K. K. T., Matia-Merino, L., Mawson, J., & Brennan, C. (2015). Extraction and characterisation of pomace pectin from gold kiwifruit (Actinidia chinensis). *Food Chemistry*, *187*, 290–296.

Zanoni, F., Primiterra, M., Angeli, N., & Zoccatelli, G. (2020). Microencapsulation by spray-drying of polyphenols extracted from red chicory and red cabbage: Effects on stability and color properties. *Food Chemistry*, *307*, 125535.
Zhang, D., & Hamauzu, Y. (2004). Phenolics, ascorbic acid, carotenoids and antioxidant activity of broccoli and their changes during conventional and microwave cooking. *Food Chemistry*, *88*, 503–509.
Ziberna, L., Lunder, M., Moze, S., Vanzo, A., Tramer, F., Passamonti, S., et al. (2010). Acute cardioprotective and cardiotoxic effects of bilberry anthocyanins in 429 ischemia–reperfusion injury: beyond concentration-dependent antioxidant 430 activity. *Cardiovascular Toxicology*, *10*, 283–294.

CHAPTER FIVE

Strategies to increase the shelf life of meat and meat products with phenolic compounds

Paulo E.S. Munekata[a], Mirian Pateiro[a], Elisa Rafaela Bonadio Bellucci[b], Rubén Domínguez[a], Andrea Carla da Silva Barretto[b], and José M. Lorenzo[a,c,*]

[a]Centro Tecnológico de la Carne de Galicia, rúa Galicia n° 4, Parque Tecnológico de Galicia, Ourense, Spain
[b]Department of Food Technology and Engineering, UNESP—São Paulo State University, São José do Rio Preto, SP, Brazil
[c]Facultad de Ciencias de Ourense, Área de Tecnología de los Alimentos, Universidad de Vigo, Ourense, Spain
*Corresponding author: e-mail address: jmlorenzo@ceteca.net

Contents

Abstract

Oxidative reactions and microbial growth are the main processes involved in the loss of quality in meat products. Although the use of additives to improve the shelf life is a common practice in the meat industry, the current trends among consumers are pushing the researchers and professionals of the meat industry to reformulate meat products. Polyphenols are compounds with antioxidant and antimicrobial activity naturally found in several plants, fruits, and vegetables that can be used in the production of extracts and components in active packaging to improve the shelf life of meat products. This chapter aims to discuss the advances in terms of (1) encapsulation techniques to protect phenolic compounds; (2) production of active and edible packages rich on phenolic compounds; (3) use of phenolic-rich additives (free or encapsulated form) with non-thermal technologies to improve the shelf life of meat products; and (4) use of

Advances in Food and Nutrition Research, Volume 98
ISSN 1043-4526
https://doi.org/10.1016/bs.afnr.2021.02.008

active packaging rich on phenolic compounds on meat products. Innovative strategies to encapsulated polyphenols and produce films are mainly centered in the use of innovative and emerging technologies (such as ultrasound and supercritical fluids). Moreover, the combined use of polyphenols and non-thermal technologies is a relevant approach to improve the shelf life of meat products, especially using high pressure processing. In terms of application of innovative films, nanomaterials have been largely explored and indicated as relevant strategy to preserve meat and meat products.

1. Introduction: Quality decay and shelf life of meat products and phenolic compounds

The quality of meat and meat products is drastically reduced by the progression of oxidative reactions and the growth of microorganisms (Horita et al., 2018; Lorenzo, Varges, et al., 2018). In terms of oxidation, this effect is mainly observed in myoglobin and lipids. The loss of characteristic bright cherry-red in meat and meat products is a consequence of myoglobin oxidation, which is directly related to the redox state of iron in the heme fraction of myoglobin molecule (Lorenzo, Pateiro et al., 2018; Ribeiro et al., 2019). Once exposed to oxidizing conditions, oxymyoglobin is gradually oxidized to metmyoglobin, which causes the shift in color of meat and meat products from the bright cherry-red to brown (Lorenzo, Pateiro, et al., 2018). This muscle pigment is the first attribute when judging the quality of meat and meat products (Faustman & Cassens, 1990; Suman & Joseph, 2013). The oxidative stability of myoglobin is affected by many factors which include exposure to oxidizing agents (UV-radiation and atmospheric oxygen), decay of endogenous antioxidant activity, pH, processing conditions (thermal treatment and refrigeration), and ingredients (food-grade antioxidants, food components, and nitrate salts) (Faustman, Sun, Mancini, & Suman, 2010).

The lipid oxidation has a central role in the quality decay of meat products. The importance of protecting lipids from the action of oxidizing agents is related to the progression of the reactions involved in the degradation of lipids (especially unsaturated fatty acids) (Lorenzo, Pateiro, et al., 2018). The auto-oxidation of fatty acids play an central role in this scenario and is known to occur in three stages: initiation (formation of the first alkyl radical radicals by the action of oxidizing agents), propagation (chain reaction that disseminated the formation of radical species through lipid phase), and termination (formation of stable oxidation products) (Domínguez et al., 2019). The eventual accumulation of oxidation products is associated with changes in the perception and acceptance (particularly the odor and flavor) of meat and meat products (Campo et al., 2006; Greene & Cumuze, 1982; Verma & Sahoo, 2000).

Regarding the microbial spoilage of meat products, this category of food is susceptible to growth of microorganism due to their chemical composition and physic-chemical characteristics and inadequate practices during processing and storage (temperature, packaging conditions and the presence of hurdles) (Doulgeraki, Ercolini, Villani, & Nychas, 2012; Nychas, Skandamis, Tassou, & Koutsoumanis, 2008). Once favorable conditions for microbial spoilage occurs, the growth of microorganisms is triggered and will gradually affect the characteristics of meat products. Consequently, important changes such as the unpleasant odors, abnormal discoloration, and the presence of slime cause the rejection by consumers (Nychas et al., 2008). This effect is related to a variety of microorganisms that may be pathogenic and affected consumers healthy. Some common groups of bacteria associated with meat products spoilage are *Bacillus* spp., *Clostridium* spp., *Escherichia* spp., *Pseudomonas* spp., *Salmonella* spp., *Sarcina* spp., *Streptococcus* spp., and other *Enterobacteriaceae*. In the case of molds, Geotrichum, Mucor, and Penicillium can be cited as relevant species. Relevant yeasts implicated in the quality loss of muscle foods are *Rhodotorula* spp. and *Candida* spp., for instance (Dave & Ghaly, 2011).

In this sense, compounds with antioxidant and antimicrobial activity play a central role in the preservation of meat and meat products. Polyphenols are secondary metabolites in plant tissues and non-essential dietary compounds that are characterized by, at least, one aromatic ring with hydroxyls groups. The polyphenol group comprises several compounds that can be classified according to their structure as: simple phenolics, flavonoids, phenolic acids, and lignins (Lorenzo et al., 2017; Vuolo, Lima, & Maróstica Junior, 2018).

The antioxidant mechanism comprises two reactions: H-atom and electron transfer. In both cases, a H or an electron is donated to stabilize a radical and prevent the progression of oxidative reactions (Leopoldini, Marino, Russo, & Toscano, 2004). In terms of antimicrobial mechanism, these compounds affect the membrane of microorganisms and cause depolarization, modification in membrane fluidity, and membrane disruption. Moreover, polyphenols can also influence the normal metabolism (stress response, ATP generation and citric acid cycle processes, for instance) and DNA related activity (such as DNA gyrase, helicase, and protein kinase activity) (Lima et al., 2019; Rempe, Burris, Lenaghan, & Stewart, 2017).

These characteristics support the use of phenolic compounds in the preservation of meat products either as additive or active component in packaging material (de Carvalho et al., 2019; Munekata, Rocchetti, et al., 2020). Moreover, current advances in the use of phenolic compounds are increasing the opportunities for these class of natural food and packaging additive in free and encapsulated form. Thus, this chapter aims to discuss recent advances in the: (1) encapsulation techniques to protect phenolic compounds; (2) production of active and edible packages rich on phenolic compounds; (3) use of phenolic-rich additives (free or encapsulated form) with non-thermal technologies to improve the shelf life of meat products; and (4) use of active packaging rich on phenolic compounds on meat products.

2. Innovative encapsulation techniques to protect phenolic compounds

Many approaches to improve the stability of phenolic compounds have been proposed so far. Encapsulation consist in the production of particles composed of encapsulating material (solid or immiscible material) and entrapped compounds (bioactive compounds) (Assadpour & Jafari, 2019). Encapsulation by spray-drying is the most widely spread technology to obtain polyphenol-rich additives for the food industry. The main reason to the popularity of spray-drying is due to the simplicity, economic and fast drying stage as well as facilitating the transportation and use (powder form) (Assadpour & Jafari, 2019; Vinceković et al., 2017).

Moreover, other technologies and equipment have been proposed and explored in scientific literature in order to produce encapsulated materials. Some relevant approaches that can be cited are the use of freeze-drying, ultrasound, and supercritical fluids. Freeze-drying or lyophilization is technology that facilitates the removal of a solvent from a solution by applying low pressure. In order to achieve the sublimation of solvent, the solution containing the phenolic compounds and encapsulating material are properly homogenized and frozen (below $-18\,^{\circ}C$). Additionally, it is relevant to mention that encapsulating polyphenols with this technology can take few hours. Once the solvent is removed, particles containing the phenolic compounds are formed (Sharif, Khoshnoudi-Nia, & Jafari, 2020).

Another relevant technology to encapsulate polyphenols consist in the use of supercritical fluids. This technology consists in the encapsulation of active compounds (polyphenols, for instance) using a supercritical fluid. Due to the thermodynamic properties of CO_2, this gas is usually applied in this approach (Cocero, Martín, Mattea, & Varona, 2009). Once the gas flow, pressure, and temperature are stabilized to achieve a supercritical condition, the solution containing the encapsulating material and polyphenols is pumped to the precipitation vessel along with the flow of pressurized CO_2. In this stage, the atomization of polyphenol-rich solution takes place and the droplets formed are surrounded by supercritical CO_2 and precipitate. The configuration of the equipment can be modified due to the composition and characteristics of encapsulating material (Trucillo, Campardelli, Aliakbarian, Perego, & Reverchon, 2018; Visentin, Rodríguez-Rojo, Navarrete, Maestri, & Cocero, 2012).

The use of ultrasound to produce encapsulated materials has been proposed to protect polyphenols during storage. This fundament of this technology can be explained by the passing of ultrasound wave (above 16 kHz) that causes cycles of compression and decompression in the liquid media. In the negative pressure stage of the cycle, bubbles are formed and grow in size. Following these events, these bubbles are forced to collapse and produce shear forces (Yasui, 2018). In the case of ultrasound encapsulation, this process and the extraction of compounds occurs simultaneously: source of high-added value compounds, solvent (usually water or an aqueous solution) and the encapsulating material are sonicated together in a two-step process. Initially, a low ultrasound frequency is applied to favor the extraction of high-added value compounds. The following stage consist in the use of high frequency to favor the formation of microspheres, which entrap the extracted compounds (Leong, Martin, & Ashokkumar, 2017; Vilkhu, Mawson, Simons, & Bates, 2008).

Regarding the studies carried out to produce encapsulated polyphenols with these innovative methods, selecting appropriate encapsulating conditions and encapsulating materials play a central role in the production of encapsulated extracts with high polyphenol content and storage stability (Fig. 1).

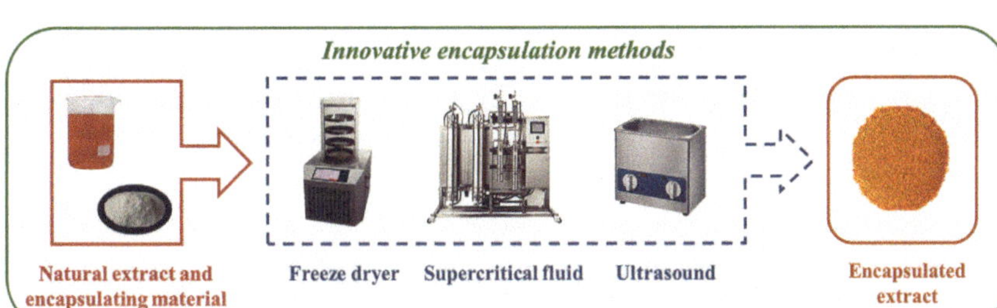

Fig. 1 Innovative encapsulation methods to protect polyphenols.

The Table 1 displays studies carried out with freeze-drying, ultrasound, and supercritical fluids. In the case of freeze-drying, recent studies explored both influence of drying stage (Ballesteros et al., 2017; Kuck & Noreña, 2016; Papoutsis et al., 2018; Ravichai & Muangrat, 2019; Saikia et al., 2015) and storage stability (Antigo et al., 2018; Ramírez et al., 2015; Rocha-Parra et al., 2016; Šaponjac et al., 2017) of encapsulated polyphenols. A relevant study regarding the encapsulation of polyphenols using this technology was carried out with star fruit (*Averrhoa carambola*) pomace (Saikia et al., 2015). In this case, the encapsulation efficiency was influenced by the ratio of encapsulating material and extract. According to the authors, the increasing the proportion of encapsulating material from 10:1 to 20:1 was associated with higher encapsulation efficiency, especially in the core of particles. Similarly, the encapsulation of polyphenols from fermented tea leaf wastewater was influenced by this ratio (Ravichai & Muangrat, 2019). In the case of tea polyphenols, reducing the proportion of encapsulating material from 80:1 to 40:1 improved the encapsulating efficiency as well as the antioxidant activity of particles obtained from freeze-drying.

The composition of encapsulating material can also influence the phenolic content in the particles obtained from freeze-drying. For instance, the experiment carried out with spent coffee grounds explored the effect encapsulating material composition (maltodextrin, gum arabic and their mixture (1:1)) in the production of encapsulated polyphenols (Ballesteros et al., 2017). The authors indicated that using gum arabic or its combination with maltodextrin produced particles with lower polyphenol contents than that obtained using maltodextrin. A similar outcome was reported for the encapsulation of polyphenols from lemon by-products (Papoutsis et al., 2018). In this case, the combination of maltodextrin and soybean protein produced particles with higher polyphenol content than using maltodextrin alone or the combination of maltodextrin with ι-carrageenan. In the same line of thought, a recent study explored the encapsulation of polyphenols found in grape skin using different encapsulating materials (Kuck & Noreña, 2016). Among the materials tested (gum arabic, partially hydrolyzed guar gum, and polydextrose alone or combined), the combination of gum arabic and polydextrose produced particles with the highest polyphenol content and antioxidant activity.

Considering the influence of storage conditions is also another important aspect in the production of encapsulated polyphenols using freeze-drying approach. In this sense, Ramírez et al. (2015) studied the effect of encapsulating material composition and drying conditions (using a response

Table 1 Innovative encapsulation techniques to improve the preservation of phenolic-rich natural extracts.

Source of polyphenols	Encapsulation conditions	Effect	References
Freeze-drying			
Star fruit (*Averrhoa carambola*) pomace	EM:extract ratio (10:1, 15:1, and 20:1 w/w): maltodextrin; T: −55 °C; and t: 24 h	EM:extract of 20:1 had the highest encapsulating efficiency (97%)	Saikia, Mahnot, and Mahanta (2015)
Fermented tea leaf wastewater	EM:extract ratio (1:10, 1:40 and 1:80 *w*/w): maltodextrin, gum arabic, and modified starch; P: 0.133 mbar; and t: 72 h	Gum arabic (EM: extract, 10:1) provided the highest encapsulation efficiency of polyphenols and preservation of AA	Ravichai and Muangrat (2019)
Spent coffee grounds	EM:extract ratio (2:10 *w*/*v*): maltodextrin, gum arabic and their mixture (1:1); T: −60 °C; P: 0.05 bar; and t: 48 h	Maltodextrin had lower losses in comparison to gum Arabic and their combination	Ballesteros, Ramirez, Orrego, Teixeira, and Mussatto (2017)
Lemon by-products	EM:extract ratio (3:10 w/v): maltodextrin and its combinations with soybean protein (5:1) and ι-carrageenan (9:1); liquid nitrogen as freezing agent; and t: 48 h	Maltodextrin with soybean protein was the most efficient EM to protect polyphenols and AA; particles with low moisture and water activity	Papoutsis et al. (2018)
Grape skin	EM:extract ratio (1:10 w/v): gum arabic, partially hydrolyzed guar gum, and polydextrose alone or combined (1:1); T: −57 °C; P:< 20 μmHg; and t: 48 h	Combining gum arabic and polydextrose was the most efficient EM to protect polyphenols and AA	Kuck and Noreña (2016)

Continued

Table 1 Innovative encapsulation techniques to improve the preservation of phenolic-rich natural extracts.—cont'd

Source of polyphenols	Encapsulation conditions	Effect	References
Model fruit juice (citrus pectin, sucrose and gallic acid)	EM:extract ratio (1:10, 2:10, and 3:10 w/v): maltodextrin, gum arabic and their mixture (1:1); P: 300 to 500 mTorr; t: 16 h; and freezing rate: from 0.3 °C/min to 0.7 °C/min	EM in juice: 10%–12%; 80%–100% maltodextrin in EM; P: 475–500 mTorr; 87% of polyphenols retention after 200 days at 22 °C	Ramírez, Giraldo, and Orrego (2015)
Red wine (*Cabernet Sauvignon*)	EM:extract ratio (9:91 w/w): maltodextrin with gum arabic (65:35); T: RT; P: 100 mmHg; and t: 40 h	Polyphenols and AA were preserved throughout the storage (145 days at 38 °C); main reduction was observed in anthocyanins (aw > 0.33)	Rocha-Parra, Lanari, Zamora, and Chirife (2016)
Red beetroot *(Beta vulgaris* l.)	EM:extract ratio (3:7 w/w): maltodextrin alone and a mixture with xanthan gum (99.5:0.5); pH: 3–6; T: −36 °C; P: 1.09 Pa; and t: 2 days	The pH had a major effect; polyphenol retention between 80% and 91% after 7 days at 30 °C and pH 4, regardless of EM composition	Antigo, de Bergamasco, and Madrona (2018)
Sour cherry pomace	EM:extract ratio (2:1 *v/w*): whey or soy protein; T: −40 °C, and t: 24 h	Soy protein protected the polyphenols; non-significant differences in AA between EM after 6 weeks at 25 °C	Šaponjac et al. (2017)
Supercritical fluids			
Rosemary leaves	EM:extract ratio (2.5:1 w/w): Pluronic F88; CO_2 flow rate:0.7 kg/h; extract flow: 1 mL/min; T: 25–50 °C; and P: 8–12 MPa	No clear effect in terms of P or T; all treatments had higher phenolic content than concentrated extract	Visentin et al. (2012)

Table 1 Innovative encapsulation techniques to improve the preservation of phenolic-rich natural extracts.—cont'd

Source of polyphenols	Encapsulation conditions	Effect	References
Olive pomace	EM:polyphenol ratio (95:5, 90:10, 85:15, and 80:20 w/w): phosphatidylcholine; CO_2 flow rate: 6.5 g/min; extract flow: 0.7 mL/min and 5 mL/min; pressure 130 bar and 170 bar; and T: 40 °C	Encapsulation efficiency increased with polyphenol content; encapsulation efficiency was higher than conventional method	Trucillo et al. (2018)
Ultrasound			
Polygonum cuspidatum	EM:extract ratio (7.5:10 v/w): β-cyclodextrin; t: 5 min; I: 100 W; and s/s: 20 g/L	Preservation of AA in relation to fresh extract	Mantegna et al. (2012)

AA: antioxidant activity; TPC: total phenolic content; EM: encapsulating material; I: intensity of treatment; P: pressure; RT: room temperature; s/s: solid/solvent ratio; T: temperature; and t: time.

surface design) in the polyphenol content of particles obtained by freeze-dried from a model juice (composed of citrus pectin, sucrose and gallic acid). The authors indicated that optimum conditions to improve the phenolic content in freeze-dried particles were 10%–12% of maltodextrin/100 mL model juice, 80%–100% maltodextrin in encapsulating material (up to 20% of gum arabic in the mixture) and pressure in the range of 475–500 mTorr. Moreover, around 87% of polyphenols were retained in the particles (around 87%) after 200 days at 22 °C. Another interesting experiment was carried out with red wine (Rocha-Parra et al., 2016). In this study, the wine was freeze-dried with a combination of maltodextrin and gum arabic as encapsulating material. According to the authors, encapsulated polyphenols and the antioxidant activity were preserved during the storage period (145 days at 38 °C). However, it is also relevant to comment that a drastic reduction in anthocyanins (especially for malvidin-3-glucoside chloride) was observed during storage while other selected polyphenols (such as gallic acid, catechin and epicatechin) remained stable in the freeze-dried encapsulated.

A recent study characterized the influence of encapsulating material composition and pH of solution prior to freeze-drying red beetroot extract (Antigo et al., 2018). In this case, the pH of solution containing the extract and encapsulating material had an important effect in the stability of polyphenols during storage (7 days at 30 °C). At pH 4, non-significant loss was observed after the storage period (regardless of the composition of the encapsulating material). A similar experiment was carried out with sour cherry pomace where the effect of encapsulating material composition was evaluated (Šaponjac et al., 2017). The use of soy protein was indicated as more appropriated to preserve the phenolic compounds (especially the anthocyanins) found in this residue of food industry in comparison to whey protein. Additionally, non-significant effect was observed between soy protein and whey in terms of antioxidant activity during storage (6 weeks at 25 °C).

Regarding the use of supercritical fluids in the encapsulation of phenolic compounds, a study with rosemary leaves explored the effect of processing temperature and pressure using supercritical CO_2 (Visentin et al., 2012). Although no clear effect was reported in terms of temperature or pressure, the content of phenolic compounds in the encapsulated extract was superior to that reported for the extract prior to encapsulation (up to 230 vs 110 mg GAE/g extract dry matter). Another related experiment using supercritical CO_2 to encapsulate polyphenol found in olive pomace (Trucillo et al., 2018). In this case, the proportion of polyphenol to encapsulating material was studied and indicated that increasing the proportion of polyphenols in this ratio increase the encapsulation efficiency. Moreover, the authors also indicated that the supercritical encapsulation method proposed by the authors had higher encapsulation efficiency than the traditional encapsulating method (using an ethanolic solution containing lipids mixed with aqueous solution with polyphenols dried under vacuum).

Finally, a study using β-cyclodextrin explored the influence of ultrasound-assisted encapsulation in the antioxidant compounds found in *Polygonum cuspidatum* (Mantegna et al., 2012). According to the authors, non-significant differences in relation to non-encapsulated extract were observed in terms of antioxidant activity. Collectively, the use of emerging and environmentally friendly technologies as innovative approach to encapsulated polyphenols from different sources support the importance of studying and improving the knowledge of these technologies. Freeze-drying seems to play an important role in the encapsulation of polyphenols due to lower impact on polyphenol content during both processing and storage.

3. Novel approaches to produce active and edible packages rich on phenolic compounds

The production of packaging materials for foods has been traditionally been carried out by casting, extrusion and thermo compression. Although these methods are widely used, the demand for packaging materials with modern consumer concerns such as the use of synthetic food additives, sustainable packaging materials and fresh-like properties and also preserving quality and safety, improving the characteristics of films gained more importance (Han, Ruiz-Garcia, Qian, & Yang, 2018; Umaraw et al., 2020).

More recent approaches to produce food packaging materials with active properties have explored the use of: (1) sub- and supercritical fluids for impregnation of films and (2) grafting of active compounds in the surface of films. Particularly for sub- and supercritical fluids impregnation methods, this method consists in the exposure of films to phenolic compounds (extracts or pure compounds) under sub- and supercritical conditions. The process initiates with the loading a vessel with both film/food and active compounds. Once the temperature and pressure to achieve the sub- or supercritical conditions are achieved, the flow of fluid and agitation take place. In this condition, the active compounds migrate toward the film/food and impregnate it. When the selected impregnation time is completed, the vessel is depressurized and active films/impregnated food are obtained. The use of CO_2 is interesting due to low technical requirements (in terms of pressure and temperature) to achieve supercritical state, low toxicity, and availability at low price. Moreover, the films are free of solvent and the removed CO_2 can be recycled in the system (Champeau, Thomassin, Tassaing, & Jérôme, 2015; Rojas, Torres, José Galotto, Guarda, & Julio, 2020).

Grafting a polyphenol to a structural component of film consist in a production of covalent bond (polyphenol-g-structural component) between these two compounds. The reaction can take place by different methods: grafting by carbodiimide reaction, enzyme catalyzed coupling, free radical based coupling and electrochemical technique. Details regarding the reaction mechanism can be found elsewhere (Liu, Pu, Liu, Kan, & Jin, 2017). Once the reaction is completed, the films are dialyzed to remove reagents and non-grafted polyphenols and active films are obtained. The use of these two techniques to produce food packages with enhanced properties is displayed in Fig. 2.

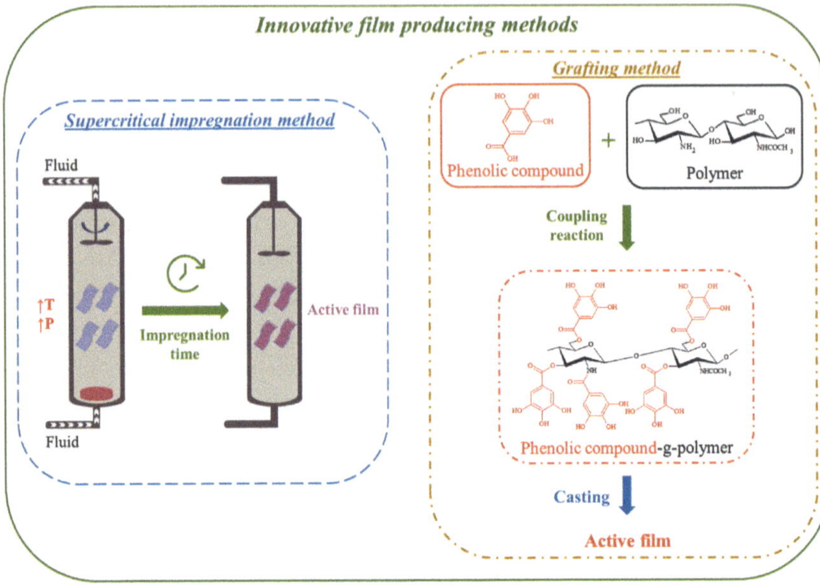

Fig. 2 Innovative methods to produce films with polyphenol-rich extract.

Regarding the use of impregnation of films with sub- and supercritical fluids, many studied explored the influence of producing conditions and the effect in foods (Table 2). For instance, a recent study studied the influence of pressure and impregnation time in the production of a film rich in polyphenols extracted from *Olea europea* leaves (Bastante et al., 2018). According to the authors, the combination of 40 MPa and 1 h produced films with highest impregnation yield in comparison to other combinations. Additionally, the film produced with the optimized condition also delayed the progression of lipid oxidation in coatless sunflower seeds. In a similar way, the development of a film using mango leaves polyphenols characterized the effect of pressure, temperature and impregnation time (Belizón et al., 2018). In this case, the optimum conditions to improve the impregnation of polyphenols were 10 MPa at 45 °C for 22 h, which resulted in a film with high antioxidant capacity. Moreover, the application in fresh lettuce and tangerine improve the shelf life up to 14 and 39 days, respectively. Similarly, an experiment with polyphenols extracted from black walnut husk indicated that increasing the proportion of this residue in the husk: ethanol ratio (used for simultaneous extraction and impregnation) enhanced the impregnation yield in LDPE film (Wenzel et al., 2020).

Table 2 Innovative strategies to produce films rich in polyphenols.

Source or active compound	Film components and forming conditions	Film properties and optimum conditions	References
Impregnation with sub- and supercritical fluids			
Olea europea leaves	P (10–40 MPa); T (35 °C); t (5 min, 1, 2 and 22 h); agitation (40 rpm); CO_2; and PET/PP film	The highest impregnation yield was obtained using 40 MPa for 1 h; reduced lipid oxidation in sunflower seeds (15 days at RT)	Bastante, Cardoso, Ponce, Serrano, and de la Ossa-Fernández (2018)
Mango leaves	P (10–20 MPa); T (35–55 °C); t $(3-22h)$; agitation (40 rpm); CO_2; and PET/PP film	The highest impregnation yield was obtained using 10 MPa at 45 °C for 22 h; film with high antioxidant capacity; preservation of lettuce (14 days) and tangerine (39 days) at RT	Belizón, Fernández-Ponce, Casas, Mantell, and La Ossa-Fernández (2018)
Black walnut husk	P (19.4 MPa); T (60 °C); CO_2; ethanol (15%, as modifier); CO_2; husk: ethanol ratio (0.4:1–1:1) and LDPE film	Impregnation yield enhanced as the proportion of husk increased	Wenzel et al. (2020)
Eugenol	P (10, 12 and 15 MPa); T (45 °C); t (4 h); CO_2; depressurization rate (0.5, 1 and 5 MPa/min); and LLDPE film	Impregnation yield was maximum using 15 MPa and depressurization rate of 0.5 MPa/min; film with high antioxidant capacity	Goñi, Gañán, Strumia, and Martini (2016)

Continued

Table 2 Innovative strategies to produce films rich in polyphenols.—cont'd

Source or active compound	Film components and forming conditions	Film properties and optimum conditions	References
Gallic acid	Potato peel/potato cull ratio (0:1–1.3:1); glycerol/starch ratio (0:1–2:1); gallic acid/starch ratio (0:1–0.3:1); H_20; P (5–19 MPa); and T (100–150 °C)	Optimum conditions were: 125 °C, 12 MPa, 1.3 g peel/g cull, 1 g glycerol/g cull, and 0.2 g gallic acid/g cull; film with high antioxidant capacity	Zhao and Saldaña (2019)
Grafting method			
Gallic acid (20 g/L)	Radical initiation (chitosan, H_2O_2 and ascorbic acid; 12 h at RT; and dialysis), carbodiimide coupling (chitosan, EDC, NHS, and MES buffer; ice bath for 1 h followed by 12 h at 20 °C and a dialysis) and tyrosinase catalysis (chitosan, phosphate buffer, and tyrosinase; 30 °C for 18 h; and dialysis)	Carbodiimide coupling method produced the highest grafting ratio, tensile strength and antioxidant activity; this method also reduced the water vapor permeability	Zhang, Liu, Qian, Kan, and Jin (2019)
Protocatechuic acid (0.15; 0.45 and 0.75 M)	Chitosan, EDC, NHS, and MES buffer; reactions carried out in ice bath for 1 h followed by 12 h at RT and a dialysis step	Increasing protocatechuic acid proportion enhanced grafting ratio and tensile strength; decreased moisture and water vapor permeability	Liu, Meng, Liu, Kan, and Jin (2017)

Table 2 Innovative strategies to produce films rich in polyphenols.—cont'd

Source or active compound	Film components and forming conditions	Film properties and optimum conditions	References
EGCG (Chitosan: EGCG ratio 9:1, 8:2, 7:3, 6:4 and 5:5)	Chitosan, glycerol, H_2O_2 and ascorbic acid; reaction was 40 °C for 24 h followed by dialysis	Increasing levels of EGCG enhanced darkness, redness and yellowness as well as antioxidant and antimicrobial activity (*Pseudomonas aeruginosa* and *Listeria monocytogenes*)	Mittal et al. (2021)
Gallic acid, gentisic acid, protocatechuic acid, syringic acid and vanillic acid (3 M)	Chitosan, EDC, NHS, and MES buffer reaction was carried out in ice bath for 1 h followed by 12 h at RT and a dialysis step	GLA had the highest ratio, darker, redder and yellower film among all phenolic acids; GLA also reduced water vapor permeability and increased tensile strength	Liu, Liu, et al. (2017)
Caffeic acid, gallic acid, gentisate acid and sinapic acid (0.16 M)	Chitosan, EDC, and HOBt; reaction carried out in the dark for 24 h followed by dialysis	Films with caffeic acid displayed higher antioxidant and antimicrobial (*Pseudomonas aeruginosa*, *Escherichia coli*, *Bacillus subtilis*, and *Staphylococcus aureus*) activity than films produced with other polyphenols	Wang et al. (2019)

EDC: 1-ethyl-3-(3-dimethylaminopropyl) carbodiimide hydrochloride; EGCG: epigallocatechin gallate; HOBt: hydroxybenzotriazole; IBHB: isobutyl-4-hydroxybenzoate; LDPE: low-density polyethylene; LLDPE: linear low density polyethylene; MES: 2-(*N*-morpholino)ethanesulfonic acid; NHS: *N*-Hydroxysuccinimide; P: pressure; PET/PP: polyethylene terephthalate/polypropylene; and RT: room temperature.

The use of supercritical CO_2 was also tested to produce active packaging with single polyphenols. A relevant example of this approach was the production of active linear low density polyethylene (LLDPE) film rich in eugenol (Goñi et al., 2016). In this experiment, the pressure and depressurization rate were indicated was relevant variables that affect the impregnation rate. The film with the highest impregnation rate and antioxidant activity was produced using 15 MPa and depressurization rate of 0.5 MPa/min. Another relevant example of the application of impregnation technique in production of films with a single active component was the study carried out by Zhao and Saldaña (2019) with potato cull, peel, glycerol and gallic acid. In this case, the authors explored the influence of different ratios in the impregnation rate using water as subcritical fluid. The authors observed that using 1.3 g peel/g cull, 1 g glycerol/g cull, and 0.2 g gallic acid/g cull was the most appropriate condition to produce the film with high antioxidant capacity. Moreover, temperature and pressure were also indicated as key factors to improve the antioxidant activity of active film (optimum conditions: 125 °C and 12 MPa).

The use of grafting technique to produce active packaging was also recent explored with different polyphenols. A recent experiment explored the influence of grafting methods (radical initiation, carbodiimide coupling, and tyrosinase catalysis) in the grafting ratio of gallic acid as well as the properties of chitosan film (Zhang, Liu, et al., 2019). Although all methods grafted gallic acid in the film, carbodiimide coupling method produced films with the highest grafting ratio and antioxidant activity. However, significant differences were observed in the water vapor permeability of films.

Using a different phenolic acid, the study carried out by Liu, Meng, et al. (2017) explored the effect of different concentrations of protocatechuic acid (0.15; 0.45 and 0.75 M) in grafting ratio of chitosan films. In this case, an increase in grafting ratio was observed as the concentration of this phenolic acid was augmented from 0.15 to 0.75 M. Moreover, this change in protocatechuic acid content also influenced the properties of film, particularly the tensile strength, moisture and water vapor permeability. A related study also explored the proportion of epigallocatechin gallate in the properties of a chitosan film using the radical initiation method (Mittal et al., 2021). The highest levels of this polyphenol were associated with changes in color (films became darker, redder and yellower). Moreover, the authors also observed that a significant increase in the antioxidant and antimicrobial activity (against *Pseudomonas aeruginosa* and *Listeria monocytogenes*) of film in relation to control (chitosan film without epigallocatechin gallate).

Differences in terms of phenolic acid can also influence the grafting ratio of active films. For instance, a recent experiment explored the differences in the grafting ratio and properties of films produce with different phenolic acids: gallic acid, gentisic acid, protocatechuic acid, syringic acid and vanillic acid (Liu, Liu, et al., 2017). Among all the films produced with these compounds, the use of gallic acid produced the highest grafting ratio, reduced lightness and water vapor permeability, and increased the redness, yellowness and tensile strength in comparison to other films. A similar experiment with phenolic acids (caffeic acid, gallic acid, gentisate acid and sinapic acid) in chitosan films also indicated significant differences (Wang et al., 2019). In this case, the use of caffeic acid was associated with highest antioxidant and antimicrobial.

Collectively, seems reasonable to consider that optimizing the producing conditions as well as the structure of polyphenol have a central role in the production of active films. This consideration is particularly relevant for the grafting method whereas for impregnation with sub- and supercritical fluids method, more studies are necessary to clarify the influence of structure composition in the properties of active films.

4. Application of phenolic-rich additives (free or encapsulated form) with non-thermal technologies to improve the shelf life of meat and meat products

The preservation of meat and meat using non-thermal technologies is an innovative that fits in the emerging sector of food market that is composed of consumers interested in consuming microbiologically safe foods with fresh-like characteristics (Park, Park, & Shimanouchi, 2019). Although this category is composed by different technologies, the common characteristics is the absence or reduced temperatures (below pasteurization conditions). The main advantage of this aspect is the preservation of sensorial attributes associated with fresh foods in comparison to pasteurization and sterilization (Alirezalu et al., 2020; Huang, Wu, et al., 2017; Pankaj, Wan, & Keener, 2018). Some of the most relevant non-thermal technologies are high-pressure processing and cold plasma. Additionally, supercritical CO_2 (SC-CO_2) has been proposed to treat food (González-Alonso et al., 2020).

High pressure processing (HPP) is a technology that consist in the application of pressure in a packaged sample immersed a liquid media. The HPP treatment can affect both microorganisms and enzymes and ultimately leads to enhanced shelf life with reduced effect in the fresh-like characteristics of treated sample. Moreover, HPP can influence the bioaccessibility of functional compounds and nutrients as well as prevent the formation of harmful compounds (Chemat et al., 2020).

The mechanism of action that explain the inactivation of microorganisms is centered in the modifications of microbial membrane (increased permeability leading to loss of internal components) and in the intracellular components (structural modification of vital proteins, organelles and genetic material) (Sehrawat, Kaur, Nema, Tewari, & Kumar, 2020). In terms of application in meat and meat products preservation, HPP *per se* can be applied as processing to improve the preservation of meat and meat products but an important concern related to the potential effect in sensorial active compounds as well as the development of oxidative must be consider to avoid quality loss during storage (Bolumar et al., 2020). Regarding the effect in polyphenols (particularly for food rich in these compounds), the use of HPP can improve the retention of phenolic compounds in treated products (Tadapaneni, Daryaei, Krishnamurthy, Edirisinghe, & Burton-Freeman, 2014).

Cold plasma is another interesting and relevant approach to inactivate spoilage and pathogenic microorganisms in fresh food products. The fundament of cold plasma treatment is centered in the application of gas (in plasma state) in the surface of a food that will generate reactive nitrogen and oxygen species (Munekata, Domínguez, Pateiro, & Lorenzo, 2020) that damage microbial membrane (by etching, causing lesioning and eventually disrupting the membrane) and induce the oxidation or disruption of chemical bonds in proteins, lipids, and DNA. Another important aspect is the generation of cause electrostatic stress due to the impact of charged particles with eventual accumulation in the surface of the treated food (Bourke, Ziuzina, Han, Cullen, & Gilmore, 2017). It is worth to mention that solo application of cold plasma treatment *per se* can reduce the microbial load in meat and meat products, which increase the safety with minimal impact in the characteristics of meat and meat products (Misra & Jo, 2017). In a similar way, its application can, under optimized conditions, keep and even increase the phenolic content in foods rich in polyphenols (Munekata, Domínguez, et al., 2020).

The use of supercritical fluids is mainly associated with the extraction of high-added value compounds from several foods, by-products of industry and other sources. This aspect is mainly related to the favorable characteristics that a fluid has when appropriate pressure and temperature are applied in it. Once achieving the supercritical condition (above the critical pressure and temperature), the fluid displays gaseous and liquid properties (in terms of solvation, density, and viscosity) that improve its capacity to separate compounds from a matrix (Ameer, Shahbaz, & Kwon, 2017).

Interestingly, this condition can also lead to microbial inactivation of spoilage and pathogenic bacteria, yeasts and molds (González-Alonso et al., 2020). The mechanism associated with the inactivation of microorganisms is initiated by the solubilization of supercritical fluid in the free water followed by the diffusion through the cellular membrane with eventual cellular internalization. Once these events take place, a rapid drop on pH and the disruption of vital metabolic process occurs inside the microorganism. Moreover, the passage of supercritical fluid through the membrane also causes its disruption and lead to loss of intracellular components (Giulitti, Cinquemani, & Spilimbergo, 2011; Spilimbergo, Mantoan, Quaranta, & Mea, 2009).

Collectively, the use of more than one technology or strategy to improve the preservation of food is a valid and important concept known as hurdle technology (Singh & Shalini, 2016). In this sense, combining the aforementioned technologies and polyphenolic-rich extracts can be seen as a major up-to-date strategy to preserve muscle foods (Fig. 3).

In this sense, recent studies explored the use of high-pressure processing (HPP), cold plasma and SC-CO_2 (Table 3). Regarding the combined use of HPP with natural extracts rich in phenolic compounds, three strategies have been explored to preserve meat and meat products: free, encapsulated and incorporated into films. For instance, Alves et al. (2012) studied the

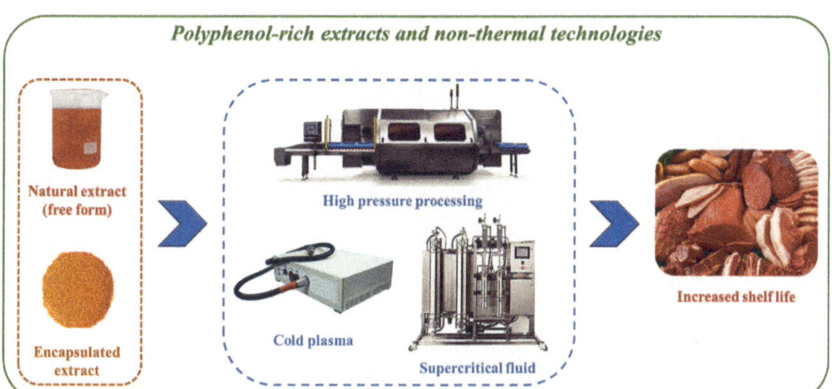

Fig. 3 Use of non-thermal technologies with polyphenols to improve the shelf life of meat products.

Table 3 Preservation of meat and meat products combining non-thermal technologies and phenolic-rich additives.

Technology (treatment conditions)	Source (concentration and/or TPC); strategy	Sample	Storage conditions	Effect	References
High pressure processing					
HPP: 300, 600, or 800 MPa for 10 min	Tomato: in natura (0.1%, 435 mg GAE/100 g DW), waste (skin and seed, 0.3%, 173 mg GAE/100 g DW), peeled tomato (0.1%, 497 mg GAE/100 g DW), and tomato paste (0.1%, 500 mg GAE/100 g DW); free form	Minced raw chicken breast	15 days at 5 °C	Tomato paste and waste reduced LO; no effect in color	Alves, Bragagnolo, Da Silva, Skibsted, and Orlien (2012)
HPP: 600 MPa/180 s at 25 °C	Carvacrol (200 ppm); free form	Cooked turkey breast ham	60 days at 4 °C	Reduced LO; delayed MG; increased syneresis; no effect in color, pH, and texture	de Oliveira et al. (2015)
HPP: 100 and 200 MPa, at 15 °C for 5 min	Sea tangle (1.5% and 3%); free form	Emulsion-type sausage	14 days at 4 °C	No effect in pH, cooking loss, and WHC; reduced texture, PO, LO, and MG	Lee et al. (2018)
HPP: 500 MPa for 10 min	Pitaya peel (100 and 1000 ppm, 49.5 mg GAE/g); microencapsulated	Pork patties	9 days at 4 °C	Reduced pH; increased color, texture, LO and PO	Cunha et al. (2018)

HPP: 800 MPa, 10 min, 5 °C	Rosemary (10%); film	Chicken breast and thigh patties	25 days at 5 °C	Reduce LO in both surface and inner section of breast and thigh patties	Bolumar, Andersen, and Orlien (2011)
HPP: 700 MPa, 10 min, 5 °C	Rosemary (10%); film	Pork patties	60 days at 5 °C	Reduce LO in both surface and inner section of breast and thigh patties	Bolumar, LaPeña, Skibsted, and Orlien (2016)
HPP: 600 MPa for 7 min	Olive leaf (986 μg CAE/1091.5 cm^2); film	Sliced dry-cured Iberian pig shoulders	150 days at 5 °C	Low MG with HPP alone or combined with film; no effect in color, LO, PO, and sensory analysis	Amaro-Blanco, Delgado-Adámez, Martín, and Ramírez (2018)
Cold plasma processing					
Cold plasma: 70 kV for 180 s with gas (65% O_2 and 30% CO_2)	Rosemary (1%); free	Chicken patties	5 days at 4 °C	Reduced LO, MG and L*; increased pH, a*, and b*	Gao, Zhuang, Yeh, Bowker, and Zhang (2019)
Supercritical fluid					
SC-CO_2: 13.8 MPa at 35 °C for 2 h	Rosemary (2.5 and 5%); free	Ground pork	7 days at 4 °C	Reduced MG, pH, TVB-N; increased LO, L* and b*; no effect in a*	Huang, Liu, Ge, & Dai (2017)

CAE: caffeic acid equivalent, GAE: gallic acid equivalent, HPP: high pressure processing, LO: lipid oxidation, MG: microbial growth, PO: protein oxidation, SC-CO_2: supercritical CO_2, TPC: total phenolic content, TVB-N: total volatile basic nitrogen, and WHC: water holding capacity.

influence of different products and by-products from tomato paste processing (in natura, waste, peeled tomato, and tomato paste; all in free form) in minced raw chicken breast. The most efficient extracts were tomato paste and waste in terms of lipid oxidation regardless of the applied pressure. Additionally, no effect in color was reported by the authors.

In the same line of research (using extracts in free form), other studies explored the this combinatory approach in the production of meat products reformulated with health- and consumer-oriented strategies (de Oliveira et al., 2015; Lee et al., 2018). Regarding the application in healthier meat products, de Oliveira et al. (2015) explored the combination of HPP with carvacrol in turkey breast ham with reduce salt content. Due to the antioxidant and antimicrobial potential of this natural polyphenol, lipid oxidation was inhibited and the growth of lactic acid and psychrotrophic bacteria, *Staphylococcus* sp. and *Listeria innocua* was delayed by at least 30 days during storage, respectively. Moreover, the treated products also displayed the same color, pH, and texture but higher syneresis than control samples (neither extract nor HPP treatment). In a similar way, by Lee et al. (2018) studied the combination of HPP with sea tangle powder to reduce the content of phosphate in emulsion-type sausage. According to the authors, the use of sea tangle powder at 3% and 100 MPa was the most efficient treatment to preserve the pH, cooking loss, and WHC of sausage but reduced the texture parameters, lipid oxidation, protein oxidation, and microbial growth during storage.

Regarding the use of microencapsulated extracts with HPP, Cunha et al. (2018) explored the effect of pitaya extract (100 and 1000 ppm) in the preservation of pork patties. The combination of microencapsulated extracts with HPP caused a reduction in pH and increased the formation of lipid and protein oxidation products in relation to non-treated samples. Moreover, texture and color were also affected by the combination of treatments. It is relevant to mention that a protective effect against oxidative reactions was reported between HPP-treated samples and the combination of HPP and pitaya extract, supporting the role of this combinatory approach to extend the shelf life, especially using 1000 ppm of extract.

In terms of incorporation of phenolic-rich extract into films, Bolumar et al. (2011) explored the combination of active film incorporated of rosemary extract with HPP in the preservation of patties produced with either chicken breast or thigh meat. in both patties, the combinatory approach delayed the formation of lipid oxidation products in the surface and internal sections. A further experiment carried out by the same research group using the same approach indicated the same protective effect in pork patties (Bolumar et al., 2016).

In an experiment with sliced dry-cured Iberian pig shoulders, the combination of HPP and active film (added of olive leaf extract) displayed protective effect against microbial growth during storage Additionally, the samples packaged with the active film and treated with HPP did not cause significant changes in other physic-chemical and oxidative properties during the storage of this meat products (Amaro-Blanco et al., 2018).

Another relevant technology that can be applied with phenolic-rich extracts to preserve meat and meat products is cold plasma. In this regard, a recent study indicated that treatment with either rosemary extract (in free form) or cold plasma improved the preservation of chicken patties but the combination did not show significant interaction (Gao et al., 2019). In both treatments, microbial growth, lipid oxidation, and lightness were significantly reduced in relation to untreated control. Moreover, redness was better preserved in these two treatments.

The use of supercritical fluids with natural extracts was also explored in scientific literature. In this case, the treatment is carried out with simultaneous extraction and treatment in a pressurized chamber where the SC-CO_2 flows (Huang, Liu, et al., 2017). According to the authors, the samples treated with this combinatory strategy displayed lower values of microbial growth, total volatile basic nitrogen and pH. However, the treatment with rosemary and SC-CO_2 caused a significant increase in lipid oxidation, lightness and yellowness whereas no significant effect was reported for redness during storage. Collectively, the main advances in the combination of non-thermal technologies with phenolic-rich extracts are centered in the use of HPP using rosemary extract, which is particularly relevant to improve oxidative and microbiological stability.

5. Advances in the use of active packaging rich on phenolic compounds on meat and meat products

Packaging plays a central role in the preservation in food. The initial concept of packaging food (protecting from contamination and spoilage as well as facilitating the transport and informing) has been changing due to factors that encompass different factors as technological advances, consumer trends, and concerns about the environment, for instance. In this sense, concept such as "active," "smart," "edible," and "biodegradable" gained more importance to improve the preservation of foods as well as facilitating the communication with consumers in terms of product freshness (Umaraw et al., 2020; Umaraw & Verma, 2017). Moreover, the use of nanoscale materials and real time monitoring are also important aspects related to development of new films and coatings to improve the preservation of meat and meat products (Holman, Kerry, & Hopkins, 2018; Umaraw & Verma, 2017). The real time monitoring of quality in fresh meat and the development of films using nanoscale materials are relevant current strategies in the development of intelligent and active packaging material for meat and meat products. A schematic representation of these strategies is shown in Fig. 4.

The fundamental approach of real time monitoring packaging materials is to facilitate the indication of quality from acceptable to unacceptable condition. In this sense, recent studies have explored the production of indicators of quality that uses different colors to indicate the freshness of a meat cut (Table 4). For instance, Alizadeh-Sani et al. (2021) explored the use of saffron flower anthocyanins in a film composed of chitosan and nanofibers in methyl cellulose to monitor the loss of quality in lamb meat during refrigerated storage. According to authors, this film was sensible to pH changes as the storage time progressed and also displayed a change from violet (fresh) to gray/green (spoiled). Another experiment carried out by the same research group explored the use of barberry anthocyanins in methylcellulose/chitosan nanofiber film in lamb meat (Alizadeh-Sani et al., 2020). This new film was efficient to indicate the progression of quality loss and indicate that the cut was not suitable for consumption (pH from 5.8 to 7.5 with a color change from reddish to pale peach). In a similar way, roselle anthocyanins were incorporated in starch and polyvinyl alcohol film to improve the preservation of pork meat (Zhang, Zou, et al., 2019). In a similar way to reported in the aforementioned studies, a change in pH was accompanied by a change in the color of film (from pinkish to pale peach), which also a relevant indication of the increase of total volatile basic nitrogen (TVB-N) content in the samples.

Aiming to improve the preservation of fresh meat and meat products, the production of active films using nanoscale materials was explored in many studies. One relevant example of the potential to prevent loss of quality in muscle foods is the study carried out by Khan et al. (2020) with a film composed of bovine gelatin, carrageenan, glycerol and nanoemulsions with polyphenols in chicken meat. Among the tested compounds, the film contained curcumin nanoemulsion displayed the highest potential increase the

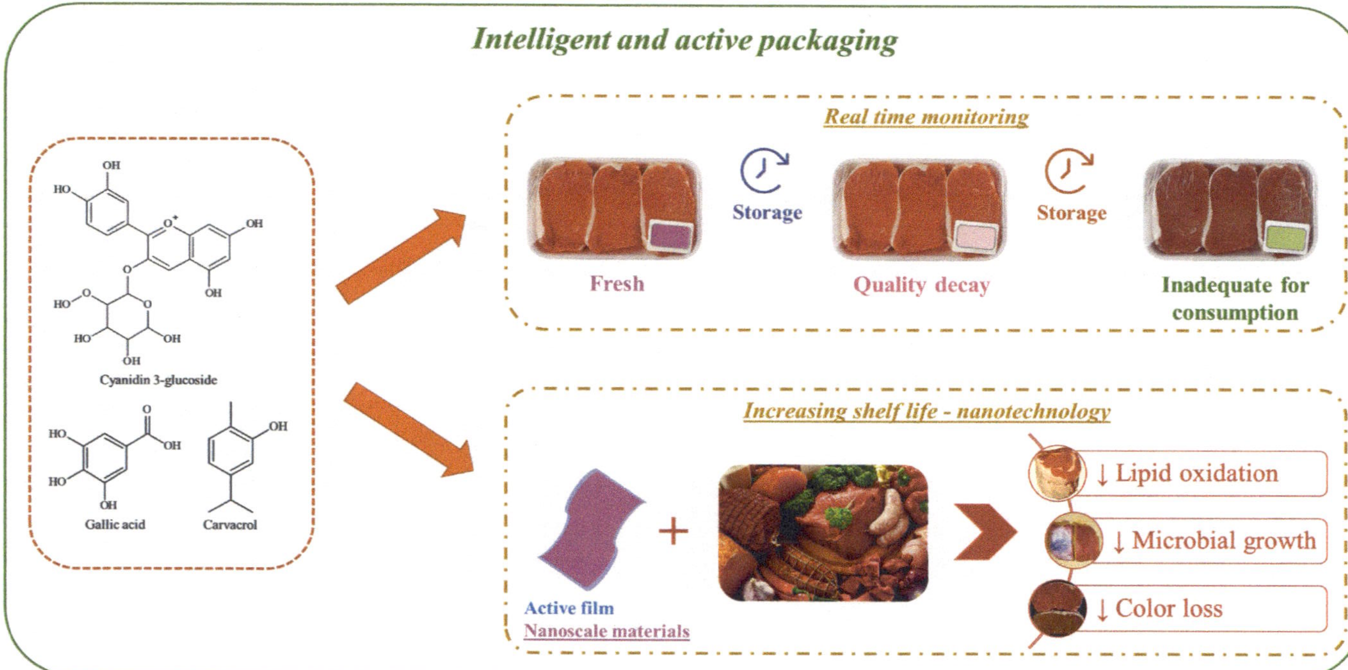

Fig. 4 Use of intelligent and active packaging with polyphenols in meat and meat products.

Table 4 Advances in intelligent and active packaging for meat and meat products.

Sample	Source (concentration and/or TPC in film/coating solution)	Packaging forming components	Storage conditions	Effect	References
Real time monitoring					
Lamb meat	Saffron flower anthocyanins (3% in film solution)	Chitosan and nanofibers and methyl cellulose	72 h at 25 °C	Visible color change in film corresponded with changes in pH (final value: 7.3)	Alizadeh-Sani, Tavassoli, McClements, and Hamishehkar (2021)
Lamb meat	Barberry anthocyanins (2.2 mg/mL in film solution)	Methylcellulose/chitosan nanofiber	72 h at 25 °C	Visible color change in film corresponded with changes in pH (final value: 7.5)	Alizadeh-Sani et al. (2020)
Pork meat	Roselle anthocyanins (2.5% in film solution)	Starch and polyvinyl alcohol	72 h at 25 °C	Visible color change in film corresponded with changes in TVB-N values (final value >15 mg/100 g)	Zhang, Zou, et al. (2019)
Preservation					
Broiler meat	Curcumin, gallic acid and quercetin (10–30 mg/100 mL in nanoemulsion) in nanoemulsion	Bovine gelatin, carrageenan, and glycerol	18 days at 4 °C	Curcumin nanoemulsion film was the most efficient compounds to inhibit microbial growth	Khan, Sadiq, and Mehmood (2020)
Pork	Olive leaves (2%, 5%, 10%, and 15%)	Polyethylene	15 days at 4 °C	Reduced loss of redness and the formation of lipid oxidation products	Moudache, Nerín, Colon, and Zaidi (2017)

Pork	Resveratrol (3%, 6%, and 9% in gelatin/zein mixture) in mats	Gelatin and zein	12 days at 4 °C	Delayed color changes and the increase of TVB-N and microbial growth; optimum resveratrol concentration: 6%	Li et al. (2020)
Chicken breasts	*Porphyra dioica* (1% and 5%) in nanofiber	Polypropylene, polyvinyl alcohol, gelatine	4 days at 4 °C	Both concentrations inhibited microbial growth and color changes; no effect in lipid oxidation	Reboleira et al. (2020)
Chicken sausage	*Eucalyptus* leaf with Ag nanoparticles (chitosan:Ag with *Eucalyptus* extract 1:1 and 3:1)	Chitosan	15 days at 20 °C	1:1 was the most efficient to inhibit fungi growth	Nwabor, Singh, Paosen, Vongkamjan, and Voravuthikunchai (2020)
Beef	Pomegranate peel (20 mg/mL)	Chitosan/ polyethylene oxide	10 days at 4 °C and 7 days at 25 ° C	Reduced *Escherichia coli* growth and preserved sensory properties in both storage conditions	Surendhiran, Li, Cui, and Lin (2020)
Pork meat	Pomegranate peel extract (1% in film solution) in chitosan nanoparticles	Zein, chitosan nanoparticles and glycerol	24 h at 37 °C	Inhibition of *L. monocytogenes* growth	Cui, Surendhiran, Li, and Lin (2020)

TPC: total phenolic content and TVB-N: total volatile basic nitrogen.

shelf life of fresh chicken meat by 7 days in comparison to samples without curcumin nanoemulsion. An interesting film was produced by Moudache et al. (2017). These authors explored the inclusion of olive leaves in a multilayer polyethylene film to improve the preservation of fresh pork meat. The authors observed that all tested concentrations (from 2% to 15%) improved the preservation of red color and reduced the formation of lipid oxidation products throughout the storage period.

Li et al. (2020) explored the influence of resveratrol concentration (3%–9%) in the capacity to preserve pork meat using a gelatin and zein film. According to authors, using the film with 6% of resveratrol was the most propriate to delay the discoloration, microbial growth and accumulation of TVB-N. An increase of 3 days in the shelf life of pork meat was obtained using this concentration of resveratrol. A similar experiment was recently carried out with *Porphyra dioica* algae extract (1% and 5%) in a film produced with polypropylene, polyvinyl alcohol, and gelatin by electrospinning (Reboleira et al., 2020). Both concentrations inhibited the discoloration and growth of spoilage microorganisms during storage (4 days at 4 °C). However, a non-significant effect against lipid oxidation was reported in this study.

A recent experiment with chitosan explored the influence of *Eucalyptus* leaf extract with Ag nanoparticles proportion to preserve chicken sausage (Nwabor et al., 2020). The authors observed that proportion of 1:1 (chitosan:Ag with Eucalyptus extract) was the more efficient to preserve this meat product against fungi growth than using a proportion of 3:1.

In addition, the antimicrobial effect of films containing nanomaterials has also been reported against specific pathogens. For instance, the mat composed of chitosan/polyethylene oxide nanofibers with pomegranate peel extract affected the growth of *Escherichia coli* in beef (Surendhiran et al., 2020). Moreover, the authors also indicated that samples with active mat received higher scores in sensory evaluation of appearance, color, odor, and overall acceptability than control film in the two tested storage conditions (10 days at 4 °C and 7 days at 25 °C). In a similar experiment, Cui et al. (2020) combined zein, chitosan nanoparticles and glycerol with pomegranate peel extract and observed an inhibition in the growth of *Listeria monocytogenes* in fresh pork meat during a more challenging condition: 12 days at 37 °C. In both cases the, the active films displayed important inhibitory effect against the growth of pathogenic bacteria, which support the strategy of active films to preserve fresh meat against pathogenic bacteria.

6. Conclusion

Polyphenols have a key role in the preservation of meat and meat products and goes beyond the use as additives (in free form or encapsulated by spray-drying). The branches considered in this review comprises the main advances in terms of phenolic compounds. An important aspect in these advances is the use of non-thermal and emerging technologies (sub- and supercritical fluids, ultrasound, high pressure processing and cold plasma) that are being used in combination with phenolic-rich extracts and in an innovative way that is beyond its currently and well-known use. In terms of combined use, high pressure processing can be seen as one of the main options. In the case of innovative use, supercritical fluids (a technology widely applied in the extraction of high-added value compounds) has been used to encapsulate polyphenol-rich extracts and impregnate films with polyphenols and also as treatment technology.

It is also relevant to notice that nanotechnology has also been explored and present major advances to advance the knowledge about the use of polyphenols in the preservation of meat and meat products. Further applications are suggested to explored the of less studied technologies and approaches in the production of additives and packaging materials such as the use of ultrasound for simultaneous extraction and encapsulation and grafting polyphenols in films. Another relevant advance in terms of application that can be explored in the preservation of healthier and functional meat products. This fast-growing share of food market is composed of products in line with health-oriented reformulation (in terms of fat, sodium chloride, dietary fiber, for instance) that require protection against the oxidative reaction and growth of microorganisms.

Acknowledgments

Thanks to GAIN (Axencia Galega de Innovación) for supporting this study (grant number IN607A2019/01). Paulo E. S. Munekata acknowledges postdoctoral fellowship support from the Ministry of Economy and Competitiveness (MINECO, Spain) "Juan de la Cierva" program (FJCI-2016-29486). Authors are members of the HealthyMeat network, funded by CYTED (ref. 119RT0568).

References

Alirezalu, K., Munekata, P. E. S., Parniakov, O., Barba, F. J., Witt, J., Toepfl, S., et al. (2020). Pulsed electric field and mild heating for milk processing: A review on recent advances. *Journal of the Science of Food and Agriculture, 100*, 16–24.

Alizadeh-Sani, M., Tavassoli, M., McClements, D. J., & Hamishehkar, H. (2021). Multifunctional halochromic packaging materials: Saffron petal anthocyanin loaded-chitosan nanofiber/methyl cellulose matrices. *Food Hydrocolloids*, *111*, 106237.

Alizadeh-Sani, M., Tavassoli, M., Mohammadian, E., Ehsani, A., Khaniki, G. J., Priyadarshi, R., et al. (2020). pH-responsive color indicator films based on methylcellulose/chitosan nanofiber and barberry anthocyanins for real-time monitoring of meat freshness. *International Journal of Biological Macromolecules*, *166*, 741–750 (in press).

Alves, A. B., Bragagnolo, N., Da Silva, M. G., Skibsted, L. H., & Orlien, V. (2012). Antioxidant protection of high-pressure processed minced chicken meat by industrial tomato products. *Food and Bioproducts Processing*, *90*, 499–505.

Amaro-Blanco, G., Delgado-Adámez, J., Martín, M. J., & Ramírez, R. (2018). Active packaging using an olive leaf extract and high pressure processing for the preservation of sliced dry-cured shoulders from Iberian pigs. *Innovative Food Science and Emerging Technologies*, *45*, 1–9.

Ameer, K., Shahbaz, H. M., & Kwon, J.-H. (2017). Green extraction methods for polyphenols from plant matrices and their byproducts: A review. *Comprehensive Reviews in Food Science and Food Safety*, *16*, 295–315.

Antigo, J. L. D., de Bergamasco, R. C., & Madrona, G. S. (2018). Effect of ph on the stability of red beet extract (*Beta vulgaris* l.) microcapsules produced by spray drying or freeze drying. *Food Science and Technology*, *38*, 72–77.

Assadpour, E., & Jafari, S. M. (2019). Advances in spray-drying encapsulation of food bioactive ingredients: From microcapsules to nanocapsules. *Annual Review of Food Science and Technology*, *10*, 103–131.

Ballesteros, L. F., Ramirez, M. J., Orrego, C. E., Teixeira, J. A., & Mussatto, S. I. (2017). Encapsulation of antioxidant phenolic compounds extracted from spent coffee grounds by freeze-drying and spray-drying using different coating materials. *Food Chemistry*, *237*, 623–631.

Bastante, C. C., Cardoso, L. C., Ponce, M. T. F., Serrano, C. M., & de la Ossa-Fernández, E. J. M. (2018). Characterization of olive leaf extract polyphenols loaded by supercritical solvent impregnation into PET/PP food packaging films. *Journal of Supercritical Fluids*, *140*, 196–206.

Belizón, M., Fernández-Ponce, M. T., Casas, L., Mantell, C., & La Ossa-Fernández, E. J. M. D. (2018). Supercritical impregnation of antioxidant mango polyphenols into a multilayer PET/PP food-grade film. *Journal of CO_2 Utilization*, *25*, 56–67.

Bolumar, T., Andersen, M. L., & Orlien, V. (2011). Antioxidant active packaging for chicken meat processed by high pressure treatment. *Food Chemistry*, *129*, 1406–1412.

Bolumar, T., LaPeña, D., Skibsted, L. H., & Orlien, V. (2016). Rosemary and oxygen scavenger in active packaging for prevention of high-pressure induced lipid oxidation in pork patties. *Food Packaging and Shelf Life*, *7*, 26–33.

Bolumar, T., Orlien, V., Sikes, A., Aganovic, K., Bak, K. H., Guyon, C., et al. (2020). High-pressure processing of meat: Molecular impacts and industrial applications. *Comprehensive Reviews in Food Science and Food Safety*, *1541-4337*, 12670.

Bourke, P., Ziuzina, D., Han, L., Cullen, P. J., & Gilmore, B. F. (2017). Microbiological interactions with cold plasma. *Journal of Applied Microbiology*, *123*, 308–324.

Campo, M. M., Nute, G. R., Hughes, S. I., Enser, M., Wood, J. D., & Richardson, R. I. (2006). Flavour perception of oxidation in beef. *Meat Science*, *72*, 303–311.

Champeau, M., Thomassin, J. M., Tassaing, T., & Jérôme, C. (2015). Drug loading of polymer implants by supercritical CO_2 assisted impregnation: A review. *Journal of Controlled Release*, *209*, 248–259.

Chemat, F., Abert Vian, M., Fabiano-Tixier, A.-S. S., Nutrizio, M., Režek Jambrak, A., Munekata, P. E. S. S., et al. (2020). A review of sustainable and intensified techniques for extraction of food and natural products. *Green Chemistry*, *22*, 2325–2353.

Cocero, M. J., Martín, Á., Mattea, F., & Varona, S. (2009). Encapsulation and co-precipitation processes with supercritical fluids: Fundamentals and applications. *Journal of Supercritical Fluids*, *47*, 546–555.
Cui, H., Surendhiran, D., Li, C., & Lin, L. (2020). Biodegradable zein active film containing chitosan nanoparticle encapsulated with pomegranate peel extract for food packaging. *Food Packaging and Shelf Life*, *24*, 100511.
Cunha, L. C. M., Monteiro, M. L. G., Costa-Lima, B. R. C., Guedes-Oliveira, J. M., Alves, V. H. M., Almeida, A. L., et al. (2018). Effect of microencapsulated extract of pitaya (Hylocereus costaricensis) peel on color, texture and oxidative stability of refrigerated ground pork patties submitted to high pressure processing. *Innovative Food Science & Emerging Technologies*, *49*, 136–145.
Dave, D., & Ghaly, A. E. (2011). Meat spoilage mechanisms and preservation techniques: A critical review. *American Journal of Agricultural and Biological Sciences*, *6*, 486–510.
de Carvalho, F. A. L. F. A. L., Lorenzo, J. M. J. M., Pateiro, M., Bermúdez, R., Purriños, L., & Trindade, M. A. (2019). Effect of guarana (Paullinia cupana) seed and Pitanga (Eugenia uniflora L.) leaf extracts on lamb burgers with fat replacement by chia oil emulsion during shelf life storage at 2 °C. *Food Research International*, *125*, 108554.
de Oliveira, T. L. C., de Castro Leite, B. R., Ramos, A. L. S., Ramos, E. M., Piccoli, R. H., & Cristianini, M. (2015). Phenolic carvacrol as a natural additive to improve the preservative effects of high pressure processing of low-sodium sliced vacuum-packed Turkey breast ham. *LWT - Food Science and Technology*, *64*, 1297–1308.
Domínguez, R., Pateiro, M., Gagaoua, M., Barba, F. J., Zhang, W., & Lorenzo, J. M. (2019). A comprehensive review on lipid oxidation in meat and meat products. *Antioxidants*, *8*, 429.
Doulgeraki, A. I., Ercolini, D., Villani, F., & Nychas, G. J. E. (2012). Spoilage microbiota associated to the storage of raw meat in different conditions. *International Journal of Food Microbiology*, *157*, 130–141.
Faustman, C., & Cassens, R. G. (1990). The biochemical basis for discoloration in fresh meat: A review. *Journal of Muscle Foods*, *1*, 217–243.
Faustman, C., Sun, Q., Mancini, R., & Suman, S. P. (2010). Myoglobin and lipid oxidation interactions: Mechanistic bases and control. *Meat Science*, *86*, 86–94.
Gao, Y., Zhuang, H., Yeh, H. Y., Bowker, B., & Zhang, J. (2019). Effect of rosemary extract on microbial growth, pH, color, and lipid oxidation in cold plasma-processed ground chicken patties. *Innovative Food Science and Emerging Technologies*, *57*, 102168.
Giulitti, S., Cinquemani, C., & Spilimbergo, S. (2011). High pressure gases: Role of dynamic intracellular pH in pasteurization. *Biotechnology and Bioengineering*, *108*, 1211–1214.
Goñi, M. L., Gañán, N. A., Strumia, M. C., & Martini, R. E. (2016). Eugenol-loaded LLDPE films with antioxidant activity by supercritical carbon dioxide impregnation. *Journal of Supercritical Fluids*, *111*, 28–35.
González-Alonso, V., Cappelletti, M., Bertolini, F. M., Lomolino, G., Zambon, A., & Spilimbergo, S. (2020). Microbial inactivation of raw chicken meat by supercritical carbon dioxide treatment alone and in combination with fresh culinary herbs. *Poultry Science*, *99*, 536–545.
Greene, B. E., & Cumuze, T. H. (1982). Relationship between TBA numbers and inexperienced panelists' assessments of oxidized flavor in cooked beef. *Journal of Food Science*, *47*, 52–54.
Han, J. W., Ruiz-Garcia, L., Qian, J. P., & Yang, X. T. (2018). Food packaging: A comprehensive review and future trends. *Comprehensive Reviews in Food Science and Food Safety*, *17*, 860–877.
Holman, B. W. B., Kerry, J. P., & Hopkins, D. L. (2018). Meat packaging solutions to current industry challenges: A review. *Meat Science*, *144*, 159–168.

Horita, C. N., Baptista, R. C., Caturla, M. Y. R., Lorenzo, J. M., Barba, F. J., & Sant'Ana, A. S. (2018). Combining reformulation, active packaging and non-thermal post-packaging decontamination technologies to increase the microbiological quality and safety of cooked ready-to-eat meat products. *Trends in Food Science and Technology*, *72*, 45–61.

Huang, S., Liu, B., Ge, D., & Dai, J. (2017). Effect of combined treatment with supercritical CO_2 and rosemary on microbiological and physicochemical properties of ground pork stored at 4 °C. *Meat Science*, *125*, 114–120.

Huang, H. W., Wu, S. J., Lu, J. K., Shyu, Y. T., & Wang, C. Y. (2017). Current status and future trends of high-pressure processing in food industry. *Food Control*, *72*, 1–8.

Khan, M. R., Sadiq, M. B., & Mehmood, Z. (2020). Development of edible gelatin composite films enriched with polyphenol loaded nanoemulsions as chicken meat packaging material. *CyTA Journal of Food*, *18*, 137–146.

Kuck, L. S., & Noreña, C. P. Z. (2016). Microencapsulation of grape (*Vitis labrusca* var. Bordo) skin phenolic extract using gum Arabic, polydextrose, and partially hydrolyzed guar gum as encapsulating agents. *Food Chemistry*, *194*, 569–576.

Lee, H., Choe, J., Yong, H. I., Lee, H. J., Kim, H. J., & Jo, C. (2018). Combination of sea tangle powder and high-pressure treatment as an alternative to phosphate in emulsion-type sausage. *Journal of Food Processing and Preservation*, *42*, e13712.

Leong, T. S. H., Martin, G. J. O., & Ashokkumar, M. (2017). Ultrasonic encapsulation—A review. *Ultrasonics Sonochemistry*, *35*, 605–614.

Leopoldini, M., Marino, T., Russo, N., & Toscano, M. (2004). Antioxidant properties of phenolic compounds: H-atom versus electron transfer mechanism. *Journal of Physical Chemistry A*, *108*, 4916–4922.

Li, L., Wang, H., Chen, M., Jiang, S., Cheng, J., Li, X., et al. (2020). Gelatin/zein fiber mats encapsulated with resveratrol: Kinetics, antibacterial activity and application for pork preservation. *Food Hydrocolloids*, *101*, 105577.

Lima, M. C., Paiva de Sousa, C., Fernandez-Prada, C., Harel, J., Dubreuil, J. D., & de Souza, E. L. (2019). A review of the current evidence of fruit phenolic compounds as potential antimicrobials against pathogenic bacteria. *Microbial Pathogenesis*, *130*, 259–270.

Liu, J., Liu, S., Chen, Y., Zhang, L., Kan, J., & Jin, C. (2017). Physical, mechanical and antioxidant properties of chitosan films grafted with different hydroxybenzoic acids. *Food Hydrocolloids*, *71*, 176–186.

Liu, J., Meng, C. G., Liu, S., Kan, J., & Jin, C. H. (2017). Preparation and characterization of protocatechuic acid grafted chitosan films with antioxidant activity. *Food Hydrocolloids*, *63*, 457–466.

Liu, J., Pu, H., Liu, S., Kan, J., & Jin, C. (2017). Synthesis, characterization, bioactivity and potential application of phenolic acid grafted chitosan: A review. *Carbohydrate Polymers*, *174*, 999–1017.

Lorenzo, J. M., Munekata, P. E. S., Baldin, J. C., Franco, D., Domínguez, R., & Trindade, M. A. (2017). The use of natural antioxidants to replace chemical antioxidants in foods. In J. M. Lorenzo, & F. J. Carballo (Eds.), *Strategies for obtaining healthier foods* (pp. 205–228). New York: Nova Science Publishers.

Lorenzo, J. M., Pateiro, M., Domínguez, R., Barba, F. J., Putnik, P., Kovačević, D. B., et al. (2018). Berries extracts as natural antioxidants in meat products: A review. *Food Research International*, *106*, 1095–1104.

Lorenzo, J. M., Vargas, F. C., Strozzi, I., Pateiro, M., Furtado, M. M., Sant'Ana, A. S., et al. (2018). Influence of Pitanga leaf extracts on lipid and protein oxidation of pork burger during shelf-life. *Food Research International*, *114*, 47–54.

Mantegna, S., Binello, A., Boffa, L., Giorgis, M., Cena, C., & Cravotto, G. (2012). A one-pot ultrasound-assisted water extraction/cyclodextrin encapsulation of resveratrol from *Polygonum cuspidatum*. *Food Chemistry*, *130*, 746–750.

Misra, N. N., & Jo, C. (2017). Applications of cold plasma technology for microbiological safety in meat industry. *Trends in Food Science and Technology*, *64*, 74–86.
Mittal, A., Singh, A., Benjakul, S., Prodpran, T., Nilsuwan, K., Huda, N., et al. (2021). Composite films based on chitosan and epigallocatechin gallate grafted chitosan: Characterization, antioxidant and antimicrobial activities. *Food Hydrocolloids*, *111*, 106384.
Moudache, M., Nerín, C., Colon, M., & Zaidi, F. (2017). Antioxidant effect of an innovative active plastic film containing olive leaves extract on fresh pork meat and its evaluation by Raman spectroscopy. *Food Chemistry*, *229*, 98–103.
Munekata, P. E. S., Domínguez, R., Pateiro, M., & Lorenzo, J. M. (2020). Influence of plasma treatment on the polyphenols of food products—A review. *Food*, *9*, 929.
Munekata, P. E. S., Rocchetti, G., Pateiro, M., Lucini, L., Domínguez, R., & Lorenzo, J. M. (2020). Addition of plant extracts to meat and meat products to extend shelf-life and health-promoting attributes: An overview. *Current Opinion in Food Science*, *31*, 81–87.
Nwabor, O. F., Singh, S., Paosen, S., Vongkamjan, K., & Voravuthikunchai, S. P. (2020). Enhancement of food shelf life with polyvinyl alcohol-chitosan nanocomposite films from bioactive Eucalyptus leaf extracts. *Food Bioscience*, *36*, 100609.
Nychas, G. J. E., Skandamis, P. N., Tassou, C. C., & Koutsoumanis, K. P. (2008). Meat spoilage during distribution. *Meat Science*, *78*, 77–89.
Pankaj, S. K., Wan, Z., & Keener, K. M. (2018). Effects of cold plasma on food quality: A review. *Food*, *7*, 4.
Papoutsis, K., Golding, J. B., Vuong, Q., Pristijono, P., Stathopoulos, C. E., Scarlett, C. J., et al. (2018). Encapsulation of citrus by-product extracts by spray-drying and freeze-drying using combinations of maltodextrin with soybean protein and ι-carrageenan. *Food*, *7*, 115.
Park, K. M., Park, S. H., & Shimanouchi, T. (2019). Innovative strategies and emerging technologies for food safety. *Journal of Chemistry*, *2019*, 1–2.
Ramírez, M. J., Giraldo, G. I., & Orrego, C. E. (2015). Modeling and stability of polyphenol in spray-dried and freeze-dried fruit encapsulates. *Powder Technology*, *277*, 89–96.
Ravichai, K., & Muangrat, R. (2019). Effect of different coating materials on freeze-drying encapsulation of bioactive compounds from fermented tea leaf wastewater. *Journal of Food Processing and Preservation*, *43*, 1–15.
Reboleira, J., Adão, P., Guerreiro, S. F. C., Dias, J. R., Ganhão, R., Mendes, S., et al. (2020). Poultry shelf-life enhancing potential of nanofibers and nanoparticles containing porphyra dioica extracts. *Coatings*, *10*, 315.
Rempe, C. S., Burris, K. P., Lenaghan, S. C., & Stewart, C. N. (2017). The potential of systems biology to discover antibacterial mechanisms of plant phenolics. *Frontiers in Microbiology*, *8*, 422.
Ribeiro, J. S., Santos, M. J. M. C., Silva, L. K. R., Pereira, L. C. L., Santos, I. A., da Silva Lannes, S. C., et al. (2019). Natural antioxidants used in meat products: A brief review. *Meat Science*, *148*, 181–188.
Rocha-Parra, D. F., Lanari, M. C., Zamora, M. C., & Chirife, J. (2016). Influence of storage conditions on phenolic compounds stability, antioxidant capacity and colour of freeze-dried encapsulated red wine. *LWT - Food Science and Technology*, *70*, 162–170.
Rojas, A., Torres, A., José Galotto, M., Guarda, A., & Julio, R. (2020). Supercritical impregnation for food applications: A review of the effect of the operational variables on the active compound loading. *Critical Reviews in Food Science and Nutrition*, *60*, 1290–1301.
Saikia, S., Mahnot, N. K., & Mahanta, C. L. (2015). Optimisation of phenolic extraction from Averrhoa carambola pomace by response surface methodology and its microencapsulation by spray and freeze drying. *Food Chemistry*, *171*, 144–152.
Šaponjac, V. T., Ćetković, G., Čanadanović-Brunet, J., Dilas, S., Pajin, B., Petrović, J., et al. (2017). Encapsulation of sour cherry pomace extract by freeze drying: Characterization and storage stability. *Acta Chimica Slovenica*, *64*, 283–289.

Sehrawat, R., Kaur, B. P., Nema, P. K., Tewari, S., & Kumar, L. (2020). Microbial inactivation by high pressure processing: Principle, mechanism and factors responsible. *Food Science and Biotechnology*, *30*, 1–17.

Sharif, N., Khoshnoudi-Nia, S., & Jafari, S. M. (2020). Nano/microencapsulation of anthocyanins; a systematic review and meta-analysis. *Food Research International*, *132*, 109077.

Singh, S., & Shalini, R. (2016). Effect of hurdle technology in food preservation: A review. *Critical Reviews in Food Science and Nutrition*, *56*, 641–649.

Spilimbergo, S., Mantoan, D., Quaranta, A., & Mea, G. D. (2009). Real-time monitoring of cell membrane modification during supercritical CO_2 pasteurization. *Journal of Supercritical Fluids*, *48*, 93–97.

Suman, S. P., & Joseph, P. (2013). Myoglobin chemistry and meat color. *Annual Review of Food Science and Technology*, *4*, 79–99.

Surendhiran, D., Li, C., Cui, H., & Lin, L. (2020). Fabrication of high stability active nanofibers encapsulated with pomegranate peel extract using chitosan/PEO for meat preservation. *Food Packaging and Shelf Life*, *23*, 100439.

Tadapaneni, R. K., Daryaei, H., Krishnamurthy, K., Edirisinghe, I., & Burton-Freeman, B. M. (2014). High-pressure processing of berry and other fruit products: Implications for bioactive compounds and food safety. *Journal of Agricultural and Food Chemistry*, *62*, 3877–3885.

Trucillo, P., Campardelli, R., Aliakbarian, B., Perego, P., & Reverchon, E. (2018). Supercritical assisted process for the encapsulation of olive pomace extract into liposomes. *Journal of Supercritical Fluids*, *135*, 152–159.

Umaraw, P., Munekata, P. E. S., Verma, A. K., Barba, F. J., Singh, V. P., Kumar, P., et al. (2020). Edible films/coating with tailored properties for active packaging of meat, fish and derived products. *Trends in Food Science and Technology*, *98*, 10–24.

Umaraw, P., & Verma, A. K. (2017). Comprehensive review on application of edible film on meat and meat products: An eco-friendly approach. *Critical Reviews in Food Science and Nutrition*, *57*, 1270–1279.

Verma, S. P., & Sahoo, J. (2000). Improvement in the quality of ground chevon during refrigerated storage by tocopherol acetate preblending. *Meat Science*, *56*, 403–413.

Vilkhu, K., Mawson, R., Simons, L., & Bates, D. (2008). Applications and opportunities for ultrasound assisted extraction in the food industry—A review. *Innovative Food Science and Emerging Technologies*, *9*, 161–169.

Vinceković, M., Viskić, M., Jurić, S., Giacometti, J., Bursać Kovačević, D., Putnik, P., et al. (2017). Innovative technologies for encapsulation of mediterranean plants extracts. *Trends in Food Science and Technology*, *69*, 1–12.

Visentin, A., Rodríguez-Rojo, S., Navarrete, A., Maestri, D., & Cocero, M. J. (2012). Precipitation and encapsulation of rosemary antioxidants by supercritical antisolvent process. *Journal of Food Engineering*, *109*, 9–15.

Vuolo, M. M., Lima, V. S., & Maróstica Junior, M. R. (2018). Phenolic compounds: Structure, classification, and antioxidant power. In *Bioactive compounds: Health benefits and potential applications* (pp. 33–50). Elsevier.

Wang, Y., Xie, M., Ma, G., Fang, Y., Yang, W., Ma, N., et al. (2019). The antioxidant and antimicrobial activities of different phenolic acids grafted onto chitosan. *Carbohydrate Polymers*, *225*, 115238.

Wenzel, J. E., Moorman, V., Wang, L., Spencer-Williams, I., Hall, M., Samaniego, C. S., et al. (2020). In-situ extraction and impregnation of black walnut husk into polyethylene film using supercritical carbon dioxide with an ethanol modifier. *Food Science & Nutrition*, *8*, 612–619.

Yasui, K. (2018). Acoustic cavitation. In K. Yasui (Ed.), *Acoustic cavitation and bubble dynamics* (pp. 1–35). Springer.

Zhang, X., Liu, J., Qian, C., Kan, J., & Jin, C. (2019). Effect of grafting method on the physical property and antioxidant potential of chitosan film functionalized with gallic acid. *Food Hydrocolloids*, *89*, 1–10.
Zhang, J., Zou, X., Zhai, X., Huang, X. W., Jiang, C., & Holmes, M. (2019). Preparation of an intelligent pH film based on biodegradable polymers and roselle anthocyanins for monitoring pork freshness. *Food Chemistry*, *272*, 306–312.
Zhao, Y., & Saldaña, M. D. A. (2019). Use of potato by-products and gallic acid for development of bioactive film packaging by subcritical water technology. *Journal of Supercritical Fluids*, *143*, 97–106.

CHAPTER SIX

Incorporation of polyphenols in baked products

Juanying Ou*
Institute of Food Safety & Nutrition, Jinan University, Guangzhou, China
*Corresponding author: e-mail address: ojy_26@163.com

Contents

Abstract

Bakery foods, including breads, cakes, cookies, muffins, rolls, buns, crumpets, pancakes, doughnuts, waffles, and bagels, etc., have been an important diet of humans for thousands of years. As the nutraceuticals with various biological activities, polyphenols, especially polyphenol-enriched products are widely used in bakery foods.

Advances in Food and Nutrition Research, Volume 98
ISSN 1043-4526
https://doi.org/10.1016/bs.afnr.2021.02.009

The polyphenol-enriched products are mainly from fruits and vegetables, including fruits in whole, juice, puree, jam, and the powder of dried fruits, pomace, and peels. Incorporation of these products not only provide polyphenols, but also supply other nutrients, especially dietary fibers for bakery products. This chapter discussed the thermal stability of different types of polyphenols during baking, and the effect of polyphenols on the sensory attributes of baked foods. Moreover, their role in mitigation of reactive carbonyl species and the subsequent formation of advanced glycation end products, antioxidant and antimicrobial activities have been also discussed. Since polyphenols are subjected to high temperature for dozens of minutes during baking, future works need to focus on the chemical interactions of polyphenols and their oxidized products (quinones) with other food components, and the safety consequence of these interactions.

1. Introduction

Baked foods, with wheat flour as the main ingredient, have been an important part of a balanced diet for thousands of years. Baked foods can be classified as unsweetened, sweetened, or filled goods, including bread, cakes, cookies, muffins, rolls, buns, crumpets, pancakes, doughnuts, waffles, and bagels, etc. (Smith, Daifas, El-Khoury, Koukoutsis, & El-Khoury, 2004). These products are largely consumed worldwide and in high quantities for individuals, and have an important role in human nutrition (Martins, Pinho, & Ferreira, 2017). Moreover, for nutrition and sensory enhancement purposes, a wide variety of filled baked products are sold in markets nowadays, such as pastries, sandwiches, cream, cakes, fruit and meat pies, sausage rolls, pizza, and quiche.

However, baked foods have encountered several problems in modern society. First of all, they are high in sugar and fats. In addition to sugar-sweetened beverages, baked foods are the main sources of sugar in processed foods (Sahin, Zannini, Coffey, & Arendt, 2019). For example, the sugar content in cakes ranges from 27% to 45%, and in biscuits, the fats and sugar level up to 30%, respectively (Sahin et al., 2019). Excessive sugar consumption increases body weight and fat gain, and promotes the development of non-communicable diseases, such as obesity, diabetes and cardiovascular disease via upregulation of hepatic uptake and fructose metabolism, which results in liver lipid accumulation, dyslipidemia, decreased insulin sensitivity and increased uric acid levels (Stanhope, 2016). The second problem is formation of toxins during baking. Various food-borne toxins, including acrylamide, heterocyclic amines, reactive carbonyl species (e.g., acrolein, glyoxal, methylglyoxal), advanced glycation end products (AGEs), furans and hydroxymethylfurfural have been detected in baked foods (Capuano & Fogliano, 2011; Koszucka & Nowak, 2019; Ou, Zheng, Huang, Ho, & Ou, 2020; Zhou et al., 2020). The third one is the spoilage, including physical spoilage (moisture loss, staling), chemical spoilage (hydrolytic and oxidative rancidity of fats), and microbiological spoilage (yeast, mold, bacterial growth) (Smith et al., 2004). Fourthly, using refined flour as raw material will lead to nutritional deficiency to a certain extent, especially dietary fibers, polyphenols and vitamin C.

In order to produce bakery products with high nutritional value, enrichment with vitamins, minerals, fibers and polyphenols is considered as an effective way (Salehi & Aghajanzadeh, 2020). Fruits and vegetables contribute 90% of vitamin C, 50% of β-carotene (vitamin A), most of dietary fiber as well as polyphenols in human diets (Aditi, Shashirekha, & Sudha, 2017). Therefore, polyphenols as a type of nutraceuticals and nutrition promoters, can be used in bakery foods through the incorporation of fruit and vegetable products, or direct supplementation of polyphenols.

Since polyphenols are antioxidants with high reactivity and sensitivity to thermal processing of foods, in this chapter, the changes of polyphenols during baking, the influence of polyphenols on the quality and safety, and antioxidant activity of baked products are discussed.

2. Incorporation methods of polyphenols in baked foods

Two major methods are applied in the incorporation of polyphenols in baked foods, one is addition of foods rich in polyphenols to bakery products, such as the products of fruits and vegetables, the other is directly mixing polyphenols standards into the formular. The products of fruits and vegetables are composed of their raw materials or their processed ones, including jams, puree or pulp, jellies, marmalade, fruit chunks, dried fruits, canned products, fruit powder, fruit seed oil, fruit peel, fruit pomace, etc. (Aditi et al., 2017).

2.1 Incorporation of fruit and vegetable pomace

Fruit and vegetable juicing generate a large amount of waste in the form of pomace, which is the by-product of the processing industry composed of cell wall compounds, stems, and seeds (Quiles, Campbell, Struck, Rohm, & Hernando, 2018). The pomace accounts for 10%–35% of raw mass but are not fully used in food industry (Majerska, Michalska, & Figiel, 2019).

Introducing pomace into bakery goods is a way out for application of them to partially replace flour, sugar, or fat so as to reduce energy load. Moreover, this is the main pathway for fortification of polyphenols and dietary fibers.

Fiber enrichment by fruit and vegetable pomace incorporation has great influence and better functional properties on the overall characteristics of bakery products, such as oil absorption capacity and low caloric value, water retention and holding property, and good colonic fermentability (Sahni & Shere, 2018). Although dietary fibers from cereals and grains have similar chemical and physical properties to those of wheat fiber, this type of dietary source contains some antinutritional factors, such as phytic acid, which remarkably decreases minerals availability. By comparison, fruit and vegetable pomace exhibits a better nutritional quality, since it has the balanced composition of water-soluble and water-insoluble dietary fibers (Table 1) and contains large amounts of bioactive compounds, such polyphenols, carotenoids, etc. (Quiles et al., 2018). Thus, the by-products from the processing of fruits and vegetables have been widely used in bakery products, such as the pomace and peels from grape, melon, pomegranate, plantain, carrot, beetroot, apple, rowanberry, blackcurrant, elderberry, passion fruit, pumpkin, guava, peach, orange, raspberry, cranberry, potato, tomato (Majzoobi, Ghavi, Farahnaky, Jamalian, & Mesbahi, 2011; Trigo, Alexandre, Saraiva, & Pintado, 2020).

Due to the high content of soluble dietary fiber, fruit and vegetable pomace can mimic techno-functional properties of fat in bakery foods, such as leavening the product, extending cake softness, tenderizing the crumb, stabilizing the gas bubbles, etc. (Quiles et al., 2018). Up to 10% of fat could be replaced by peach dietary fiber in muffins (Grigelmo-Miguel, Carreras-Boladeras, & Martin-Belloso, 2001), and up to 60% mango pulp fiber has been used to replace wheat flour in muffins (Sudha et al., 2015), leading to significant energy-lowering effect in baked products.

Moreover, fruit and vegetable pomace contain high quantities of polyphenols (Table 1), which can increase the antioxidant activity and offer the foods with various biological activities that polyphenols exhibit (Majerska et al., 2019). However, there are different types of polyphenols with various structures in fruit and vegetable pomace, which may undergo degradation and complicated reactions with food components during thermal processing. When using fruit and vegetable pomace in cereal-baked product formulations, it is important to consider their chemical changes and the resulting influence on food properties and human health.

Table 1 Contents of dietary fiber and polyphenols in different fruit and vegetable pomace.

Sources	Water-insoluble fiber (%)	Water-soluble fiber (%)	Phenolic compounds (mg GAE/g)	References
Chokeberry	43.8	13.9	55	Schmid et al. (2020)
Apple	42.6	8.3	21.9–45.9	Rana, Rana, Gupta, and Bhushan (2020)
Orange	29.9	23.3	41.7	Gutiérrez Barrutia, Curutchet, Arcia, and Cozzano (2019)
Lemon	51.4	12.7	20.7	Chang, Li, and Shiau (2015); Papoutsis et al. (2017)
Pear	75.3	0.1	2.25	Aguedo et al. (2012)
Pineapple	50.9	9.8	8.92	Montalvo-González et al. (2018)
Grape	61.2	4.1	75.4–138.4	Bender et al. (2020); Mohamed Ahmed et al. (2020)
Grapefruit	37.8	6.4		Figuerola et al. (2005)
Mango	38.9	19.2	1.05	Sudha et al. (2015)
Broccoli	49.6		7.3	Shi et al. (2019)

2.2 Incorporation of fruit and vegetable juices

Fruit and vegetable juices play important roles in the modern diets of people because they provide energy (sugars), nutrients and functional ingredients, such as vitamins, minerals, carotenes, and polyphenols (Bhardwaj, Nandal, Pal, & Jain, 2014; Shishir & Chen, 2017; Zheng et al., 2017). For example, orange juice has over 170 different phytochemicals, including more than 60 flavonoids, most of them show antioxidant and health effects (Bhardwaj et al., 2014). It is well recognized that fruit and vegetable juices may offer protection against a number of diseases, including cardiovascular disease, mental depression, diabetes, high blood pressure, chronic inflammation, arthritis, heart disease, gastric ulcers, sprains, poor digestion, arteriosclerosis, and blood vessel problems, and shows beneficial effects for cognition in healthy older adults (Bhardwaj et al., 2014; Lamport, Saunders, Butler, & Spencer, 2014).

Based on the nutrition and health-promoting effect of fruit and vegetable juices, they have been widely incorporated in bakery foods to raise nutritional value of the baked product and increase health benefits for the consumers. Moreover, due to the seasonal availability of fresh fruits and vegetables as well as very short shelf life of some special fruits, such as berries, they are chosen to be sold in the form of frozen fruits, jams and sauces (Rosales-Soto, Powers, & Alldredge, 2012). This property of fresh fruits and vegetables also impels food industry to employ fruit and vegetable juices in baked products. Thus, the fruit and vegetable juices, fruit pulp, and jam are widely used in bakery goods, such as red raspberry juice (10% of addition) in muffins (Rosales-Soto et al., 2012), peach pulp (28 Brix, 10.5% of addition) in cookies (Canalis, Baroni, León, & Ribotta, 2020), bergamot juice (5.4%–21.6% of addition) in steamed pumpkin cake (Mahidsanan & Lertnirundon, 2020), lime juice (10%–20% of addition) in pan bread (Scarton, Ganancio, de Avelar, Clerici, & Steel, 2020), sea-buckthorn pulp (5% and 10% of addition) in bread (Guo et al., 2019), and orange juice (16.5% of addition) in cakes (da SILVA, Lopes, & Valente-Mesquita, 2006). These research results showed that addition of juices not only increased the antioxidant activity of bakery products, but also decreased the energy intake via replacing fat. For example, although avocado puree was applied as a substitution of 50% fat, the low-fat muffins still got satisfactory acceptability (Othman, Abdul Manaf, Harith, & Wan Ishak, 2018).

Compared with fruit and vegetable pomace, fruit juice contains much more compounds, such as carbohydrates, water-soluble vitamins (vitamin B, C), minerals (potassium, calcium and magnesium), organic acids (lactic acid, succinic acid, and pyruvic acid, etc.), amino acids, fibers, aroma compounds, carotenoids (lycopene, β-carotene, etc.), and phenolic compounds (Bhardwaj et al., 2014). Phenolic compounds include hydroxybenzoic and hydroxycinnamic acid derivatives, anthocyanins, flavonols, flavanols, condensed tannins (proanthocyanidins) (Rosales-Soto et al., 2012).

When fruit and vegetable juices are used in bakery foods, the changes of these components and their interaction should be cautiously investigated during thermal treatment. Moreover, the juices, especially fruit juices also contain different organic acids, which markedly decrease pH value of the dough. Lower pH results in hydrolysis of sucrose and starch, leading to the release of reducing sugars (glucose and fructose). Besides, lower pH environment, divalent cations, and amino acids favor the formation of some food toxins, such as hydroxymethylfurfural produced from hexoses (Capuano & Fogliano, 2011).

2.3 Incorporation of dried fruits and vegetable powder

It has been reported that about US$310 billion and US$680 billion were lost in foods in developing and industrialized countries, respectively. In which root crops, fruits and vegetables accounts for 40%–50% (Salehi & Aghajanzadeh, 2020). Therefore, application of various preservation methods is necessary to increase their shelf life and reduce the waste of agricultural products, cost of disposal and environmental problem. After slicing of the fruits and vegetables, drying treatments are often conducted to decrease water content and preserve them. Drying technologies on fruits and vegetables include solar dryer, infrared irradiation, microwave heating, fluidized drying, flash drying, super-heated steam drying, freeze drying, and vacuum drying (Salehi & Aghajanzadeh, 2020). These technologies can also manufacture powder of dried fruits and vegetables, which act as a good source of nutrients for bakery products. Powders of different dried fruits and vegetable enriched in cakes and bread were respectively reviewed by Salehi and Aghajanzadeh (2020) and Betoret and Rosell (2020), including those from beans (faba bean, lupine, lentils and carob), berries (acai berry, blueberry, cranberry, maqui berry, raspberry and strawberry), pumpkin, carrot, melon, apple, pear, peach, apricot, grapefruit, baobab fruit, sea-buckthorn, hawthorn, rowan, guava, blackcurrant, grape, mango, kiwifruit, quince, orange, mushroom, etc. The fortification level of dried powder ranged from 2% to 50%.

Incorporation of powder of dried fruits and vegetables in bakery foods proved that they can minimize the negative effects of high temperature processing, significantly increased the total phenolic compounds and antioxidant activity, raised the fibers content and other nutrients (vitamins and minerals) (Alashi, Taiwo, Oyedele, Adebooye, & Aluko, 2019; Betoret & Rosell, 2020; Salehi & Aghajanzadeh, 2020). However, high water absorption capacity by dietary fibers affects the rheology, texture, volume, and density of the batter.

2.4 Using fruit and vegetable as fillings

The fillings, especially fruit-based fillings are widely used in bakery foods in three types, toping, sandwiching, and burying (pastries) (Fig. 1). Toping and

Fig. 1 Fruit fillings used in bakery foods: toping (up), sandwiching (middle) and burying (down).

sandwiching filled-bakery foods are usually cooked and sold in western countries, while the burying filled-bakery foods are more popular in eastern countries or regions, such as pineapple and blueberry pastries. Whatever the filling types we choose, the fillings have to be undergone two or three thermal treatment procedures, including preparation of the raw materials (juice, puree or jam), fillings preparation, and baking.

Preparation of puree and jam: Puree (or mash) is a kind of cooked food, usually are vegetables, fruits or legumes that have been ground, pressed, blended or sieved to the consistency of a creamy paste. Take apple puree preparation as an example, which includes two steps of thermal treatment, enzyme inactivation, and sterilization. The first step is to cook the diced apple at a temperature between 93 and 98 °C for about 4–5 min to soften the fruit tissue and inactivate the polyphenol oxidase that induces enzymatic browning. After pulped to remove coarse fibers, seeds, seed cells, and peel particles, purees are filled and subjected to the next heating (pasteurization) at 90 °C for 20 min (Oszmiański, Wolniak, Wojdyło, & Wawer, 2008).

Jam is semi-solid food product obtained upon cooking of fruits or vegetables pulp with sugar, citric acid and pectin, including single fruit-type and mixed fruit-type (two or more fruits pulp). Jam should contain 65% or more total soluble solids and at least 45% pulp (Ullah, Ullah, Khan, Ullah, & Badshah, 2018), thus, the mixtures (pH = 3) should be cooked until the final product contained 65% of soluble solids (Abid, Yaich, Hidouri, Attia, & Ayadi, 2018; Ahmed, Burhanuddin, Haque, & Hossain, 2011).

Preparation technology of fillings: Jam can be directly used as the fillings for bakery foods, such as jam-filled cakes (Michon, O'Sullivan, Sheehan, Delahunty, & Kerry, 2010). The fillings prepared from puree are usually mixed with other components. Cropotova, Tylewicz, Cocci, Romani, and Dalla Rosa (2016) gave a preparation technology of apple fillings that contain apple puree, sugar, inulin, pectin and gellant gum as follows: Sugar was added to apple puree, stirring and warming up to complete dissolution. Inulin, pectin and gellant gum were dissolved in hot water under continuous mixing and then simultaneously introduced into the prepared apple filling mix and homogenized under intense blending. Then the prepared fillings were immediately put into jars and sterilized at a temperature of 121 °C for 10 min. The preparation technology for other fruit puree-based fillings are similar to that of apple fillings and the pH value of the fillings is 3.0 (Aditi et al., 2017; Agudelo, Varela, & Fiszman, 2015; Agudelo, Varela, Sanz, & Fiszman, 2014). In addition to jam and puree, dried fruit powder is also used to prepare fillings, such as dehydrated chiku powder (Aditi et al., 2017).

During preparation of the fillings of pastries (bakery products in eastern regions), stir-frying is applied. The stir-frying almost takes 40 min and removes approximately 60% of the moisture (Yang et al., 2019). The temperature of this process is difficult to control and the temperature in the pot bottom is very high, which may result in a significant degradation of food components.

Lastly, the fillings are subjected to baking at high temperatures. Under this thermal treatment, various chemical reactions take place, including thermal degradation and interactions between food components, such as the Maillard reaction. The reaction may take vigorously for the toping fillings. Moreover, as mentioned before, the fillings contain a lot of thermally sensitive components (vitamins, polyphenols), organic acids, and high moisture. During baking, the liquid along with these components will run into the product, thus influences the chemical and physical properties of the bake products.

2.5 Incorporation of polyphenol

Incorporation of single polyphenol cannot provide other nutrients and produce the related health effects related to them. The main aim of polyphenols incorporation in bakery foods is to increase the antioxidant activity, anti-microbial activity, and mostly, to reduce or scavenge the toxins produced during thermal processing, such as acrylamide, heterocyclic amines, reactive carbonyl species and the resulting advanced glycation end products (Jin, Wu, & Zhang, 2013; Navarro & Morales, 2017; Teng, Hu, Tao, & Wang, 2018). Various polyphenols have been incorporated into different bakery foods, such as caffeic acid, gallic acid, ferulic acid, chlorogenic acid, rosmarinic acid, catechin, epigallocatechin gallate, naringenin, and quercetin in breads (Fu et al., 2018; Goh et al., 2015; Lin, Teo, Leong, & Zhou, 2019; Mildner-Szkudlarz, Różańska, Piechowska, Waśkiewicz, & Zawirska-Wojtasiak, 2019; Zhang, Chen, & Wang, 2014), cyanidin-3-O-glucoside in wafers (López et al., 2019), hydroxytyrosol, hesperidin and naringenin in biscuits (Mayneris-Perxachs et al., 2019; Navarro & Morales, 2017), rosmarinic acid and epigallocatechin-3-gallate in rolls (Favreau-Farhadi, Pecukonis, & Barrett, 2015), resveratrol, epicatechin and rosmarinic acid in cookies (Ou, Teng, El-Nezami, & Wang, 2018). The incorporation level ranges from 0.1% to 5%.

3. Effect of baking on the stability of polyphenols

The retention of quercetin glycosides, catechins, chlorogenic acid, phloridzin and cyanidin galactoside in apple skin were 61%, 57%, 53%, 44% and 20% after incorporation in muffins (Rupasinghe, Wang, Huber, & Pitts, 2008), and baking caused a significant reduction of lutein up to 62% in wholegrain muffins (Abdel-Aal, Young, Akhtar, & Rabalski, 2010), indicating that polyphenols, including phenolic acid, anthocyanins, and flavonoids, show low thermal stability and undergo thermal degradation during baking when incorporated in bakery products.

3.1 Phenolic acids

Ferulic and coumaric acids are the main hydroxycinnamic acid in cereals, especially in wholegrain products and exist in free or bound structures via esterifying to oligosaccharides and polysaccharides in the kernel. Baking increases the free phenolic acids while decreases the bound ones in bread, cookies, and muffins, etc. (Abdel-Aal & Rabalski, 2013), suggesting that baking can transform phenolic acids from bound to free ones. Increase of free ferulic acid from maize feruloyl oligosaccharides in biscuits confirmed that baking cleaves the ester bond of feruloyl oligosaccharides (Huang et al., 2018). In this research, approximately 52% of ferulic acid was released from feruloyl oligosaccharides when at addition concentrations of 10–40 mg/g. Alkaline pH environment (pH 8.0) by adding leaving agent $NaHCO_3$ may be responsible for the release of ferulic acid during baking.

Autoclaving 51 mM of ferulic and p-coumaric acids at pH 7.0 in different length of time (10, 20, 30, 40, 50 and 60 min) indicated that they underwent decarboxylation and formed two dimeric products through their corresponding radical intermediates, namely dimer of 4-vinylguaiacol and dimer of 4-vinylphenol (Arrieta-Baez et al., 2012). Thermal treatment degrades ferulic acid to form aroma compounds, including guaiacol, 4-methylguaiacol, 4-ethylguaiacol, 4-vinylguaiacol, and vanillin (Ezekiel, Singh, Sharma, & Kaur, 2013; Schieber & Wüst, 2020). 4-Vinylguaiacol participates in the Maillard reaction and form two adducts (Jiang, Chiaro, Maddali, Prabhu, & Peterson, 2009; Fig. 2). Adduct 2 can produce an aroma compound, a pyrazinone-type Maillard product, 2-(5-(furan-2-yl)-6-methyl-2-oxopyrazin-1(2H)-yl) acetic acid, though ferulic acid significantly

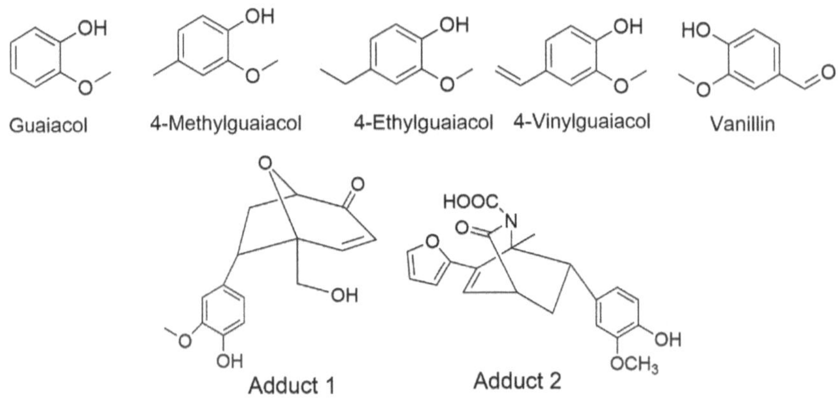

Fig. 2 Thermal degradation products from ferulic acid (top) and the two adducts (bottom) formed from the moiety ferulic acid (blue) in glucose-glycine model.

reduced the generation of other Maillard-type aroma compounds in most cases, such as methylpyrazines, furfurals, 2-acetylpyridine, 2-acetylpyrrole, 2-acetylfuran, and cyclotene (Jiang et al., 2009).

Chlorogenic acid (5-O-caffeoylquinic acid) is the main phenolic acid in potato and coffee bean. It accounts for more than 90% of the phenolics in potato (1000–4000 mg/kg) and contain up to 10% (dry matter) in green coffee (Ezekiel et al., 2013; Świeca, Gawlik-Dziki, Sęczyk, Dziki, & Sikora, 2018). Heat treatment induces isomerization and transformation of chlorogenic acid. Treatment of chlorogenic acid at 100–200 °C in aqueous solution generated 3- and 4-caffeoylquinic acid isomers, cis-5-caffeoylquinic acid, hydroxylated 5-caffeoylquinic acid derivatives, hydrolysis products (quinic acid and caffeic acid), and di-caffeoylquinic acid (Dawidowicz & Typek, 2017). Degradation of chlorogenic acid depends on the processing conditions. It was found that microwaving, steaming, and boiling does not degrade chlorogenic acid (Navarre, Shakya, Holden, & Kumar, 2010), whereas complete destruction was observed during baking of potatoes at 212 °C for 46 min (Dao & Friedman, 1992). Coffee-roasting significantly decreases the amount of chlorogenic acid; 8%–10% of its amount is lost for every 1% loss of dry matter (Nunes & Coimbra, 2010). During coffee-roasting, chlorogenic acid undergoes rapid decarboxylation and produces simple phenols, including catechol, 4-methylcatechol, 4-ethylcatechol, and 4-vinylcatechol (Moon & Shibamoto, 2010; Nunes & Coimbra, 2010). Chlorogenic acid and its hydrolyzed product, caffeic acid, also participate in the Maillard reaction to form melanoidins (Nunes & Coimbra, 2010).

3.2 Anthocyanins

Anthocyanins are vacuolar pigments existing in flowers, fruits, vegetables, beans, tubers, and grains; six major anthocyanin aglycones (anthocyans) present in nature are cyanidin, delphinidin, pelargonidin, peonidin, petunidin, and malvidin (Sadilova, Stintzing, & Carle, 2006). The glycosylation and acylation of anthocyans produce more than 500 anthocyanins in the plant kingdom (Sadilova et al., 2006).

In acidic aqueous solution ($pH < 2$), anthocyanins are presented in their type of flavylium cation. Under different pH conditions prevailing in plants, foods and in the digestive tract (from $pH = 2$ to 8), anthocyanins change to a mixture of colored and colorless forms in equilibrium between different chemical species (Fig. 3, top). Each chemical species displays specific characteristics (charge, electronic distribution, planarity, and shape) modulating its reactivity and interactions with plant or food components, such as electron-rich C–C double bonds (4-hydroxystyrenes, acrolein), the enol forms of various aldehydes and ketones (acetaldehyde, glyoxal, methylglyoxal) (Dangles & Fenger, 2018).

Due to their structural characteristics, anthocyanins are the most sensitive phenolic compounds to thermal treatment among polyphenols. Cyanidin-3-glucoside is the predominant anthocyanin in black soybeans, Slavin, Lu, Kaplan, and Yu (2013) incorporated 24% of whole fresh soybean flour into dough to prepare the crackers and found that baking at 177 °C for 10 min or at 149 °C for 16 min resulted in 74% and 68% reduction of cyanidin-3-glucoside, respectively. Berries contain various anthocyanins at high levels. Górnaś et al. (2016) incorporated fruit pomaces from strawberry, black currant, and raspberry into muffins and determined the content of anthocyanins after baking in conventional oven or halogen oven at three different temperatures (140, 180 and 220 °C); the baking timing started when the inside temperature of the muffins reached 107 °C (for example, 35 min for oven baking at 140 °C, and 24 min for halogen baking at 220 °C). The results in Table 2 indicates that anthocyanins are very sensitive to baking, most of which are lost up to 50% after long time baking (Górnaś et al., 2016). The same situation occurred in wild blueberry-incorporated bread and blueberry-filled pastries (Rodriguez-Mateos, Cifuentes-Gomez, George, & Spencer, 2014; Yang et al., 2019). The anthocyanin content in blueberry decreased extensively after cooking (stir-frying) and baking. During preparation (cooking) of the bread filling, malvidin, delphinidin, and petunidin lost 18%, 25% and 21%, respectively (Rodriguez-Mateos et al., 2014). While stir-frying degraded much more anthocyanin

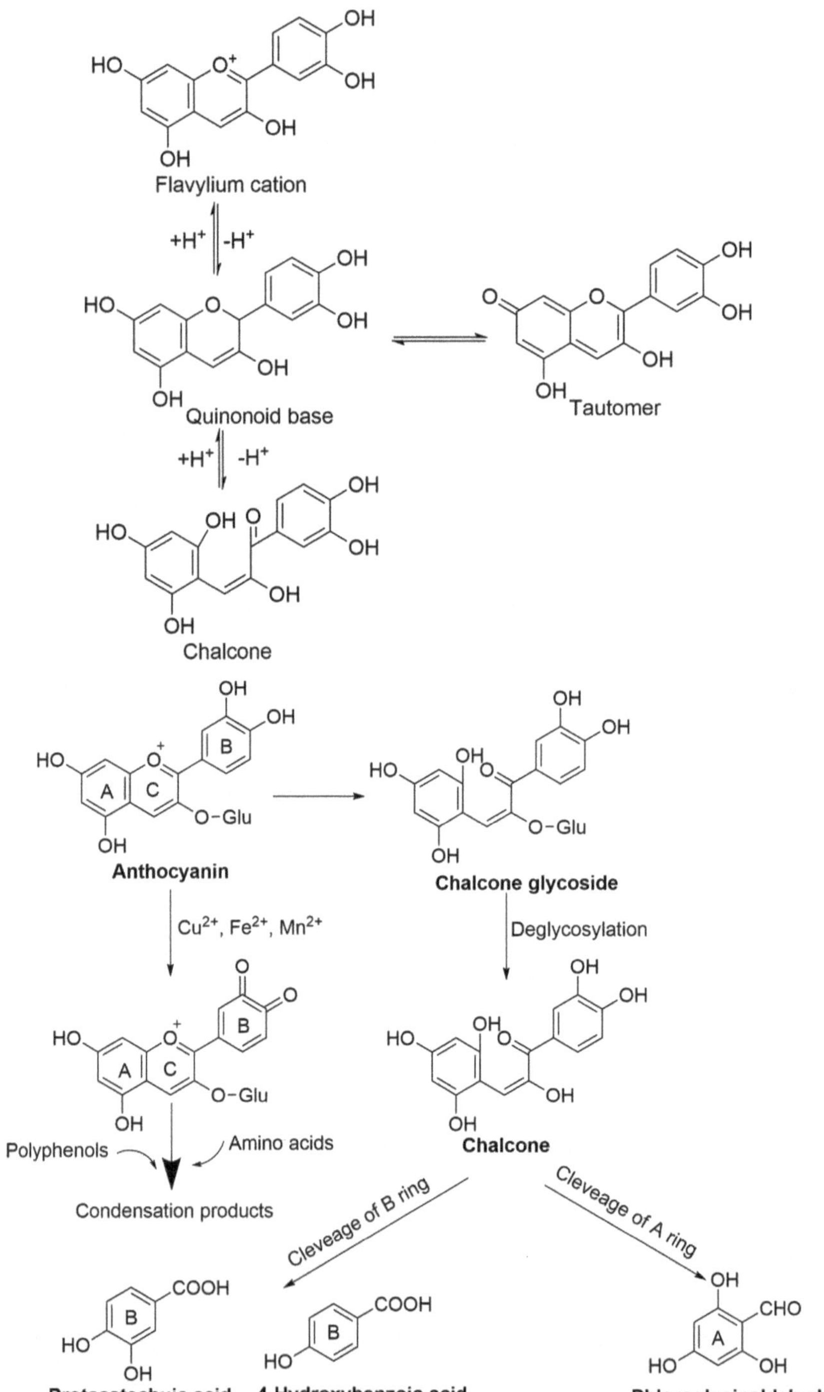

Fig. 3 Effect of pH value on structure transformation of anthocyanins (top) and their typical thermal degradation pathway and the resulting products (bottom).

Table 2 Average recovery (%) of polyphenols in muffins enriched with fruit pomaces from strawberry, black currant, and raspberry.

		In conventional oven			In halogen oven		
Baking temperatures(°C)		140	180	220	140	180	220
Strawberry pomace	Pelargonidin-3-O-glucoside	24	37	39	7	14	22
Black currant pomace	Delphinidin-3-O-glucoside	93	43	54	26	32	39
	Delphinidin-3-O-rutinoside	92	51	60	29	39	46
	Cyanidin-3-O-glucoside	44	54	57	27	31	41
	Cyanidin-3-O-rutinoside	53	63	64	30	40	49
	Myricetin-3-O-glucoside	83	84	86	79	80	81
	Myricetin-3-O-rutinoside	85	85	85	82	83	83
Raspberry pomace	Cyanidin-3-O-sophoroside	26	46	53	8	14	25
	Cyanidin-3-O-glucosylrutinoside	32	53	55	12	21	32
	Cyanidin-3-O-glucoside	29	49	53	8	20	31
	Cyanidin-3-O-rutinoside	35	57	61	9	20	36

(48%–53% of reduction in total anthocyanins), delphinidin glycosides by 61%–67%, malvidin by 52%–58%, petunidin by 40%–45%, and cyanidin by 38%–41%, respectively (Yang et al., 2019). Baking further destroyed anthocyanins, malvidin by 24%, petunidin by 26% and peonidin by 24% in breads (Rodriguez-Mateos et al., 2014).

The typical thermal degradation pathway of anthocyanins is illustrated in Fig. 3 (bottom). Under thermal treatment, anthocyanins are transformed to chalcone glycosides and then deglycosylated to release chalcones (Sadilova et al., 2006). Then chalcone undergoes cleavage to form various products, including aldehydes and acids (Patras, Brunton, O'Donnell, & Tiwari, 2010; Sadilova, Carle, & Stintzing, 2007). Sinela et al. (2017) identified various of degradation products after thermal treatment of pigments from black carrot concentrates, including phenolic acids attached to anthocyanins (Table 3). Sadilova et al. (2006) obtained various numbers of degradation compounds from heated pigment isolates of strawberry (9), elderberry (9), and black carrot (18) concentrates respectively. The typical degradation substances

Table 3 Anthocyanins and their degradation products identified from black carrot concentrates heated at pH 3.5, 95 °C for 6 h.

Anthocyanins before heating	Degradation products
Cyd-3-gal-xyl-glc	Protocatechuic acid
Cyd-3-gal-xyl-glc-sin	p-Coumaric acid derivative
Cyd-3-gal-xyl-glc-fer	Ferulic acid derivative
Cyd-3-gal-xyl-glc-coum	Phloroglucinaldehyde
	Sinapic acid derivative
	Cyd-3-gal-glc-fer
	Cyd-3-gal-glc-coum
	Ferulic acid

Coum, coumaric acid; Cyd, cyanidin; fer, ferulic acid; gal, galactose; glc, glucose; Pel, pelargonidin; Sin, sinapic acid; xyl, xylose.

include protocatechuic acid, 4-hydroxybezoic acid, coumaroyl-galactose-glucotose, phloroglucinaldehyde, cyanidin-3-galactose-glucoside-coumaric acid, coumaric acid, cyanidin, ferulic acid, sinapic acid, pelargonidin, and three tentatively assigned adducts. These adducts are possible the dimers from protocatechuic acids or protocatechuic acid and phloroglucinaldehyde, condensation products of pelargonidin-3-glucose, and its cleavage products 4-hydroxybenzoic acid. Moreover, metal ions (iron, copper, and manganese) in food materials can catalyze anthocyanins to form quinones, which can react with proteins and other polyphenols to form condensation products and contribute to food browning (Fig. 3, bottom; Sinela et al., 2017).

3.3 Catechins

Catechins are flavonoids and belong to the subgroup of flavanol. The most important catechins include catechin (C), catechin gallate (CG), epicatechin (EC), epicatechin gallate (ECG), gallocatechin (GC), epigallocatechin (EGC) and epigallocatechin gallate (EGCG). Apart from teas, these compounds are ubiquitously existing in a wide variety of fruits and vegetables, including wines, black grapes, apricots, strawberries, raspberry, apricot, apple, blackberry, orange, lemon, beans, cherry, and peach (Gadkari & Balaraman, 2015; Pedro, Maciel, Rampazzo Ribeiro, & Haminiuk, 2020).

Like anthocyanins, catechins are thermally sensitive and undergo epimerization and degradation during thermal processing of foods, and their thermal stability is influenced by food components and pH value. Heat treatment induced degradation of catechin and its epimerization to epicatechin. A 36% of reduction occurred to catechin when heating 50 μg/mL of catechin at 100 °C for 30 min, in which 9.7% was transformed to epicatechin. Sugars influenced the degradation of catechin. Trehalose inhibited while sucrose promoted the degradation. When 10% of sucrose was added, 48% of catechin were lost with 22% being epimerized to epicatechin (Lončarić, Pablo Lamas, Guerra, Kopjar, & Lores, 2018). The pH value also influenced thermal stability of tea polyphenols. At pH 7.4, (-)-epigallocatechin and (-)-epigallocatechin gallate were almost completely degraded when incubated at 100 °C, for 3 and 6 h, respectively; whereas lower pH value (5–6.5) significantly decreased the degradation of these polyphenols (Su, Leung, Huang, & Chen, 2003). The degradation of tea polyphenols occurred after fortification in bakery products, including breads, biscuits, cakes. When the biscuits were fortified with green tea extract at a concentration of 0.3%, 70% and 79% of epigallocatechin and epigallocatechin gallate were degraded, respectively (Ananingsih, Sharma, & Zhou, 2013). In breads, only 18.4% of tea catechins were left after incorporation of them at 150 mg/100 g and baking at 215 °C for 11 min (Wang & Zhou, 2004).

During thermal treatment, catechin gallate, epicatechin gallate, gallocatechin, epigallocatechin and epigallocatechin gallate can be hydrolyzed to generate gallic acid, catechin or epicatechin. Catechin or epicatechin can form dimers or degrade to 3,4-dihydroxybenzaldehyde and protocatechuic acid (Fan et al., 2016; Lončarić et al., 2018).

3.4 Flavanols and flavones

Different flavonoids show different thermal stability. When heating naringin (belong to flavanone, 0.29 mM), rutin (flavanol, 0.041 mM), mesquitol (flavanol, 0.93 mM), eriodictyol (flavanone, 0.059 mM), luteolin (flavone, 0.0087 mM), and luteolin-7-O-glucoside (flavanone, 0.014 mM) at 130 °C for 120 min, their half life time ($t_{1/2}$) were 6.08, 0.07, 0.49, 0.25, 0.25, and 0.16 h, respectively (Chaaban et al., 2017). Rutin is the most sensitive one to thermal treatment, which is mainly hydrolyzed to release quercetin.

Quercetin (belong to flavanol) is a typical flavonoid compound present in many fruits, vegetables, seeds, and medicinal plants. It is relatively stable during heat treatment. When heating quercetin alone at pH 5.0 in air or pH 8.0 in nitrogen, 100 °C for 240 min, no degradation of quercetin occurred. However, when it was heated at pH 8.0 in the air, approximately

20% of quercetin disappeared due to oxidation (Buchner, Krumbein, Rohn, & Kroh, 2006). Chassy et al. (2015) found that sauteing and direct baking of onion decreased the quercetin content by 12.6% and 1.3%, respectively.

Vogrincic, Timoracka, Melichacova, Vollmannova, and Kreft (2010) used tartary buckwheat flour to prepare bread, where the dough was initially baked at 200 °C for 10 min and then baked at 180 °C for another 30 min. They found that rutin (quercetin-3-rutinoside) content in the crust decreased by 53.5% after baking, whereas 94.2% of quercetin remained (Vogrincic et al., 2010). This circumstance also occurs in bread enriched with onion or onion peel. In which the average recovery of rutin decreased by 58% while quercetin increased by 181% (Bedrníček et al., 2020). Other quercetin glycosides also convert to quercetin during baking. When apple skin powder was incorporated in muffins, quercetin-3-galactoside, quercetin-3-glucoside, quercetin-3-rhamnoside, and quercetin-3-rutinoside decreased by 55%–73.4%, while quercetin increased by 12- to 45-fold (Rupasinghe et al., 2008).

Hydrolysis of rutin may occur during thermal treatment (Buchner et al., 2006; Chassy et al., 2015) or with the aid of rutin-degrading enzymes existing in food stuffs, such as tartary buckwheat (Kalinová, Vrchotová, & Tříska, 2018). These enzymes are quite stable during heat treatment and remain active at temperature up to 140 °C, which catalyzes the deglycosylation of kaempferol-3-O-rutinoside, kaempferol-3-O-β-D-glucoside, and piceid. The transformation of rutin or other quercetin glycosides to quercetin may counteract the decrease in quercetin content.

Quercetin is also stable in model reaction system. Buchner et al. (2006) heated quercetin at 100 °C in an aqueous solution and found no decrease at 120 min (pH 5.0), and 75% was left when heated for 300 min. At pH 8.0, quercetin was depleted to approximately 20% in 240 min (Buchner et al., 2006). Buchner et al. detected the degradation products with HPLC–MS/MS and proposed the degradation pathway as shown in Fig. 4.

Under more violent heating condition, quercetin produces more thermal degradation products. Ravber et al. (2016) heated rutin at 220 °C, 400 bar, and detected several degradation compounds from quercetin, including 3, 4-dihydroxybenzoic acid, catechol, 2-hydroxy-1-(2, 4, 6-trihydroxyphenyl) ethanone, and 4, 6-dihydroxy-3-benzofuranone. Chaaban et al. (2017) heated 1 mM of quercetin at 130 °C for 2 h and detected several new degradation products, including 2-(3′,4′-dihydroxybenzoyl)-2,4,6-trihydroxybenzofuran-3(2H)-one, (3′,4′,5′,6′-tetrahydroxybenzoyl)-2,4,6-trihydroxybenzofuran-3

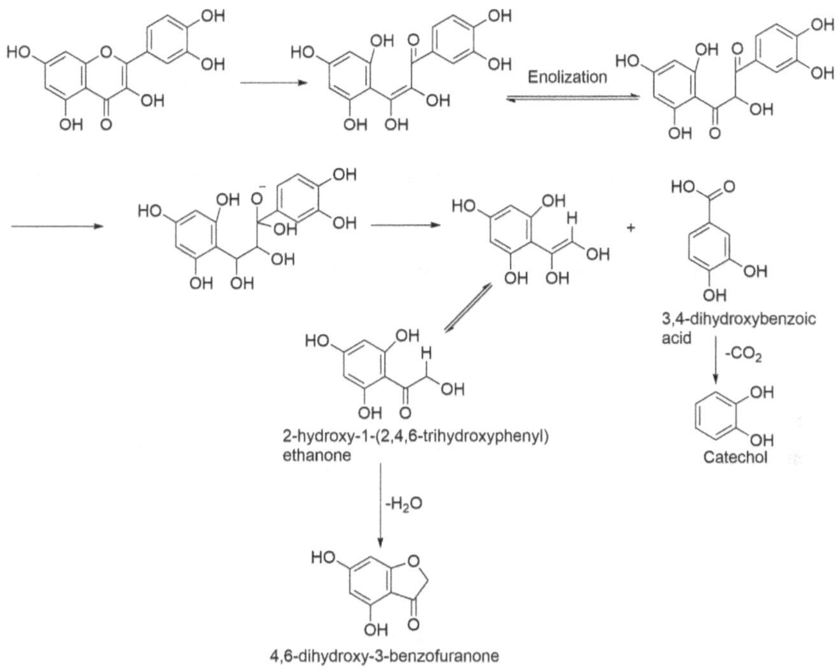

Fig. 4 Thermal degradation pathway of quercetin.

(2H)-one, (3′,4′,5′-trihydroxybenzoyl)-2,4,6-trihydroxybenzofuran-3(2H)-one, 2-(3,4-dihydroxyphenyl)-2-oxoacetic acid, and 2,4,6-trihydroxybenzoic acid.

4. Effect of polyphenols on flavor, color, and texture of baked products

4.1 Effect of polyphenols on the flavor of baked products

Due to the bitterness and astringency, phenolic compounds display a significant role in the flavors in such foods and drinks as beer, wine, coffee, tea, chocolate, tree nuts, fruit-based products, and soy products (Delgado, Zamora, & Hidalgo, 2015). Polyphenols contributed to aroma compounds formation in thermally processed foods via thermal degradation and reaction with amino acids after their transformation to quinones. For example, 4-vinylguaiacol can be generated from ferulic acid during baking of alkaline dough due to the added leaving agent, $NaHCO_3$ (Fig. 1). 4-Vinylguaiacol, guaiacol and phenol are produced via decarboxylation of ferulic acid during coffee roasting (Fig. 1; Schieber & Wüst, 2020). Delgado et al. (2015) used

20 phenolic compounds with different structures (including o-diphenols, m-diphenols, p-diphenols, trihydroxy derivatives, and complex phenols) to react with phenylalanine at 120–180 °C, pH 3–6 for different length of time. Compared with the control, catechol, 4-methylcatechol, 3,4-dihydroxybenzoic acid, caffeic acid, 2,6-dihydroxybenzoic acid, hydroquinone, 2,5-dihydroxybenzoic acid, benzoquinone, pyrogallol, gallic acid, 1,2,4-trihydroxybenzene significantly increased the formation of phenylacetaldehyde, an aroma compound with hyacinth smell (Delgado et al., 2015). Obviously, polyphenols with two hydroxyl groups located in ortho or para positions increased the Strecker degradation of the amino acid to form high level of phenylalanine, suggesting that conversion of polyphenols to their corresponding quinones is responsible for the flavor formation. Benzoquinone and hydroquinone increased phenylacetaldehyde formation with 6.2- and 2.8-fold (Delgado et al., 2015), confirming the role of quinone in flavor formation.

However, incorporation of polyphenol-enriched by-products generally reduced the aroma of baked foods. It was found that the addition of fruit pomace (20%) from blackcurrant fruit, rowan, rosehip, and elderberry decreased the flavor score of shortbread cookies (Tańska et al., 2016). Fortification of grape pomace powder (10%) decreased the acceptability of bread compared with those added with 2% and 5% (Hayta, Özuğur, Etgü, & Şeker, 2014), and addition of 5% grounded green coffee bean powder significantly decreased the bread aroma (Zain, Baba, & Shori, 2018). Similar aroma-inhibition circumstances occurred in quince fruit enriched in biscuits (Antoniewska, Rutkowska, & Pineda, 2019).

Addition of single phenolic compound also proved that phenolic compounds decreased the flavor of baked foods. At the incorporation level of 2.0%, phenolic compounds in dough, including caffeic acid, gallic acid, ferulic acid, catechin, and quercetin suppressed Maillard-type volatiles by 75.9%, 74.3%, 65.6%, 62.4%, and 59.3%, respectively. Interestingly, these compounds also decreased the level of pleasant components ranging from 18% to 26% during the fermentation process of breads (Mildner-Szkudlarz et al., 2019).

Peterson's research group carried out intensive work to investigate the effect of polyphenols on the inhibition of baked aromas and the underlying mechanism. The early purpose was to control the unwanted flavors in milk after ultrahigh-temperature processing (UHT). At the beginning of this century, Perkins and Deeth (2001) took a survey on UHT milk consumption and found that it only accounted for 9% of the drinking milk market in

Australia and 5%–13% in UK. UHT milk products have a stronger unpleasant flavor than traditionally pasteurized milk due to the UHT treatment and the subsequent aging. Methional, 2-acetyl-1-pyrroline, and 2-acetyl-2-thiazoline are off-flavor compounds produced from the Maillard reaction in UHT milk (Kokkinidou & Peterson, 2014). Polyphenols were found to decrease the amount of these off-flavor compounds, 0.1% of epicatechin addition resulted in 32-fold reduction of methional in flavor occurred in UHT milk (Colahan-Sederstrom & Peterson, 2005). In 2014, Peterson's group used five phenolic compounds (catechin, genistein, daidzein, 1,2,3-trihydroxybenzene, and 1,3,5-trihydroxybenzene) to investigate their inhibition effect on the generation of cooked off-flavor during UHT treatment (day 0) and after storage (30 days at 30 °C), finding that only catechin and daidzein slightly inhibited methional formation during UHT processing, while the others showed no substantial effect (Kokkinidou & Peterson, 2014). However, all phenolic compounds suppressed the generation of off-flavor markers 2-acetyl-1-pyrroline, methional, and 2-acetyl-2-thiazoline during storage. Specifically, catechin and genistein reduced the methional content by 80% and 75% during storage (Kokkinidou & Peterson, 2014).

Polyphenols also markedly inhibited the formation of other aroma compounds during thermal treatment. In the glycine-glucose reaction system, addition of epicatechin significantly inhibited Maillard-type volatile compounds, such as 2, 3-butanedione, pyrazine, methylpyrazine, 2, 5-methylpyrazine, 2, 3, 5-trimethylpyrazine, acetol, and cyclotene. Among them, acetol and 2,5-dimethylpyrazine were decreased by 8.5- and 113.4-fold, respectively (Totlani & Peterson, 2005). In the glycine-fructose reaction model, epicatechin significantly reduced the formation of furfural, 5-methylfurfural, furfuryl alcohol, 2-acetylfuran, 2 (5H)-furanone, acetol, 2,3,5-trimethylpyrazine, and 2-acetylpyrrole. The highest reduction occurred for 5-methylfurfural (16.9-fold) and acetol (12-fold) (Totlani & Peterson, 2007).

Similar to epicatechin, other polyphenols, such as ferulic acid, chlorogenic acid, caffeic acid, gallic acid, tannic acid, protocatechuic acid, coumaric acid, rosmarinic acid, TBHQ, BHA, epigallocatechin-3-gallate, catechin, genistein, daidzein, phloroglucinol, pyrogallol, and cyanidins inhibited flavor formation (Favreau-Farhadi et al., 2015; Jiang et al., 2009; Mildner-Szkudlarz et al., 2019; Moskowitz, Bin, Elias, & Peterson, 2012; Pasqualone et al., 2014; Porter et al., 2006; Zhao, Yao, & Ou, 2017; Zheng, Chung, & Kim, 2015).

Three possible mechanisms were proposed for their inhibition effects against flavor formation. The aroma compounds formed during thermal processing of foods are mainly generated from the Maillard reaction, caramelization, lipid oxidation, as well as yeast fermentation as a minor source (Moskowitz et al., 2012). In the crust of the bread, the major aroma compounds from the Maillard reaction are 6-acetyltetrahydropyridine, 5-ethyl-3-hydroxy-4-methyl-2(5H)-furanone, methional, 3-methylbutanal, 2-acetyl-1-pyrroline, 3-hydroxy-4,5-dimethyl-2(5H)-furanone, 2-ethyl-3, 5-dimethylpyrazine, and 4-hydroxy-2,5-dimethyl-3(2H)-furanone. Whereas those from lipid oxidation are (E)-2-nonenal and (E,E)-2,4-decadienal (Jiang et al., 2009). Wang and Ho (2012) found that di-carbonyl compounds, especially methylglyoxal and glyoxal produced from hexoses degradation, are playing important roles in the formation of flavor compounds, such as pyrazines, pyrrolines, pyrazinones, pyridines, furanones, phenylacetaldehyde, 3-methylbutanal, thiazoles, and thiazolines. Many polyphenols are potential trappers of di-carbonyl compounds as discussed later. Hence, the first mechanism for the inhibition of flavor formation by polyphenols is their scavenging properties of di-carbonyl compounds generated from sugar fragmentation. Secondly, polyphenols or their degradation products can react with the flavor compounds, such as reaction of the degradation product between ferulic acid (2-methoxy-4-vinyl-phenol) and pyrazinone (Jiang et al., 2009). Lastly, polyphenols are efficient scavengers for free radicals involved in the formation of flavors from Maillard reaction and lipid oxidation (Rizzi, 2003).

4.2 Effect of polyphenols on the color and texture of the baked foods

Mechanical properties and color are qualitative characteristics of bakery products since they influence the crispness, mouthfeel, and acceptance by the consumer (De Pilli, Lopriore, Montemitro, & Alessandrino, 2019).

Incorporation of different types and levels of polyphenols in dough has different effects on the color and texture of baked foods. Li et al. (2012) incorporated three polyphenols (tea polyphenols, antioxidants of bamboo leaves, and *tert*-butyl hydroquinone) in cookies at 0.1–0.5 g/kg to inhibit acrylamide formation. Among the three polyphenol products, antioxidants of bamboo leaves showed no obvious effects on the color of samples at all addition levels from 0.1 to 0.5 g/kg, while tea polyphenols and *tert*-butyl hydroquinone affected the color of the cookies while without influence on the texture (Li et al., 2012). Addition of rosmarinic acid, epicatechin, and resveratrol at 0.2% and 0.02% changed the color of cookies depending on the phenolic type, the addition levels, and baking temperatures (Ou et al., 2018). Compared with the control, cookies fortified with 0.02% of rosmarinic acid and resveratrol showed no difference in chromatic value, whereas their addition at 0.2% significantly increased the chromatic value of the cookies. However, polyphenols had no influence on the hardness or fracturability of the cookies at the given conditions (Ou et al., 2018). With the addition levels increased, the color and the texture of bakery products were influenced. Fortification of 0.25% (w/w) of epicatechin, quercetin, naringenin, chlorogenic acid, and rosmarinic acid in cookies significantly increased chromatic value and hardness of cookies compared with the control (Zhang et al., 2014). Quercetin also significantly influenced the color and texture in breads. In comparison with the control, addition of 0.2% quercetin increased the hardness and chewiness by 23.8% and 30.5%, respectively, while decreased the specific volume by 9.7% (Lin & Zhou, 2018). Addition of feruloylated oligosaccharides from maize bran (1.25–40 mg/g) increased the browning and hardness of biscuits in a dose-dependent manner (Huang et al., 2018). The released ferulic acid and oligosaccharides may play important roles in browning, and the high-water absorption capacity of oligosaccharides increased the hardness.

Compared with the addition of the single polyphenol, food by-products (such as bran, fruit and vegetable pomace, and fruit peel powder) enriched with both polyphenols and dietary fibers are used more prevalently in baked foods as a source of antioxidant and functional ingredients. According to a review by Dziki, Różyło, Gawlik-Dziki, and Świeca (2014), these plant materials could enhance the antioxidant activity of wheat bread, while most of them adversely influenced the sensory properties (Dziki et al., 2014). Salehi and Aghajanzadeh (2020) summarized the effect of dozens of dried powders from fruits and vegetables on the physiological properties in cakes, and concluded that fortification of these products influences volume, density, color, texture and rheology properties of cakes. The same circumstances also occur in other cereal-based baked products, such as breads, muffins, cookies, and biscuits (Quiles et al., 2018). In recent years, there are many reports on the effect of polyphenols or their the polyphenols-enriched products on the color and texture of cookies, biscuits, muffins, cakes, and steam bread (Almaski, Coe, Lightowler, & Thondre, 2017; Gbenga-Fabusiwa, Oladele, Oboh, Adefegha, & Oshodi,

2018; Jayalaxmi, Vijayalakshmi, & Maruthesha, 2018; Pasukamonset et al., 2018; Rumiyati, James, & Jayasena, 2015; Tańska et al., 2016; Walker, Tseng, Cavender, Ross, & Zhao, 2014; Wang & Zhu, 2018; Zain et al., 2018; Zhu & Sun, 2019).

Interaction of polyphenols with wheat proteins and starch may affect the texture of baked foods (Amoako & Awika, 2016; Gammoh et al., 2018). The main protein in wheat flour is gluten, which is composed of gliadin and glutenin proteins. The gliadin is responsible for dough viscosity, while the glutenin is responsible for dough strength and elasticity (Xu, Wang, & Li, 2019). Polyphenols acting as the reducing agents, may disrupt the glutens. When cinnamic, p-coumaric, caffeic, ferulic and chlorogenic acids (0.05%, 0.1% and 0.2%) were added into dough, the breakdown of dough, a decrease in disulfide content, and an increase in the content of free SH groups were observed (Krekora, Szymańska-Chargot, Niewiadomski, Miś, & Nawrocka, 2020). Interactions between protein and phenolic compounds in dough are influenced by various factors, including temperature, pH value, ionic strength, protein characteristics, and the structure of phenolic compounds (Xu et al., 2019). Phenolic compounds may inhibit starch granules from absorbing water to gelatinize. Hydrophilic phenolics can compete with starch for water. They also form non-covalent bonds with starch and change the pH value to retard water absorption of starch granules, starch gelatinization, and pasting properties (Xu et al., 2019). In addition to influence starch gelatinization, polyphenols, such as green tea powder also promoted starch retrogradation during the cooling stage (Ning, Hou, Sun, Wan, & Dubat, 2017).

The browning effect of polyphenols on color may attribute to the oxidized products of their corresponding quinones, and quinones reaction with amino acids. Polyphenol oxidase, metals, heat, or high pH conditions induces the oxidation of polyphenols into quinones, which polymerize or react with amino acids and proteins through the Maillard reaction, Michael addition or Strecker reaction (Fig. 5), resulting in the formation of brown polymeric pigments (Bittner, 2006). For example, the quinones produced from catechin and epigallocatechin affect the color of black tea; the reaction of the quinone from caffeic acid with glycine contributes to the green color in coffee; and the polymerization of quinones produced from chlorogenic acid or their reaction with proteins forms the green pigment during baking of sunflower seed (Liang & Were, 2018; Wildermuth, Young, & Were, 2016).

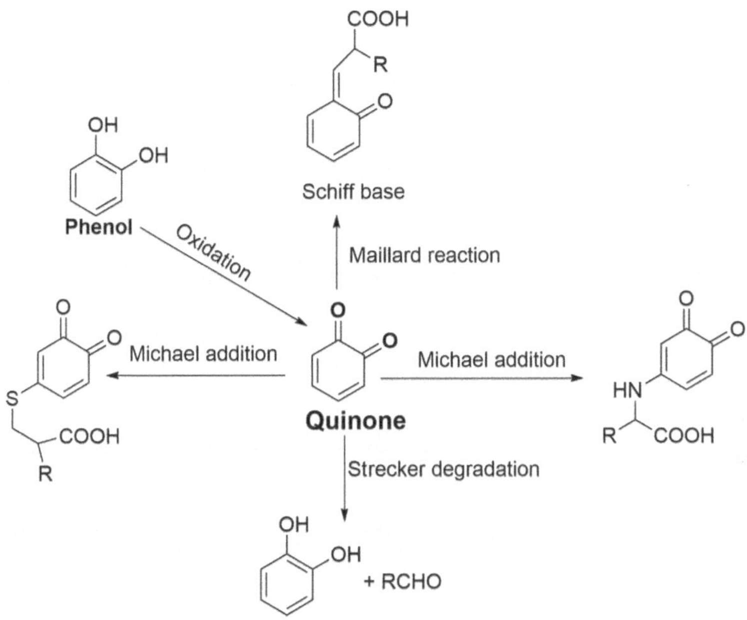

Fig. 5 Reaction of quinones of polyphenol with amino acids.

5. Effect of polyphenols on reactive carbonyls and advanced glycation end products in baked products

During baking treatments, some unpleasant toxic substances are produced, including reactive carbonyl species, advanced glycation end products (AGEs), furans, acrylamide, and heterocyclic amines, etc. Polyphenols are found to influence their exposure level in baked foods (Teng et al., 2018). For the role of polyphenols in the mitigation of furans, acrylamide, and heterocyclic amines, readers can refer to the recently published review articles (Batool et al., 2020; Dong, Xian, Li, Bai, & Zeng, 2020; Jin et al., 2013; Teng et al., 2018). In this section, we focus on their reducing effects and the underlying mechanisms for reactive carbonyl species and AGEs.

5.1 Reactive carbonyl species

Reactive carbonyl species are a type of toxic aldehydes, such as glyoxal, methylglyoxal, 3-deoxyglucosone, malondialdehyde, acrolein, 4-hydroxynonenal, 4-hydroxynonenal, 4-hydroxy-2-alkenals, 4-oxononenal, and hydroxymethylfurfural (Amoroso, Maga, & Daglia, 2013; Wang et al., 2017;

Yalcinkaya, Uzilday, Ozgur, Turkan, & Mano, 2019). They are mainly produced during sugar degradation and lipid oxidation (Fig. 6).

Polyphenols decrease the level of reactive carbonyl species via inhibition of their formation and trapping of the formed ones. Polyphenols can inhibit their formation through scavenging of free radicals (Khan, Liu, Wang, & Sun, 2020). Reactive carbonyl species from sugars involve the participation of free radicals. For example, in the Maillard reaction, formation of Schiff bases, Amadori products, and the subsequent reactive carbonyl species are accompanied by a large production of free radicals (Yeh, Hsia, Lee, & Wu, 2017), including superoxide radical anions, the hydroxyl radical, and alkoxyl radicals (Hellwig, Löbmann, Orywol, & Voigt, 2014). Cereal-based bakery foods contain bountiful of sugars and fats for undergoing the Maillard reaction and lipid autoxidation, which are the main pathways to generate reactive carbonyl species (Fig. 6). Polyphenols as the antioxidants can capture free radicals and decrease the formation of reactive carbonyl species produced from the oxidative degradation of sugars and lipids.

Moreover, polyphenols can trap the reactive carbonyl species after their formation and thus decrease their contents in baked foods. Polyphenols scavenge these toxic aldehydes through adduct formation (Bourden et al., 2008; Cui et al., 2018; Es-Safi, Cheynier, & Moutounet, 2000; Hou et al., 2018; Huang et al., 2018; Huang, Zhu, Lv, & Sang, 2020; Kokkinidou & Peterson, 2014; Ou, Huang, Wang, & Ou, 2017; Szwergold, 2017; Wang et al., 2015; Wang, Chen, & Sang, 2016; Zamora, Aguilar, Granvogl, & Hidalgo, 2016; Zamora & Hidalgo, 2018; Zhou, Gong, & Wang, 2019; Zhu et al., 2012; Zhu, Poojary, Andersen, & Lund, 2020). Different types of reactive carbonyl species form different adducts. The typical adducts produced from the reaction of mono-, di-carbonyl, and α, β-unsaturated carbonyl compounds with various polyphenols being identified as illustrated in Fig. 7. The pathways for adducts formation between polyphenols and the carbonyl compounds from lipid-derived reactive carbonyls have been summarized by Zamora and Hidalgo (2018).

The trapping mechanism of polyphenols for reactive carbonyl species involves a nucleophilic addition reaction. For phenols with two or more phenyl hydroxyls, the deprotonation of OH groups in phenols increases the electron density at the adjacent C atoms, especially the C atoms adjacent to or between the OH groups in meta-phenols. The electron-rich C atoms

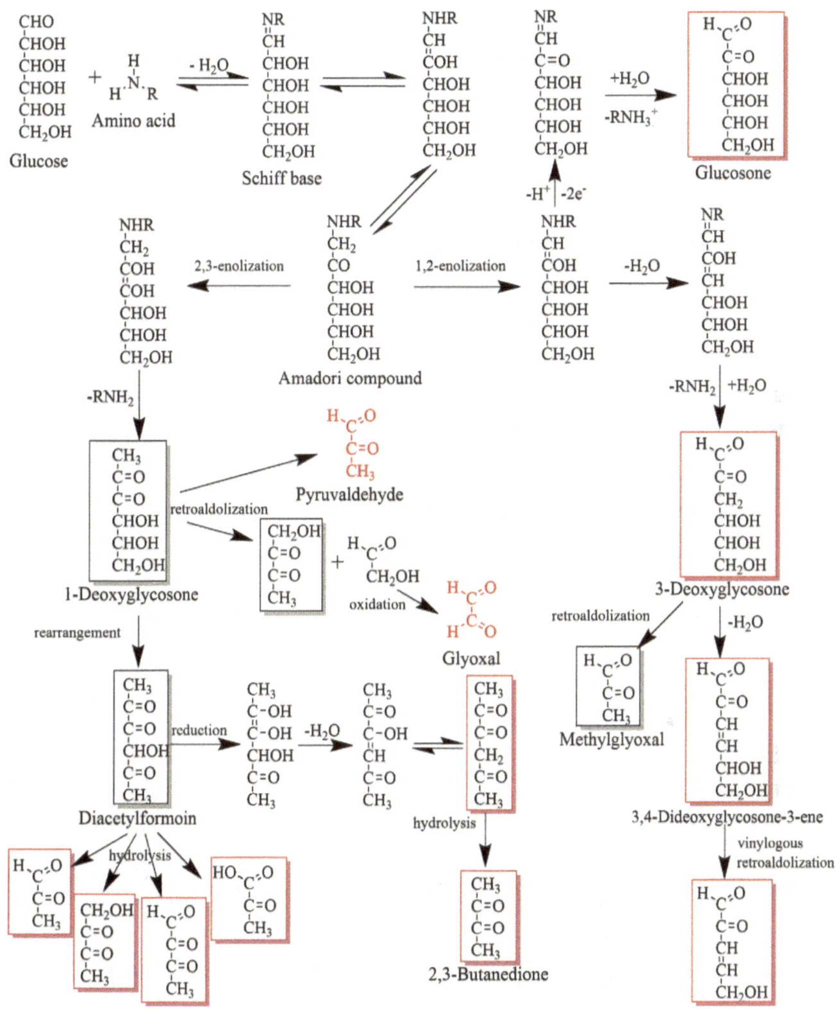

Fig. 6 The formation pathways for the reactive carbonyl species produced from the degradation of sugars (left) and lipids (right).

Continued

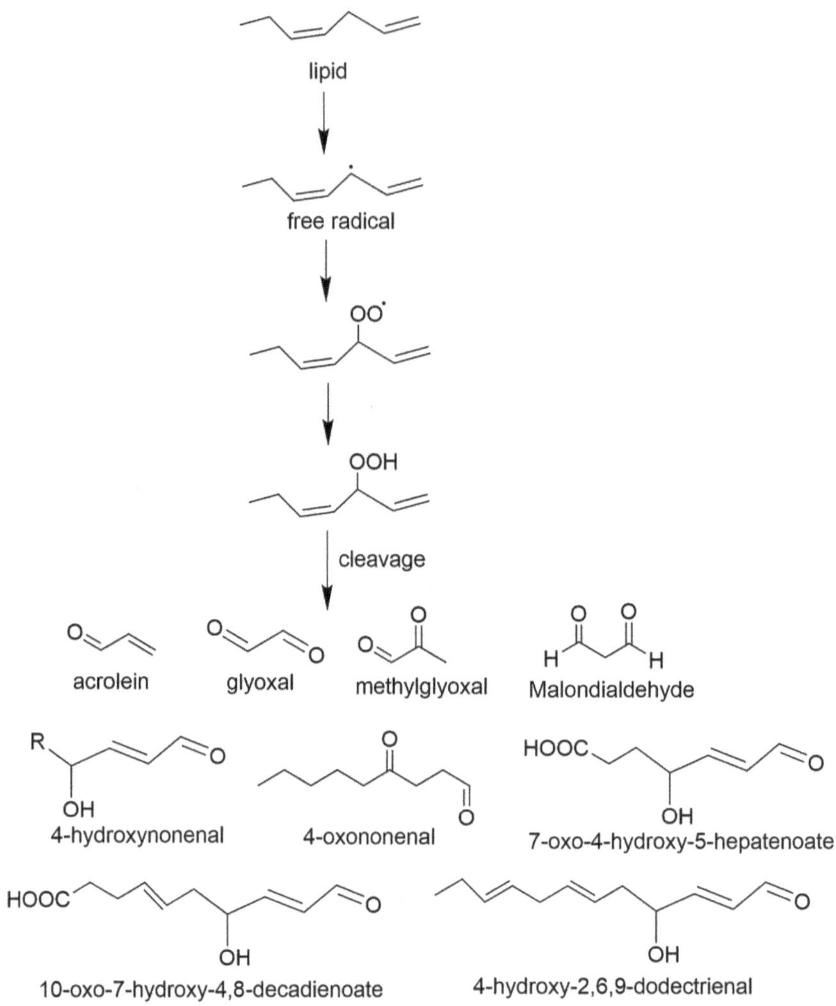

Fig. 6—Cont'd

(nucleophilic centers) attacks C atoms in the carbonyl groups of dicarbonyl compounds via nucleophilic addition reaction to generate adducts (Meng et al., 2018; Zhu, Poojary, Andersen, & Lund, 2019). For α, β-unsaturated carbonyl compounds, such as acrolein, 4-hydroxynonenal, 4-hydroxynonenal, the electron-rich C in phenols attacks β-C atom in the vinyl group via the mechanism of Michael addition reaction, and then

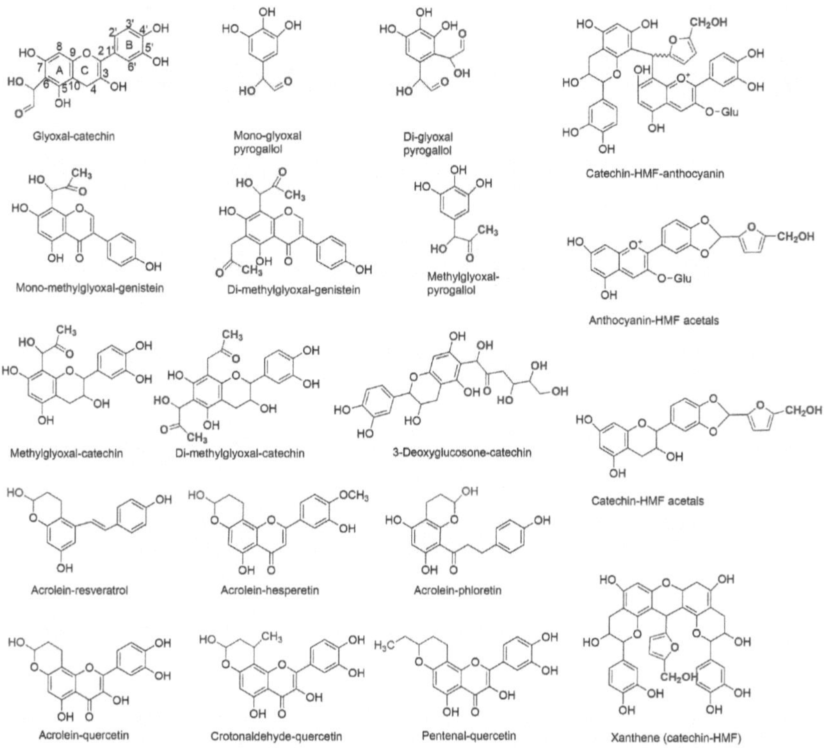

Fig. 7 Typical adducts formed between polyphenols and the toxic carbonyl compounds (blue).

the adjacent phenol hydroxyl group forms hemiacetal with the aldehyde group to form a cyclic adduct, such as acrolein-resveratrol, acrolein-hesperetin, acrolein-phloretin, and acrolein-quercetin shown in Fig. 7.

5.2 Advanced glycation end products (AGEs)

AGEs in baked foods can be generated from the reaction of dicarbonyl compounds (methylglyoxal, glyoxal, and 3-deoxyglucosone, etc.) with proteins. Dietary AGEs are absorbed and distributed in the whole body. For example, ^{14}C-labeled AGEs from ovalbumin was detected in the liver, kidney, lung, heart, and spleen after administration in Sprague Dawley rats (He, Sabol, Mitsuhashi, & Vlassara, 1999). A large fraction of dietary AGEs remains in the body and binds non-specifically to proteins in the organs, especially in the liver and kidney. The urine excreted AGEs only accounts for approximately 30% of the absorbed amount in healthy subjects, and only 5% in renal disease patients (Poulsen et al., 2013). This finding suggests that most of the dietary AGEs absorbed are retained in the body. Thus, the reduction of dietary AGEs intake shows promotion effects on human health, including improving wound healing, insulin resistance and cardiovascular diseases (Guilbaud, Niquet-Leridon, Boulanger, & Tessier, 2016; Luevano-Contreras & Chapman-Novakofski, 2010).

Yeh et al. (2017) listed several possible mechanisms that can delay or reduce AGEs formation, including reduction of the formation of dicarbonyls via scavenging of free radicals, reduction of Schiff base and Amadori products to block the dicarbonyls, inhibiting the carbonylation of proteins, chelating the metal ions that catalyze AGEs production. Polyphenols can decrease AGEs via scavenging of free radicals and the precursors (dicarbonyl compounds), chelation of the metal ions, and inhibition of the carbonylation of proteins, and the subsequent oxidation of the carbonylated products as we proposed previously (Ou et al., 2017).

Trapping (scavenging) of dicarbonyl compounds, such as glyoxal and methylglyoxal, is an important mechanism for polyphenols to inhibit AGEs formation (Yeh et al., 2017) and has been intensively investigated in recent years. Many polyphenols or their enriched products, including plant extracts (tea, tomato, garlic, rosemary, cinnamon, *Ilex paraguariensis*, and herbs) and phenolics, such as flavonoids, phenolic acids, and tannins (Peng, Ma, Chen, & Wang, 2011; Wu, Huang, Lin, & Yen, 2011) have been found to efficiently trap the dicarbonyl compounds. Incorporation of polyphenols also reduces the content of some the dicarbonyl compounds in baked foods. The content of total phenolic compounds remained in the end products has a highly negative correlation with these dicarbonyl compounds in cookies, the correlative coefficients are −0.85, −0.77, and −0.77 for glyoxal, methylglyoxal, and diacetyl, respectively, after baking at 200 °C for 7 min (Kocadağlı et al., 2016). We found that addition of 0.2% each of resveratrol, epicatechin, and rosmarinic acid reduced the content of methylglyoxal by 52.6%, 42.4%, and 23.8%, respectively (Ou et al., 2018).

However, using polyphenols to reduce the dicarbonyl compounds for AGEs inhibition should not be exaggerated as we discussed in our previous review (Ou, Wang, Zheng, & Ou, 2019): Firstly, polyphenols decrease the content of some dicarbonyl compounds while markedly increase others, such as 3-deoxyglucosone, and resulting in an elevation of other AGEs formation. 3-Deoxyglucosone formation increases by threefold after adding 50 mM of chlorogenic acid into the fructose-aspartic reaction model (Zhang et al., 2016), and a highly positive correlation between total phenolic compounds and 3-deoxyglucosone was found in cookies ($R^2=0.93$) and honey ($R^2=0.843$) (Arena, Ballistreri, Tomaselli, & Fallico, 2011; Kocadağlı et al., 2016). 3-Deoxyglucosone is also a precursor of AGE, such as 3-deoxyglucosone imidazolones (Ashraf et al., 2014; Wu et al., 2011), which promotes diabetes development by decreasing the secretion of insulin and glucagon-like peptide-1 (Zhang et al., 2016). Secondly, the phenol-dicarbonyl adducts have been detected in the urine of animals, such as in mice after the intake of genistein (Wang et al., 2016), the toxicological evaluation of the adducts formed between polyphenols and the toxic aldehydes have not been reported. Whether the adducts can really decrease the toxicity of the toxic aldehydes and how they metabolize *in vivo* are unknown. Thirdly, the detoxifying capacity of polyphenols is low compared with drugs. Navarro and Morales (2016) incubated methylglyoxal (5.551 μmol/mL) with different phenolic compounds under physiological conditions for 168 h and found that all tested phenolic compounds showed much lower trapping capacity than that of pyridoxamine, a drug for slowing the progression of diabetic neuropathy (Table 4). Fourthly, thermal degradation of polyphenols decreases their scavenging capacity for dicarbonyl compounds in baked foods. For example, scavenging capacity of epicatechin for methylglyoxal decreased by nearly 50% in baked cookies compared with that in model system at lower temperature (Noda & Peterson, 2007; Ou et al., 2018).

Table 4 Methylglyoxal trapping capacity of 10 phenolic compounds.

Compounds	IC_{50} (μmol/mL)
Hydroxytyrosol	0.195
Hydroxytyrosol acetate	0.204
Gallic acid	0.176
Caffeic acid	0.222
Chlorogenic acid	0.367
Protocatechuic acid	0.389
Sinapinic acid	0.892
Ferulic acid	1.648
Gentisic acid	3.179
p-Coumaric acid	2.985
Pyridoxamine	0.059

6. Effect of polyphenols on the antioxidant and antimicrobial activity of baked products

Food additives with antioxidant, antimicrobial, and anti-browning activities are commonly used to preserve products for longer storage time in bakery industry. However, the safety concern for synthetic additives arouses us to use natural alternatives (Antoniewska et al., 2019). Thus, polyphenol-based ingredients, including those from fruit and vegetables are incorporated in bakery products for increasing antioxidant and antimicrobial activity, in addition to control the toxins formation.

6.1 Antioxidant activity

Natural polyphenols have been used to increase the antioxidant capacity of bakery foods worldwide. Various polyphenols-rich products including powder of fruits or their pomace, vegetables powder, rice bran, wheat bran, buckwheat, barley hull, flaxseed hull, extracts of grape seed, chamomile flowers, fennel leaves, fennel seed, green tea, ginger, turmeric, and by-products from spice industry are usually used in bakery foods (Dziki et al., 2014; Quiles et al., 2018; Salehi & Aghajanzadeh, 2020; Sowbhagya, 2019). Their incorporation in dough considerably increases the content of polyphenols and antioxidant capacity of breads and biscuits (Caleja, Barros, Antonio, Oliveira, & Ferreira, 2017; Verardo et al., 2011). In cookies, when 56% of wheat flour was replaced by juice production byproduct of blueberry, the total polyphenol content and antioxidant capacity were 232.32 mg gallic acid/g and 117.22 mM TE/g, respectively, three- or fourfold of those in plain cookies (Perez, Tagliani, Arcia, Cozzano, & Curutchet, 2018).

Wheat flour-based dough contain proteins (gluten protein) and is rich in fat in the formulation of bakery products (breads, cakes, muffins, pies, etc.). The oxidation of proteins (protein carbonylation) and lipids induces deteriorations of the products during baking and storage. Due to the continuous chain reaction of lipid autoxidation after baking, the content of aldehydes and ketones increase greatly, resulting in the formation of off-flavors in bakery goods (Ma, Sun, Li, & Zhu, 2020). Polyphenols incorporation can reduce the oxidation level of lipids and proteins and produce less oxidation products of lipids and proteins than the plain ones. When 10% of the flour (by weight) was substituted with freeze dried beetroot powder, the content of thiobarbituric acid reactive substances and protein carbonyls decreased from 9.74 nmol/g and 81.1 nmol/mg protein in plain breads to 6.7 nmol/g and 34.9 nmol/mg protein in beetroot powder-fortified breads, respectively (Ranawana et al., 2016). The extracts from breads (with 0.5 mg GAE/mL) fortified with dried leafy vegetable powders from *Amaranthus viridis*, *Solanum macrocarpon*, and *Telfairia occidentalis* also showed inhibition effects on lipid peroxidation and hydroperoxides formation from linoleic acid (Alashi et al., 2019).

However, some researches indicated that the application of polyphenols or polyphenols-rich products in bakery foods would not produce satisfactory antioxidant capacity. We incorporated 0.02% of resveratrol, epicatechin and rosmarinic acid in cookies, and determined the antioxidant activity of cookies aqueous extracts with DPPH radical method. The results showed that, compared with the control, the antioxidant activity increased after fortification of three polyphenols. However, when the same amounts of polyphenols that had been incorporated in cookies was added to the extracts of the control cookies, the antioxidant capacity was significantly higher than that in the polyphenols-fortified cookies (Ou et al., 2018), suggesting that baking decreased the antioxidant capacity of polyphenols. Zhang et al. (2014) added 0.25% (w/w) of quercetin, epicatechin, naringenin, chlorogenic acid, and rosmarinic acid individually to dough to prepare cookies and determined the antioxidant capacity of the cookie extract by using Trolox equivalent antioxidant capacity assay. They found that the total antioxidant activity was much lower than the quantity of polyphenol that was originally added to the cookie recipe, especially for epicatechin, which is thermally sensitive (Zhang et al., 2014). Verardo et al. (2011) replaced 60% (w/w) of the refined wheat flour with barley coarse fraction that was rich in flavan-3-ol to prepare the biscuits. Although the flavan-3-ol content in biscuits was increased by more than seven-fold, the peroxide value, and the content of oxidized fatty acids in the enriched biscuits were higher than that in the control biscuits after storage for 10–12 months (Verardo et al., 2011). In the research of Ranawana et al. (2016), the contents of TBARS in the tomato and broccoli incorporated breads were 55.3% and 24% higher than that in the plain breads, although beetroot powder decreased its content. These findings suggest that some polyphenols may increase the antioxidant capacity or lipid stability of baked foods, while others may not.

Surprisingly, the release of polyphenols increased during gastrointestinal digestion of some polyphenol-fortified bakery foods. *In vitro* gastrointestinal digestion of white breads enriched with buckwheat, the total phenolic content had an 8- and 11-fold increase in breads with 50% addition of white buckwheat flour and roasted buckwheat groats, respectively

(Szawara-Nowak, Bączek, & Zieliński, 2016). Similar results were obtained from Hidalgo et al. (2018), who carried out an *in vitro* digestion of einkorn flour biscuits incorporated with different proportions of amaranth, quinoa, and buckwheat (Hidalgo et al., 2018). However, the underlying mechanisms needs to be further investigated.

6.2 Antimicrobial activity

Consumption of microbial contaminated food with food-borne pathogens is considered as one of the most pressing threats to human health. Many polyphenols exhibit inhibitory effects against both food spoilage microorganisms and food-borne pathogens. Polyphenols can disrupt the cell wall and cell membranes to release cellular contents and to lose chemiosmotic coupling via interaction with bacteria (Efenberger-Szmechtyk, Nowak, & Czyzowska, 2020). The interactions (via hydrogen and covalent bonding, hydrophobic and lipophilic forces) influence the microbial membranes, adhesins, enzymes, cell envelope transport proteins, and the delocalization of electrons (Bouarab-Chibane et al., 2018; Papuc, Goran, Predescu, Nicorescu, & Stefan, 2017). The electrons delocalization depolarizes bacteria membrane and thus reduces the pH gradient across the membrane and the ATP production (Bouarab Chibane, Degraeve, Ferhout, Bouajila, & Oulahal, 2019).

Many polyphenols-enriched plant extracts show antimicrobial activity for both gram-positive and gram-negative bacteria, including grape extract (pomace, seeds, peels, leaves), berry fruits, plumes, pomegranate, green tea, turmeric, ginger, etc. (Efenberger-Szmechtyk et al., 2020).

Microbial contamination seldom occurs in baked foods with low moisture content. However, for obtaining naturally chewable mouthfeel products, many wheat-based goods are provided with high-water content, such as breads, muffins, cakes, pastries, and their fillings. Which leads to a short shelf life and limits their distribution and marketing due to microbial growth (Ma et al., 2020). Seabuckthorn pulp extracts show inhibition effects for *Escherichia coli, Staphylococcus aureus, Pseudomonas aeruginosa, Salmonella, Chromobacterium violaceum, Aspergillus niger, Penicillium notatum*. After fortification of the powder of seabuckthorn pulp in breads, the bacterial infection of breads was significantly inhibited through a 3-day incubation compared with control groups (Guo et al., 2019). The freeze-dried Japanese quince fruit shows markable inhibition effects on microorganisms in cookies. It is a fruit containing 20 polyphenolic compounds with five representative compounds (chlorogenic acid, catechin, procyanidin B1, epicatechin, and procyanidin B2). When the freeze-dried fruit powder was incorporated in cookies, no yeast or molds were grown during the 16 weeks of storage, while 3.5×10^4 CFU/g (molds) were detected in the control cookies.

Apart from phenolic compounds, other components in these plant-derived products are also responsible for the antimicrobial activities. For example, aldehydes (2-hexenal and hexanal) produced from linolenic and linoleic acids by lipoxygenase during maturation of fruits, exhibit high antimicrobial activity towards pathogen microorganisms at low concentrations (Antoniewska et al., 2019). Moreover, baking can improve antimicrobial activity of some polyphenols. Speaking of tannins, which are divided into condensed tannin and hydrolysable tannin. As a hydrolysable tannin, tannic acid is composed of gallic acid esters of glucose. Baking enhanced antimicrobial capacity of tannic acid due to the formation of gallic acid and pyrogallol (Xu et al., 2019).

7. Conclusion

Baked foods are consumed in large quantities in capita and very important for human nutrition. For neutraceutical and other purposes, polyphenols and polyphenol-enriched by-products have been widely used in baked foods. Incorporation of polyphenols can increase the antioxidant and antimicrobial activity, decrease food-borne toxins such as reactive carbonyl species and AGEs, and lower the postprandial serum glucose by retarding starch digestion. However, polyphenols negatively influence the flavor, color, texture of baked foods, and the subsequent acceptance by the consumer. Moreover, most polyphenols are thermally sensitive and intensively degrade during baking, although and occasionally, the degradation releases some polyphenols with higher activity than their precursors. Since baking occurs at very high temperature and lasts for a dozen of minutes, future works need to focus on the chemical interactions of polyphenols and their oxidized products (quinones) with other food components, and the safety consequence of these interaction.

References

Abdel-Aal, E. S. M., & Rabalski, I. (2013). Effect of baking on free and bound phenolic acids in wholegrain bakery products. *Journal of Cereal Science*, *57*, 312–318.

Abdel-Aal, E. S. M., Young, J. C., Akhtar, H., & Rabalski, I. (2010). Stability of lutein in wholegrain bakery products naturally high in lutein or fortified with free lutein. *Journal of Agricultural and Food Chemistry*, *58*(18), 10109–10117.

Abid, M., Yaich, H., Hidouri, H., Attia, H., & Ayadi, M. A. (2018). Effect of substituted gelling agents from pomegranate peel on colour, textural and sensory properties of pomegranate jam. *Food Chemistry*, *239*, 1047–1054.

Aditi, M., Shashirekha, M. N., & Sudha, M. L. (2017). Characterization and bake stability of dry fruit fillings in dehydrated chiku (Manilkara zapota LP Royen) incorporated biscuits. *Journal of Texture Studies*, *48*(3), 231–240.

Agudelo, A., Varela, P., & Fiszman, S. (2015). Fruit fillings development: A multiparametric approach. *LWT-Food Science and Technology*, *61*(2), 564–572.

Agudelo, A., Varela, P., Sanz, T., & Fiszman, S. (2014). Formulating fruit fillings. Freezing and baking stability of a tapioca starch–pectin mixture model. *Food Hydrocolloids*, *40*, 203–213.

Aguedo, M., Kohnen, S., Rabetafika, N., Bossche, S. V., Sterckx, J., Blecker, C., et al. (2012). Composition of by-products from cooked fruit processing and potential use in food products. *Journal of Food Composition and Analysis*, *27*(1), 61–69.

Ahmed, T., Burhanuddin, M., Haque, M. A., & Hossain, M. A. (2011). Preparation of jam from sapota (Achras zapota). *The Agriculturists*, *9*(1–2), 1–7.

Alashi, A. M., Taiwo, K. A., Oyedele, D. J., Adebooye, O. C., & Aluko, R. E. (2019). Polyphenol composition and antioxidant properties of vegetable leaf-fortified bread. *Journal of Food Biochemistry*, *43*(6), e12625.

Almaski, A., Coe, S., Lightowler, H., & Thondre, S. (2017). Sensory evaluation of polyphenol-rich millet-based muffins and their effect on in vitro starch digestion. *Proceedings of the Nutrition Society*, *76*(OCE4). https://doi.org/10.1017/S0029665117003895.

Amoako, D., & Awika, J. M. (2016). Polyphenol interaction with food carbohydrates and consequences on availability of dietary glucose. *Current Opinion in Food Science*, *8*, 14–18.

Amoroso, A., Maga, G., & Daglia, M. (2013). Cytotoxicity of α-dicarbonyl compounds submitted to in vitro simulated digestion process. *Food Chemistry*, *140*, 654–659.

Ananingsih, V. K., Sharma, A., & Zhou, W. (2013). Green tea catechins during food processing and storage: A review on stability and detection. *Food Research International*, *50*(2), 469–479.

Antoniewska, A., Rutkowska, J., & Pineda, M. M. (2019). Antioxidative, sensory and volatile profiles of cookies enriched with freeze-dried Japanese quince (Chaenomeles japonica) fruits. *Food Chemistry*, *286*, 376–387.

Arena, E., Ballistreri, G., Tomaselli, F., & Fallico, B. (2011). Survey of 1, 2-dicarbonyl compounds in commercial honey of different floral origin. *Journal of Food Science*, *76*, C1203–C1210.

Arrieta-Baez, D., Dorantes-Álvarez, L., Martinez-Torres, R., Zepeda-Vallejo, G., Jaramillo-Flores, M. E., Ortiz-Moreno, A., et al. (2012). Effect of thermal sterilization on ferulic, coumaric and cinnamic acids: Dimerization and antioxidant activity. *Journal of the Science of Food and Agriculture*, *92*(13), 2715–2720.

Ashraf, J. M., Ahmad, S., Rabbani, G., Jan, A. T., Lee, E. J., Khan, R. H., et al. (2014). Physicochemical analysis of structural alteration and advanced glycation end products generation during glycation of H2A histone by 3-deoxyglucosone. *IUBMB Life*, *66*, 686–693.

Batool, Z., Xu, D., Zhang, X., Li, X., Li, Y., Chen, Z., et al. (2020). A review on furan: Formation, analysis, occurrence, carcinogenicity, genotoxicity and reduction methods. *Critical Reviews in Food Science and Nutrition*, *61*, 1–12.

Bedrníček, J., Jirotková, D., Kadlec, J., Laknerová, I., Vrchotová, N., Tříska, J., et al. (2020). Thermal stability and bioavailability of bioactive compounds after baking of bread enriched with different onion by-products. *Food Chemistry*, *319*, 126562.

Bender, A. B. B., Speroni, C. S., Moro, K. I. B., Morisso, F. D. P., dos Santos, D. R., da Silva, L. P., et al. (2020). Effects of micronization on dietary fiber composition, physicochemical properties, phenolic compounds, and antioxidant capacity of grape pomace and its dietary fiber concentrate. *LWT*, *117*, 108652.

Betoret, E., & Rosell, C. M. (2020). Enrichment of bread with fruits and vegetables: Trends and strategies to increase functionality. *Cereal Chemistry, 97*(1), 9–19.

Bhardwaj, R. L., Nandal, U., Pal, A., & Jain, S. (2014). Bioactive compounds and medicinal properties of fruit juices. *Fruits, 69*(5), 391–412.

Bittner, S. (2006). When quinones meet amino acids: Chemical, physical and biological consequences. *Amino Acids, 30*(3), 205–224.

Bouarab Chibane, L., Degraeve, P., Ferhout, H., Bouajila, J., & Oulahal, N. (2019). Plant antimicrobial polyphenols as potential natural food preservatives. *Journal of the Science of Food and Agriculture, 99*(4), 1457–1474.

Bouarab-Chibane, L., Oulahal, N., Dumas, E., Thanh, N. T. T., Bouajila, J., Souchard, J. P., et al. (2018). Effect of interaction with food constituents on plant extracts antibacterial activity. *Food Science and Applied Biotechnology, 1*(1), 77–85.

Bourden, M. N., Vivas, N., Absalon, C., Vitry, C., Fouquet, E., & de Gaulejac, N. V. (2008). Structural diversity of nucleophilic adducts from flavanols and oak wood aldehydes. *Food Chemistry, 107*, 1494–1505.

Buchner, N., Krumbein, A., Rohn, S., & Kroh, L. W. (2006). Effect of thermal processing on the flavonols rutin and quercetin. *Rapid Communications in Mass Spectrometry, 20*(21), 3229–3235.

Caleja, C., Barros, L., Antonio, A. L., Oliveira, M. B. P., & Ferreira, I. C. (2017). A comparative study between natural and synthetic antioxidants: Evaluation of their performance after incorporation into biscuits. *Food Chemistry, 216*, 342–346.

Canalis, M. B., Baroni, M. V., León, A. E., & Ribotta, P. D. (2020). Effect of peach puree incorporation on cookie quality and on simulated digestion of polyphenols and antioxidant properties. *Food Chemistry, 333*, 127464.

Capuano, E., & Fogliano, V. (2011). Acrylamide and 5-hydroxymethylfurfural (HMF): A review on metabolism, toxicity, occurrence in food and mitigation strategies. *LWT–Food Science and Technology, 44*(4), 793–810.

Chaaban, H., Ioannou, I., Chebil, L., Slimane, M., Gérardin, C., Paris, C., et al. (2017). Effect of heat processing on thermal stability and antioxidant activity of six flavonoids. *Journal of Food Processing and Preservation, 41*(5), e13203.

Chang, R. C., Li, C. Y., & Shiau, S. Y. (2015). Physico-chemical and sensory properties of bread enriched with lemon pomace fiber. *Czech Journal of Food Sciences, 33*(2), 180–185.

Chassy, A. W., Bueschl, C., Lee, H., Lerno, L., Oberholster, A., Barile, D., et al. (2015). Tracing flavonoid degradation in grapes by MS filtering with stable isotopes. *Food Chemistry, 166*, 448–455.

Colahan-Sederstrom, P. M., & Peterson, D. G. (2005). Inhibition of key aroma compound generated during ultrahigh-temperature processing of bovine milk via epicatechin addition. *Journal of Agricultural and Food Chemistry, 53*(2), 398–402.

Cropotova, J., Tylewicz, U., Cocci, E., Romani, S., & Dalla Rosa, M. (2016). A novel fluorescence microscopy approach to estimate quality loss of stored fruit fillings as a result of browning. *Food Chemistry, 194*, 175–183.

Cui, H., Tao, F., Hou, Y., Lu, Y., Zheng, T., Sang, S., et al. (2018). Dual effects of propyl gallate and its methylglyoxal adduct on carbonyl stress and oxidative stress. *Food Chemistry, 265*, 227–232.

da SILVA, P. T., Lopes, M. L. M., & Valente-Mesquita, V. L. (2006). Efeito de diferentes processamentos sobre o teor de ácido ascórbico em suco de laranja utilizado na elaboração de bolo, pudim e geléia. *Ciência e Tecnologia de Alimentos, 26*(3), 678–682.

Dangles, O., & Fenger, J. A. (2018). The chemical reactivity of anthocyanins and its consequences in food science and nutrition. *Molecules, 23*(8), 1970.

Dao, L., & Friedman, M. (1992). Chlorogenic acid content of fresh and processed potatoes determined by ultraviolet spectrophotometry. *Journal of Agricultural and Food Chemistry, 40*, 2152–2156.

Dawidowicz, A. L., & Typek, R. (2017). Transformation of chlorogenic acids during the coffee beans roasting process. *European Food Research and Technology*, *243*(3), 379–390.

De Pilli, T., Lopriore, G., Montemitro, M., & Alessandrino, O. (2019). Effects of two sweet cherry cultivars (Prunus avium L., cvv.'Ferrovia' and 'Lapins') on the shelf life of an innovative bakery product. *Journal of Food Science and Technology*, *56*(1), 310–320.

Delgado, R. M., Zamora, R., & Hidalgo, F. J. (2015). Contribution of phenolic compounds to food flavors: Strecker-type degradation of amines and amino acids produced by o-and p-diphenols. *Journal of Agricultural and Food Chemistry*, *63*, 312–318.

Dong, H., Xian, Y., Li, H., Bai, W., & Zeng, X. (2020). Potential carcinogenic heterocyclic aromatic amines (HAAs) in foodstuffs: Formation, extraction, analytical methods, and mitigation strategies. *Comprehensive Reviews in Food Science and Food Safety*, *19*(2), 365–404.

Dziki, D., Różyło, R., Gawlik-Dziki, U., & Świeca, M. (2014). Current trends in the enhancement of antioxidant activity of wheat bread by the addition of plant materials rich in phenolic compounds. *Trends in Food Science & Technology*, *40*(1), 48–61.

Efenberger-Szmechtyk, M., Nowak, A., & Czyzowska, A. (2020). Plant extracts rich in polyphenols: Antibacterial agents and natural preservatives for meat and meat products. *Critical Reviews in Food Science and Nutrition*, *61*, 1–30.

Es-Safi, N. E., Cheynier, V., & Moutounet, M. (2000). Study of the reactions between (+)-catechin and furfural derivatives in the presence or absence of anthocyanins and their implication in food color change. *Journal of Agricultural and Food Chemistry*, *48*, 5946–5954.

Ezekiel, R., Singh, N., Sharma, S., & Kaur, A. (2013). Beneficial phytochemicals in potato—A review. *Food Research International*, *50*(2), 487–496.

Fan, F. Y., Shi, M., Nie, Y., Zhao, Y., Ye, J. H., & Liang, Y. R. (2016). Differential behaviors of tea catechins under thermal processing: Formation of non-enzymatic oligomers. *Food Chemistry*, *196*, 347–354.

Favreau-Farhadi, N., Pecukonis, L., & Barrett, A. (2015). The inhibition of Maillard browning by different concentrations of rosmarinic acid and epigallocatechin-3-gallate in model, bakery, and fruit systems. *Journal of Food Science*, *80*, C2140–C2146.

Figuerola, F., Hurtado, M. L., Estévez, A. M., Chiffelle, I., & Asenjo, F. (2005). Fibre concentrates from apple pomace and citrus peel as potential fibre sources for food enrichment. *Food Chemistry*, *91*(3), 395–401.

Fu, Z., Yoo, M. J., Zhou, W., Zhang, L., Chen, Y., & Lu, J. (2018). Effect of (−)-epigallocatechin gallate (EGCG) extracted from green tea in reducing the formation of acrylamide during the bread baking process. *Food Chemistry*, *242*, 162–168.

Gadkari, P. V., & Balaraman, M. (2015). Catechins: Sources, extraction and encapsulation: A review. *Food and Bioproducts Processing*, *93*, 122–138.

Gammoh, S., Alu'datt, M. H., Alhamad, M. N., Rababah, T., Al-Mahasneh, M., Qasaimeh, A., et al. (2018). The effects of protein-phenolic interactions in wheat protein fractions on allergenicity, antioxidant activity and the inhibitory activity of angiotensin I-converting enzyme (ACE). *Food Bioscience*, *24*, 50–55.

Gbenga-Fabusiwa, F. J., Oladele, E. P., Oboh, G., Adefegha, S. A., & Oshodi, A. A. (2018). Nutritional properties, sensory qualities and glycemic response of biscuits produced from pigeon pea-wheat composite flour. *Journal of Food Biochemistry*, *42*, e12505.

Goh, R., Gao, J., Ananingsih, V. K., Ranawana, V., Henry, C. J., & Zhou, W. (2015). Green tea catechins reduced the glycaemic potential of bread: An in vitro digestibility study. *Food Chemistry*, *180*, 203–210.

Górnaś, P., Juhņeviča-Radenkova, K., Radenkovs, V., Mišina, I., Pugajeva, I., Soliven, A., et al. (2016). The impact of different baking conditions on the stability of the extractable polyphenols in muffins enriched by strawberry, sour cherry, raspberry or black currant pomace. *LWT–Food Science and Technology*, *65*, 946–953.

Grigelmo-Miguel, N., Carreras-Boladeras, E., & Martin-Belloso, O. (2001). Influence of the addition of peach dietary fiber in composition, physical properties and acceptability of reduced-fat muffins. *Food Science and Technology International*, 7(5), 425–431.

Guilbaud, A., Niquet-Leridon, C., Boulanger, E., & Tessier, F. J. (2016). How can diet affect the accumulation of advanced glycation end-products in the human body? *Foods*, *5*, 84.

Guo, X., Shi, L., Yang, S., Yang, R., Dai, X., Zhang, T., et al. (2019). Effect of sea-buckthorn pulp and flaxseed residues on quality and shelf life of bread. *Food & Function*, *10*(7), 4220–4230.

Gutiérrez Barrutia, M. B., Curutchet, A., Arcia, P., & Cozzano, S. (2019). New functional ingredient from orange juice byproduct through a green extraction method. *Journal of Food Processing and Preservation*, *43*(5), e13934.

Hayta, M., Özuğur, G., Etgü, H., & Şeker, İ. T. (2014). Effect of grape (V itis Vinifera L.) pomace on the quality, total phenolic content and anti-radical activity of bread. *Journal of Food Processing and Preservation*, *38*(3), 980–986.

He, C., Sabol, J., Mitsuhashi, T., & Vlassara, H. (1999). Dietary glycotoxins: Inhibition of reactive products by aminoguanidine facilitates renal clearance and reduces tissue sequestration. *Diabetes*, *48*, 1308–1315.

Hellwig, M., Löbmann, K., Orywol, T., & Voigt, A. (2014). Model studies on the oxidation of benzoyl methionine in a carbohydrate degradation system. *Journal of Agricultural and Food Chemistry*, *62*(19), 4425–4433.

Hidalgo, A., Ferraretto, A., De Noni, I., Bottani, M., Cattaneo, S., Galli, S., et al. (2018). Bioactive compounds and antioxidant properties of pseudocereals-enriched water biscuits and their in vitro digestates. *Food Chemistry*, *240*, 799–807.

Hou, Y., Xie, Z., Cui, H., Lu, Y., Zheng, T., Sang, S., et al. (2018). Trapping of glyoxal by propyl, octyl and dodecyl gallates and their mono-glyoxal adducts. *Food Chemistry*, *269*, 396–403.

Huang, J., Wang, Y., Yang, L., Peng, X., Zheng, J., & Ou, S. (2018). Effect of maize bran feruloylated oligosaccharides on the formation of endogenous contaminants and the appearance and textural properties of biscuits. *Food Chemistry*, *245*, 974–980.

Huang, Q., Zhu, Y., Lv, L., & Sang, S. (2020). Translating in vitro acrolein-trapping capacities of tea polyphenol and soy genistein to in vivo situation is mediated by the bioavailability and biotransformation of individual polyphenols. *Molecular Nutrition & Food Research*, *64*(1), 1900274.

Jayalaxmi, B., Vijayalakshmi, D., & Maruthesha, A. (2018). Application of polyphenol extract from mango peel powder as a source of natural phytonutrients into biscuits. *International Journal of Current Microbiology and Applied Sciences*, 7, 1206–1213.

Jiang, D., Chiaro, C., Maddali, P., Prabhu, K. S., & Peterson, D. G. (2009). Identification of hydroxycinnamic acid – maillard reaction products in low-moisture baking model systems. *Journal of Agricultural and Food Chemistry*, *57*, 9932–9943.

Jin, C., Wu, X., & Zhang, Y. (2013). Relationship between antioxidants and acrylamide formation: A review. *Food Research International*, *51*(2), 611–620.

Kalinová, J. P., Vrchotová, N., & Tříska, J. (2018). Contribution to the study of rutin stability in the achenes of Tartary buckwheat (Fagopyrum tataricum). *Food Chemistry*, *258*, 314–320.

Khan, M., Liu, H., Wang, J., & Sun, B. (2020). Inhibitory effect of phenolic compounds and plant extracts on the formation of advance glycation end products: A comprehensive review. *Food Research International*, *130*, 108933.

Kocadağlı, T., Žilić, S., Taş, N. G., Vančetović, J., Dodig, D., & Gökmen, V. (2016). Formation of α-dicarbonyl compounds in cookies made from wheat, hull-less barley and colored corn and its relation with phenolic compounds, free amino acids and sugars. *European Food Research and Technology*, *242*(1), 51–60.

Kokkinidou, S., & Peterson, D. G. (2014). Control of Maillard-type off-flavor development in ultrahigh-temperature-processed bovine milk by phenolic chemistry. *Journal of Agricultural and Food Chemistry, 62*, 8023–8033.

Koszucka, A., & Nowak, A. (2019). Thermal processing food-related toxicants: A review. *Critical Reviews in Food Science and Nutrition, 59*(22), 3579–3596.

Krekora, M., Szymańska-Chargot, M., Niewiadomski, Z., Miś, A., & Nawrocka, A. (2020). Effect of cinnamic acid and its derivatives on structure of gluten proteins–a study on model dough with application of FT-Raman spectroscopy. *Food Hydrocolloids, 107*, 105935.

Lamport, D. J., Saunders, C., Butler, L. T., & Spencer, J. P. (2014). Fruits, vegetables, 100% juices, and cognitive function. *Nutrition Reviews, 72*(12), 774–789.

Li, D., Chen, Y., Zhang, Y., Lu, B., Jin, C., Wu, X., et al. (2012). Study on mitigation of acrylamide formation in cookies by 5 antioxidants. *Journal of Food Science, 77*, C1144–C1149.

Liang, S., & Were, L. M. (2018). Chlorogenic acid induced colored reactions and their effect on carbonyls, phenolic content, and antioxidant capacity in sunflower butter cookies. *LWT–Food Science and Technology, 87*, 16–22.

Lin, J., & Zhou, W. (2018). Role of quercetin in the physicochemical properties, antioxidant and antiglycation activities of bread. *Journal of Functional Foods, 40*, 299–306.

Lin, J., Teo, L. M., Leong, L. P., & Zhou, W. (2019). In vitro bioaccessibility and bioavailability of quercetin from the quercetin-fortified bread products with reduced glycemic potential. *Food Chemistry, 286*, 629–635.

Lončarić, A., Pablo Lamas, J., Guerra, E., Kopjar, M., & Lores, M. (2018). Thermal stability of catechin and epicatechin upon disaccharides addition. *International Journal of Food Science & Technology, 53*(5), 1195–1202.

López, C. J., Caleja, C., Prieto, M. A., Sokovic, M., Calhelha, R. C., Barros, L., et al. (2019). Stability of a cyanidin-3-O-glucoside extract obtained from Arbutus unedo L. and incorporation into wafers for colouring purposes. *Food Chemistry, 275*, 426–438.

Luevano-Contreras, C., & Chapman-Novakofski, K. (2010). Dietary advanced glycation end products and aging. *Nutrients, 2*, 1247–1265.

Ma, M., Sun, Q. J., Li, M., & Zhu, K. X. (2020). Deterioration mechanisms of high-moisture wheat-based food—A review from physicochemical, structural, and molecular perspectives. *Food Chemistry, 318*, 126495.

Mahidsanan, T., & Lertnirundon, N. (2020). Effectiveness of Bergamot juice on the survival of bacillus cereus and quality of Thai steamed pumpkin cake. *Food and Applied Bioscience Journal, 8*(1), 33–42.

Majerska, J., Michalska, A., & Figiel, A. (2019). A review of new directions in managing fruit and vegetable processing by-products. *Trends in Food Science & Technology, 88*, 207–219.

Majzoobi, M., Ghavi, F. S., Farahnaky, A., Jamalian, J., & Mesbahi, G. (2011). Effect of tomato pomace powder on the physicochemical properties of flat bread (Barbari bread). *Journal of Food Processing and Preservation, 35*(2), 247–256.

Martins, Z. E., Pinho, O., & Ferreira, I. M. P. L. V. O. (2017). Food industry by-products used as functional ingredients of bakery products. *Trends in Food Science & Technology, 67*, 106–128.

Mayneris-Perxachs, J., Alcaide-Hidalgo, J. M., de la Hera, E., del Bas, J. M., Arola, L., & Caimari, A. (2019). Supplementation with biscuits enriched with hesperidin and naringenin is associated with an improvement of the metabolic syndrome induced by a cafeteria diet in rats. *Journal of Functional Foods, 61*, 103504.

Meng, Q., Li, S., Huang, J., Wei, C. C., Wan, X., Sang, S., et al. (2018). Importance of the nucleophilic property of tea polyphenols. *Journal of Agricultural and Food Chemistry, 67*, 5379–5383.

Michon, C., O'Sullivan, M. G., Sheehan, E., Delahunty, C. M., & Kerry, J. P. (2010). Investigation of the influence of age, gender and consumption habits on the liking of jam-filled cakes. *Food Quality and Preference, 21*(5), 553–561.

Mildner-Szkudlarz, S., Różańska, M., Piechowska, P., Waśkiewicz, A., & Zawirska-Wojtasiak, R. (2019). Effects of polyphenols on volatile profile and acrylamide formation in a model wheat bread system. *Food Chemistry, 297*, 125008.

Mohamed Ahmed, I. A., Özcan, M. M., Al Juhaimi, F., Babiker, E. F. E., Ghafoor, K., Banjanin, T., et al. (2020). Chemical composition, bioactive compounds, mineral contents, and fatty acid composition of pomace powder of different grape varieties. *Journal of Food Processing and Preservation, 44*(7), e14539.

Montalvo-González, E., Aguilar-Hernández, G., Hernández-Cázares, A. S., Ruiz-López, I. I., Pérez-Silva, A., Hernández-Torres, J., et al. (2018). Production, chemical, physical and technological properties of antioxidant dietary fiber from pineapple pomace and effect as ingredient in sausages. *CyTA Journal of Food, 16*(1), 831–839.

Moon, J. K., & Shibamoto, T. (2010). Formation of volatile chemicals from thermal degradation of less volatile coffee components: Quinic acid, caffeic acid, and chlorogenic acid. *Journal of Agricultural and Food Chemistry, 58*, 5465–5470.

Moskowitz, M. R., Bin, Q., Elias, R. J., & Peterson, D. G. (2012). Influence of endogenous ferulic acid in whole wheat flour on bread crust aroma. *Journal of Agricultural and Food Chemistry, 60*, 11245–11252.

Navarre, D. A., Shakya, R., Holden, J., & Kumar, S. (2010). The effect of different cooking methods on phenolics and vitamin C in developmentally young potato tubers. *American Journal of Potato Research, 87*, 350–359.

Navarro, M., & Morales, F. J. (2016). In vitro investigation on the antiglycative and carbonyl trapping activities of hydroxytyrosol. *European Food Research and Technology, 242*, 1101–1110.

Navarro, M., & Morales, F. J. (2017). Effect of hydroxytyrosol and olive leaf extract on 1, 2-dicarbonyl compounds, hydroxymethylfurfural and advanced glycation endproducts in a biscuit model. *Food Chemistry, 217*, 602–609.

Ning, J., Hou, G. G., Sun, J., Wan, X., & Dubat, A. (2017). Effect of green tea powder on the quality attributes and antioxidant activity of whole-wheat flour pan bread. *LWT—Food Science and Technology, 79*, 342–348.

Noda, Y., & Peterson, D. G. (2007). Structure-reactivity relationships of flavan-3-ols on product generation in aqueous glucose/glycine model systems. *Journal of Agricultural and Food Chemistry, 55*, 3686–3691.

Nunes, F. M., & Coimbra, M. A. (2010). Role of hydroxycinnamates in coffee melanoidin formation. *Phytochemistry Reviews, 9*, 171–185.

Oszmiański, J., Wolniak, M., Wojdyło, A., & Wawer, I. (2008). Influence of apple purée preparation and storage on polyphenol contents and antioxidant activity. *Food Chemistry, 107*(4), 1473–1484.

Othman, N. A., Abdul Manaf, M., Harith, S., & Wan Ishak, W. R. (2018). Influence of avocado puree as a fat replacer on nutritional, fatty acid, and organoleptic properties of low-fat muffins. *Journal of the American College of Nutrition, 37*(7), 583–588.

Ou, J., Huang, J., Wang, M., & Ou, S. (2017). Effect of rosmarinic acid and carnosic acid on AGEs formation in vitro. *Food Chemistry, 221*, 1057–1061.

Ou, J., Teng, J., El-Nezami, H. S., & Wang, M. (2018). Impact of resveratrol, epicatechin and rosmarinic acid on fluorescent AGEs and cytotoxicity of cookies. *Journal of Functional Foods, 40*, 44–50.

Ou, J., Wang, M., Zheng, J., & Ou, S. (2019). Positive and negative effects of polyphenol incorporation in baked foods. *Food Chemistry, 284*, 90–99.

Ou, J., Zheng, J., Huang, J., Ho, C. T., & Ou, S. (2020). Interaction of acrylamide, acrolein, and 5-hydroxymethylfurfural with amino acids and DNA. *Journal of Agricultural and Food Chemistry, 68*(18), 5039–5048.

Papoutsis, K., Pristijono, P., Golding, J. B., Stathopoulos, C. E., Bowyer, M. C., Scarlett, C. J., et al. (2017). Effect of vacuum-drying, hot air-drying and freeze-drying on polyphenols and antioxidant capacity of lemon (Citrus limon) pomace aqueous extracts. *International Journal of Food Science & Technology, 52*(4), 880–887.

Papuc, C., Goran, G. V., Predescu, C. N., Nicorescu, V., & Stefan, G. (2017). Plant polyphenols as antioxidant and antibacterial agents for shelf-life extension of meat and meat products: Classification, structures, sources, and action mechanisms. *Comprehensive Reviews in Food Science and Food Safety, 16*(6), 1243–1268.

Pasqualone, A., Bianco, A. M., Paradiso, V. M., Summo, C., Gambacorta, G., & Caponio, F. (2014). Physico-chemical, sensory and volatile profiles of biscuits enriched with grape marc extract. *Food Research International, 65*, 385–393.

Pasukamonset, P., Pumalee, T., Sanguansuk, N., Chumyen, C., Wongvasu, P., Adisakwattana, S., et al. (2018). Physicochemical, antioxidant and sensory characteristics of sponge cakes fortified with Clitoria ternatea extract. *Journal of Food Science and Technology, 55*, 2881–2889.

Patras, A., Brunton, N. P., O'Donnell, C., & Tiwari, B. K. (2010). Effect of thermal processing on anthocyanin stability in foods; mechanisms and kinetics of degradation. *Trends in Food Science & Technology, 21*, 3–11.

Pedro, A. C., Maciel, G. M., Rampazzo Ribeiro, V., & Haminiuk, C. W. I. (2020). Fundamental and applied aspects of catechins from different sources: A review. *International Journal of Food Science & Technology, 55*(2), 429–442.

Peng, X., Ma, J., Chen, F., & Wang, M. (2011). Naturally occurring inhibitors against the formation of advanced glycation end-products. *Food & Function, 2*(6), 289–301.

Perez, C., Tagliani, C., Arcia, P., Cozzano, S., & Curutchet, A. (2018). Blueberry by-product used as an ingredient in the development of functional cookies. *Food Science and Technology International, 24*(4), 301–308.

Perkins, M. L., & Deeth, H. C. (2001). A survey of Australian consumers' attitudes towards UHT milk. *Australian Journal of Dairy Technology, 56*, 28.

Porter, W. L., Conca, K. R., Yeomans, W. G., Diotte, S., Lynch, A., & Tate, J. (2006). Modification of maillard browning in a microwaved glucose/glycine model system by water-soluble natural antioxidants and foods containing them. *Journal of the American Oil Chemists' Society, 83*, 697–705.

Poulsen, M. W., Hedegaard, R. V., Andersen, J. M., de Courten, B., Bügel, S., Nielsen, J., et al. (2013). Advanced glycation end products in food and their effects on health. *Food and Chemical Toxicology, 60*, 10–37.

Quiles, A., Campbell, G. M., Struck, S., Rohm, H., & Hernando, I. (2018). Fiber from fruit pomace: A review of applications in cereal-based products. *Food Reviews International, 34*(2), 162–181.

Rana, S., Rana, A., Gupta, S., & Bhushan, S. (2020). Varietal influence on phenolic constituents and nutritive characteristics of pomace obtained from apples grown in western Himalayas. *Journal of Food Science and Technology, 58*, 166–174. https://doi.org/10.1007/s13197-020-04526-y.

Ranawana, V., Campbell, F., Bestwick, C., Nicol, P., Milne, L., Duthie, G., et al. (2016). Breads fortified with freeze-dried vegetables: Quality and nutritional attributes. Part II: Breads not containing oil as an ingredient. *Foods, 5*(3), 62.

Ravber, M., Pečar, D., Goršek, A., Iskra, J., Knez, Z., & Škerget, M. (2016). Hydrothermal degradation of rutin: Identification of degradation products and kinetics study. *Journal of Agricultural and Food Chemistry, 64*(48), 9196–9202.

Rizzi, G. P. (2003). Free radicals in the Maillard reaction. *Food Reviews International, 19*, 375–395.
Rodriguez-Mateos, A., Cifuentes-Gomez, T., George, T. W., & Spencer, J. P. (2014). Impact of cooking, proving, and baking on the (poly) phenol content of wild blueberry. *Journal of Agricultural and Food Chemistry, 62*(18), 3979–3986.
Rosales-Soto, M. U., Powers, J. R., & Alldredge, J. R. (2012). Effect of mixing time, freeze-drying and baking on phenolics, anthocyanins and antioxidant capacity of raspberry juice during processing of muffins. *Journal of the Science of Food and Agriculture, 92*(7), 1511–1518.
Rumiyati, R., James, A. P., & Jayasena, V. (2015). Effects of lupin incorporation on the physical properties and stability of bioactive constituents in muffins. *International Journal of Food Science & Technology, 50*(1), 103–110.
Rupasinghe, H. V., Wang, L., Huber, G. M., & Pitts, N. L. (2008). Effect of baking on dietary fibre and phenolics of muffins incorporated with apple skin powder. *Food Chemistry, 107*(3), 1217–1224.
Sadilova, E., Carle, R., & Stintzing, F. C. (2007). Thermal degradation of anthocyanins and its impact on color and in vitro antioxidant capacity. *Molecular Nutrition & Food Research, 51*, 1461–1471.
Sadilova, E., Stintzing, F. C., & Carle, R. (2006). Thermal degradation of acylated and nonacylated anthocyanins. *Journal of Food Science, 71*(8), C504–C512.
Sahin, A. W., Zannini, E., Coffey, A., & Arendt, E. K. (2019). Sugar reduction in bakery products: Current strategies and sourdough technology as a potential novel approach. *Food Research International, 126*, 108583.
Sahni, P., & Shere, D. M. (2018). Utilization of fruit and vegetable pomace as functional ingredient in bakery products: A review. *Asian Journal of Dairy and Food Research, 37*(3), 202–211.
Salehi, F., & Aghajanzadeh, S. (2020). Effect of dried fruits and vegetables powder on cakes quality: A review. *Trends in Food Science & Technology, 95*, 162–172.
Scarton, M., Ganancio, J. R. C., de Avelar, M. H. M., Clerici, M. T. P. S., & Steel, C. J. (2020). Lime juice and enzymes in clean label pan bread: Baking quality and preservative effect. *Journal of Food Science and Technology*. https://doi.org/10.1007/s13197-020-04693-y.
Schieber, A., & Wüst, M. (2020). Volatile phenols—Important contributors to the aroma of plant-derived foods. *Molecules, 25*(19), 4529.
Schmid, V., Steck, J., Mayer-Miebach, E., Behsnilian, D., Briviba, K., Bunzel, M., et al. (2020). Impact of defined thermomechanical treatment on the structure and content of dietary fiber and the stability and bioaccessibility of polyphenols of chokeberry (*Aronia melanocarpa*) pomace. *Food Research International, 134*, 109232.
Shi, M., Hlaing, M. M., Ying, D., Ye, J., Sanguansri, L., & Augustin, M. A. (2019). New food ingredients from broccoli by-products: Physical, chemical and technological properties. *International Journal of Food Science & Technology, 54*(4), 1423–1432.
Shishir, M. R. I., & Chen, W. (2017). Trends of spray drying: A critical review on drying of fruit and vegetable juices. *Trends in Food Science & Technology, 65*, 49–67.
Sinela, A. M., Mertz, C., Achir, N., Rawat, N., Vidot, K., Fulcrand, H., et al. (2017). Exploration of reaction mechanisms of anthocyanin degradation in a roselle extract through kinetic studies on formulated model media. *Food Chemistry, 235*, 67–75.
Slavin, M., Lu, Y., Kaplan, N., & Yu, L. L. (2013). Effects of baking on cyanidin-3-glucoside content and antioxidant properties of black and yellow soybean crackers. *Food Chemistry, 141*, 1166–1174.
Smith, J. P., Daifas, D. P., El-Khoury, W., Koukoutsis, J., & El-Khoury, A. (2004). Shelf life and safety concerns of bakery products—A review. *Critical Reviews in Food Science and Nutrition, 44*(1), 19–55.

Sowbhagya, H. B. (2019). Value-added processing of by-products from spice industry. *Food Quality and Safety*, *3*(2), 73–80.

Stanhope, K. L. (2016). Sugar consumption, metabolic disease and obesity: The state of the controversy. *Critical Reviews in Clinical Laboratory Sciences*, *53*(1), 52–67.

Su, Y. L., Leung, L. K., Huang, Y., & Chen, Z. Y. (2003). Stability of tea theaflavins and catechins. *Food Chemistry*, *83*(2), 189–195.

Sudha, M. L., Indumathi, K., Sumanth, M. S., Rajarathnam, S., & Shashirekha, M. N. (2015). Mango pulp fibre waste: Characterization and utilization as a bakery product ingredient. *Journal of Food Measurement and Characterization*, *9*(3), 382–388.

Świeca, M., Gawlik-Dziki, U., Sęczyk, Ł., Dziki, D., & Sikora, M. (2018). Interactions of green coffee bean phenolics with wheat bread matrix in a model of simulated in vitro digestion. *Food Chemistry*, *258*, 301–307.

Szawara-Nowak, D., Bączek, N., & Zieliński, H. (2016). Antioxidant capacity and bio-accessibility of buckwheat-enhanced wheat bread phenolics. *Journal of Food Science and Technology*, *53*, 621–630.

Szwergold, B. (2017). Reactions between methylglyoxal and its scavengers in-vivo appear to be catalyzed enzymatically. *Medical Hypotheses*, *109*, 153–155.

Tańska, M., Roszkowska, B., Czaplicki, S., Borowska, E. J., Bojarska, J., & Dąbrowska, A. (2016). Effect of fruit pomace addition on shortbread cookies to improve their physical and nutritional values. *Plant Foods for Human Nutrition*, *71*, 307–313.

Teng, J., Hu, X. Q., Tao, N. P., & Wang, M. F. (2018). Impact and inhibitory mechanism of phenolic compounds on the formation of toxic Maillard reaction products in food. *Frontiers of Agricultural Science and Engineering*, *5*(3), 321–329.

Totlani, V. M., & Peterson, D. G. (2005). Reactivity of epicatechin in aqueous glycine and glucose Maillard reaction models: Quenching of C2, C3, and C4 sugar fragments. *Journal of Agricultural and Food Chemistry*, *53*, 4130–4135.

Totlani, V. M., & Peterson, D. G. (2007). Influence of epicatechin reactions on the mechanisms of Maillard product formation in low moisture model systems. *Journal of Agricultural and Food Chemistry*, *55*, 414–420.

Trigo, J. P., Alexandre, E. M., Saraiva, J. A., & Pintado, M. E. (2020). High value-added compounds from fruit and vegetable by-products–Characterization, bioactivities, and application in the development of novel food products. *Critical Reviews in Food Science and Nutrition*, *60*(8), 1388–1416.

Ullah, N., Ullah, S., Khan, A., Ullah, I., & Badshah, S. (2018). Preparation and evaluation of carrot and apple blended jam. *Journal of Food Processing & Technology*, *9*(725), 2.

Verardo, V., Riciputi, Y., Messia, M. C., Vallicelli, M., Falasca, L., Marconi, E., et al. (2011). Dietary fiber and flavan-3-ols in shortbread biscuits enriched with barley flours co-products. *International Journal of Food Sciences and Nutrition*, *62*, 262–269.

Vogrincic, M., Timoracka, M., Melichacova, S., Vollmannova, A., & Kreft, I. (2010). Degradation of rutin and polyphenols during the preparation of tartary buckwheat bread. *Journal of Agricultural and Food Chemistry*, *58*(8), 4883–4887.

Walker, R., Tseng, A., Cavender, G., Ross, A., & Zhao, Y. (2014). Physicochemical, nutritional, and sensory qualities of wine grape pomace fortified baked goods. *Journal of Food Science*, *79*, S1811–S1822.

Wang, P., Chen, H., & Sang, S. (2016). Trapping methylglyoxal by genistein and its metabolites in mice. *Chemical Research in Toxicology*, *29*, 406–414.

Wang, R., & Zhou, W. (2004). Stability of tea catechins in the breadmaking process. *Journal of Agricultural and Food Chemistry*, *52*(26), 8224–8229.

Wang, S., & Zhu, F. (2018). Quality attributes of bread fortified with staghorn sumac extract. *Journal of Texture Studies*, *49*, 129–134.

Wang, W., Qi, Y., Rocca, J. R., Sarnoski, P. J., Jia, A., & Gu, L. (2015). Scavenging of toxic acrolein by resveratrol and hesperetin and identification of adducts. *Journal of Agricultural and Food Chemistry*, *63*, 9488–9495.

Wang, W., Yang, H., Johnson, D., Gensler, C., Decker, E., & Zhang, G. (2017). Chemistry and biology of ω-3 PUFA peroxidation-derived compounds. *Prostaglandins & Other Lipid Mediators*, *132*, 84–91.
Wang, Y., & Ho, C. T. (2012). Flavour chemistry of methylglyoxal and glyoxal. *Chemical Society Reviews*, *41*, 4140–4149.
Wildermuth, S. R., Young, E. E., & Were, L. M. (2016). Chlorogenic acid oxidation and its reaction with sunflower proteins to form green-colored complexes. *Comprehensive Reviews in Food Science and Food Safety*, *15*(5), 829–843.
Wu, C. H., Huang, S. M., Lin, J. A., & Yen, G. C. (2011). Inhibition of advanced glycation endproduct formation by foodstuffs. *Food & Function*, *2*(5), 224–234.
Xu, J., Wang, W., & Li, Y. (2019). Dough properties, bread quality, and associated interactions with added phenolic compounds: A review. *Journal of Functional Foods*, *52*, 629–639.
Yalcinkaya, T., Uzilday, B., Ozgur, R., Turkan, I., & Mano, J. I. (2019). Lipid peroxidation-derived reactive carbonyl species (RCS): Their interaction with ROS and cellular redox during environmental stresses. *Environmental and Experimental Botany*, *165*, 139–149.
Yang, N., Qiu, R., Yang, S., Zhou, K., Wang, C., Ou, S., et al. (2019). Influences of stir-frying and baking on flavonoid profile, antioxidant property, and hydroxymethylfurfural formation during preparation of blueberry-filled pastries. *Food Chemistry*, *287*, 167–175.
Yeh, W. J., Hsia, S. M., Lee, W. H., & Wu, C. H. (2017). Polyphenols with antiglycation activity and mechanisms of action: A review of recent findings. *Journal of Food and Drug Analysis*, *25*(1), 84–92.
Zain, M. Z. M., Baba, A. S., & Shori, A. B. (2018). Effect of polyphenols enriched from green coffee bean on antioxidant activity and sensory evaluation of bread. *Journal of King Saud University-Science*, *30*, 278–282.
Zamora, R., & Hidalgo, F. J. (2018). Carbonyl–phenol adducts: An alternative sink for reactive and potentially toxic lipid oxidation products. *Journal of Agricultural and Food Chemistry*, *66*, 1320–1324.
Zamora, R., Aguilar, I., Granvogl, M., & Hidalgo, F. J. (2016). Toxicologically relevant aldehydes produced during the frying process are trapped by food phenolics. *Journal of Agricultural and Food Chemistry*, *64*(27), 5583–5589.
Zhang, L., Song, X., Zhou, L., Liang, G., Xu, H., Wang, F., et al. (2016). Accumulation of intestinal tissue 3-deoxyglucosone attenuated GLP-1 secretion and its insulinotropic effect in rats. *Diabetology & Metabolic Syndrome*, *8*(1), 1–10.
Zhang, X., Chen, F., & Wang, M. (2014). Antioxidant and antiglycation activity of selected dietary polyphenols in a cookie model. *Journal of Agricultural and Food Chemistry*, *62*(7), 1643–1648.
Zhang, Z., Zou, Y., Wu, T., Huang, C., Pei, K., Zhang, G., et al. (2016). Chlorogenic acid increased 5-hydroxymethylfurfural formation when heating fructose alone or with aspartic acid at two pH levels. *Food Chemistry*, *190*, 832–835.
Zhao, Q., Yao, S., & Ou, S. Y. (2017). Maillard volatiles in baked products as affected by feruloylated oligosaccharides from maize bran. *International Journal of Food Properties*, *20*, 3266–3273.
Zheng, J., Zhou, Y., Li, S., Zhang, P., Zhou, T., Xu, D. P., et al. (2017). Effects and mechanisms of fruit and vegetable juices on cardiovascular diseases. *International Journal of Molecular Sciences*, *18*(3), 555.
Zheng, L. W., Chung, H., & Kim, Y. S. (2015). Effects of dicarbonyl trapping agents, antioxidants, and reducing agents on the formation of furan and other volatile components in canned-coffee model systems. *Food Research International*, *75*, 328–336.
Zhou, Q., Gong, J., & Wang, M. (2019). Phloretin and its methylglyoxal adduct: Implications against advanced glycation end products-induced inflammation in endothelial cells. *Food and Chemical Toxicology*, *129*, 291–300.

Zhou, X., Zhang, Z., Liu, X., Wu, D., Ding, Y., Li, G., et al. (2020). Typical reactive carbonyl compounds in food products: Formation, influence on food quality, and detection methods. *Comprehensive Reviews in Food Science and Food Safety, 19*(2), 503–529.

Zhu, F., & Sun, J. (2019). Physicochemical and sensory properties of steamed bread fortified with purple sweet potato flour. *Food Bioscience, 30*, 100411.

Zhu, H., Poojary, M. M., Andersen, M. L., & Lund, M. N. (2019). Effect of pH on the reaction between naringenin and methylglyoxal: A kinetic study. *Food Chemistry, 298*, 125086.

Zhu, H., Poojary, M. M., Andersen, M. L., & Lund, M. N. (2020). The effect of molecular structure of polyphenols on the kinetics of the trapping reactions with methylglyoxal. *Food Chemistry, 319*, 126500.

Zhu, Q., Zhang, N. Q. S., Lau, C. F., Chao, J., Sun, Z., Chang, R. C. C., et al. (2012). In vitro attenuation of acrolein-induced toxicity by phloretin, a phenolic compound from apple. *Food Chemistry, 135*, 1762–1768.

CHAPTER SEVEN

Antioxidant/pro-oxidant and antiproliferative activities of phenolic-rich foods and extracts: A cell-based point of view

Mariana Araújo Vieira do Carmo[a], Daniel Granato[b], and Luciana Azevedo[a,*]

[a]Federal University of Alfenas, Nutrition Faculty, Alfenas, MG, Brazil
[b]Department of Biological Sciences, Faculty of Science and Engineering, University of Limerick, Limerick, Ireland
*Corresponding author: e-mail address: luciana.azevedo@unifal-mg.edu.br

Contents

Abstract

Phenolic compounds have demonstrated several *in vitro* beneficial properties by acting as antioxidant and pro-oxidant agents. This chapter approaches the relationship among oxidative stress, cancer, phenolic compounds and antiproliferative activity. Moreover, it discusses *in vitro* techniques and their biological applications, regarding cell viability and intracellular measure of reactive oxygen assays. The *in vitro* methods are important tools for screening and understanding the pathways involved on antiproliferative and antioxidant/pro-oxidant effects of phenolic compounds. These findings open avenues for the development of innovative food, chemical structures, technological applications and future perspectives in this research field.

Advances in Food and Nutrition Research, Volume 98
ISSN 1043-4526
https://doi.org/10.1016/bs.afnr.2021.02.010

1. The role of reactive oxygen and nitrogen species (RONS) on oxidative stress

Reactive oxygen species (ROS) are intracellular chemical species that contain oxygen and are reactive toward lipids, proteins and DNA. ROS include the superoxide anion ($O_2^{\cdot -}$), hydrogen peroxide (H_2O_2), as well as hydroxyl radicals ($^{\cdot}OH$) (Glasauer & Chandel, 2014). On the other hand, reactive nitrogen species RNS refer to various nitrogenous products from nitric oxide synthases (NOS), such as nitric oxide (NO), nitrogen dioxide (NO_2), nitroxyl anion (NO^-), nitrite (NO_2^-) and peroxynitrite ($ONOO^-$) (do Carmo, Pressete, Marques, Granato, & Azevedo, 2018; Griendling et al., 2016; Shen & Liu, 2006). Endogenous sources of RONS include auto-oxidant reactions, xanthine oxidase, redox cycles, inflammations, oxidation in phagocytic cells and cytochrome P450 reactions. Moreover, RONS may be also delivered to plants by exogenous sources such as environmental toxins, chemicals and smog (Drzeżdżon, Jacewicz, & Chmurzyński, 2018). Both ROS and RNS may induce molecular damages to almost all organic compounds performing biological functions (nucleic acids, proteins, cells carbohydrates, and lipids, etc.) (Hu et al., 2020) and they play a role in JNK activation in oxidative or nitrosative stress-mediated cell death (Shen & Liu, 2006). In this chapter, we will focus on ROS pathways and their relationship with oxidative stress, phenolic compounds and cancer.

Under physiological conditions, ROS are essential to maintain cell homeostasis, from the regulation of signaling, proliferation and differentiation to the promotion of cellular damage and death (Harris & DeNicola, 2020). When ROS generation are below their optimal range, they lead to a disruption of cell signaling resulting in loss of homeostasis control. In contrast, in the case of oxidative stress, excessive ROS trigger cellular damage mechanisms, which can lead to DNA damage and consequently, to development of diseases (e.g., cancer) (Chen, Tian, She, Cai, & Li, 2020). When cellular ROS production enhance to more toxic levels, the JNK pathway can be activated resulting in cell apoptosis initiated by intrinsic apoptotic signaling in the mitochondria or extrinsic apoptotic signaling by death receptor pathways (do Carmo et al., 2018). Interestingly, cancer cells are characterized by an increased rate of localized ROS generation, compared to normal cells. This occurs in order to hyper-activate signaling pathways that promote cell proliferation, survival and metabolic adaptation to the tumor microenvironment (Glasauer & Chandel, 2014), through activation of many pathways including PI3K/AKT, MAPK/ERK1/2 and PKD and inactivation of their downstream targets, such as Bad, Bax, Bim, Foxo, and PTEN (do Carmo et al., 2018). In addition to high levels of ROS, cancer cells also upregulate their scavenging systems (glutathione peroxidase, catalase, superoxide dismutase, glutathione, peroxiredoxin) in order to maintain pro-tumorigenic signaling and to keep ROS levels under the cytotoxic limit (Glasauer & Chandel, 2014).

Several organelles, such as the mitochondria, peroxisomes, and endoplasmic reticulum, produce ROS during their physiological processes. In most cells, mitochondria are considered the most relevant ROS generators (Chen et al., 2020). The complexes I, II and III are localized within the mitochondrial electron transport chain and they generate superoxide by the one-electron reduction of molecular oxygen. In the cytosol, superoxide dismutase 1 (SOD) rapidly converts superoxide anion into H_2O_2 (Glasauer & Chandel, 2014).

The intracellular removal of H_2O_2 can be performed by the action of catalase (CAT), peroxiredoxin (PRX) and glutathione peroxidase (GPx), which reduces H_2O_2 to water (Fig. 1). In addition, H_2O_2 can react with ferrous or cuprous ions, forming $^{\bullet}OH$ radicals by Fenton's type reaction that can be damaging to DNA, proteins and lipids (Glasauer & Chandel, 2014; Morry, Ngamcherdtrakul, & Yantasee, 2017). Moreover, $^{\bullet}OH$ radicals are able to interact with purine and pyrimidine bases (e.g., guanine) of DNA, by oxidation reaction, producing biomarkers of oxidative stress and carcinogenesis (e.g., 8-hydroxyguanine) (Fig. 1) (Valko, Rhodes, Moncol, Izakovic, & Mazur, 2006). These events, if not repaired, can stimuli carcinogenesis processes by the tumor-initiating cell, which refers to the original cell that receives the first oncogenic hits and undergoes clonal expansion in the earliest stage of tumor progression (Gupta, Pastushenko, Skibinski, & Blanpain, 2019). Bearing it in mind, antioxidants perform an important role in preventing ROS induced oxidation of molecules and subsequent DNA damage (Harris & DeNicola, 2020). The main function of antioxidants is to scavenge or neutralize free radical formation and to inhibit the deleterious downstream effects of ROS (Morry et al., 2017). The balance of ROS is achieved by non-enzymatic molecules (glutathione, phenolic compounds, vitamin A, C and E) and enzymatic antioxidants, such as SOD, superoxide reductase, CAT, GPx, glutathione reductase (GR), PRX, and thioredoxin (TRX) (do Carmo et al., 2018).

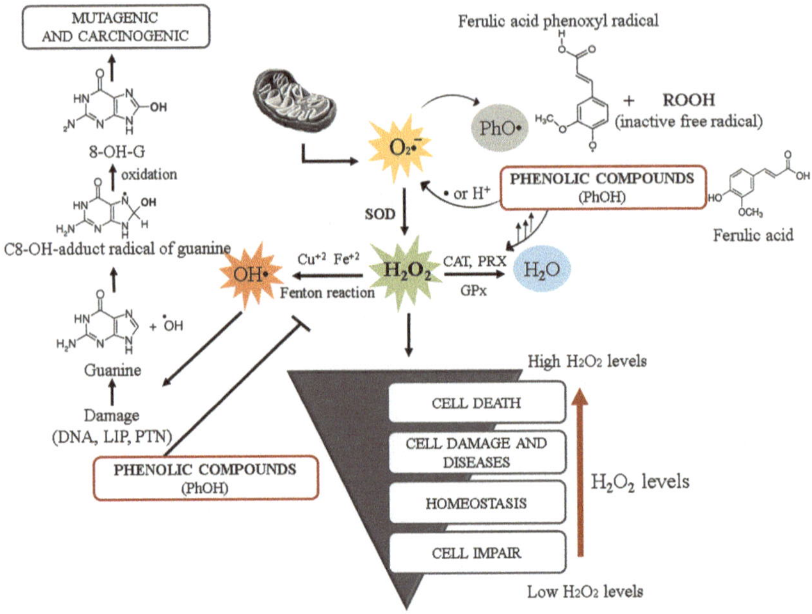

Fig. 1 Regulation of ROS balance throughout endogenous and exogenous antioxidants. Mitochondria can release either superoxide or H_2O_2. Phenolic compounds (e.g., ferulic acid) can react with superoxide anion, generating phenoxyl radicals (PhO˙) (e.g., ferulic acid phenoxyl radical). In the cytosol, the superoxide anion can be converted to H_2O_2 by SOD. Then, this H_2O_2 is transformed in water by antioxidant enzymes (CAT, GPx, PRx) and phenolic compounds stimuli their activity. H_2O_2 can also react with ferrous or cuprous ions, producing ˙OH radicals that can be damaging to DNA (e.g., reaction of guanine with hydroxyl radical), proteins and lipids. Phenolic compounds are able to chelate with these ions, preventing oxidation reactions and cell damage. Different levels of H_2O_2 lead to different cellular behavior. H_2O_2 levels below this appropriated range lead to a disruption of cell signaling resulting in loss of homeostasis. Optimal intracellular H_2O_2 concentrations in allow a permissive oxidative environment for cellular proliferation. High levels of H_2O_2 cause oxidative cell damage and aberrant cell signaling resulting in pathologies like cancer and too high levels of H_2O_2 stimuli apoptosis pathways.

2. The dual face behavior of phenolic compounds on cell proliferation

Phenolic compounds are defined as secondary plant metabolites with several thousand compounds identified from plants. Their chemical structure contains an aromatic ring bearing one or more hydroxyl groups and they range from that of a simple phenolic molecule to that of a highly polymerized compound. Considering their carbon chain, the phenolic compounds can be divided into different classes, mainly including phenolic acids, flavonoids, lignans, biflavonoids, tannins (hydrolysable and condensed), stilbenes and lignans (Ge, Li, & Lisak, 2020).

Interestingly, despite phenolic compounds are widely known for high antioxidant capacity, they are able to modulate the ROS levels, exerting multiple health benefits by acting as pro-oxidant (do Carmo et al., 2018; Eghbaliferiz & Iranshahi, 2016). In other words, phenolic compounds can prevent or induce oxidation of molecules depending on the environment circumstances (pH, phenolic compounds concentration, presence of oxygen and ions) and their behavior impact directly on cell proliferation.

2.1 Phenolic compounds as antioxidant agents

Thinking about their antioxidant capacity, phenolic compounds (PhOH) are potent H^+ donors for free radical acceptors ($ROO^{\bullet} + PhOH \rightarrow ROOH + PhO^{\bullet}$) due to electron delocalization across the molecule (Valko et al., 2006), producing less reactive and more stable radicals, preventing DNA, protein and lipid damage (Fig. 1). Moreover, phenolic compounds stimuli antioxidant enzymes activity (e.g., GPx, SOD, CAT), remove and repair damages caused by the reactive species (Olszowy, 2019) and they act as effective metal chelators (Fig. 1) (Santos, Alvarenga Brizola, & Granato, 2017). In this sense, phenolic compounds are deprotonated at physiological pH in the presence of iron and cooper, forming very stable complexes, which decreases the Fe^{+2} and Cu^{+2} availability to participate of Fenton reaction (Hider, Liu, & Khodr, 2001). Several chemical structure (Fig. 2B) features can enhance the antioxidant phenolic compounds potential, such as:

(1) The higher the numbers of hydroxyl groups (OH), the better their antioxidant properties;

(2) hydroxyl groups in *para* position (*p*-coumaric) enhances the antioxidant activity compared to *ortho* (2,5-dihydroxybenzoic acid) and *meta* (gallic acid) ones;

(3) cinnamic acids (caffeic acid, *p*-coumaric) are better antioxidants compared to benzoic acid (gallic acid, syringic acid) due to the presence in their structure of the -CH=CH-COOH group which enhances the antioxidant activity more than the -COOH group (Olszowy, 2019).

For flavonoids (Fig. 2A) (Olszowy, 2019; Valko et al., 2006) it was observed that:

(1) The presence of catechol group in ring B (e.g., cyaniding-3-glucoside) facilitates the donation of H^+ to free radicals;

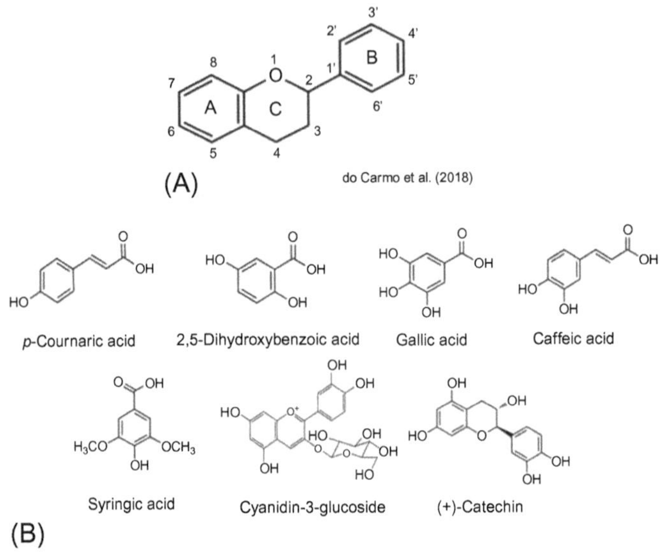

Fig. 2 Chemical structures of phenolic compounds. (A) General structure of flavonoids. (B) Examples of phenolic compounds.

(2) the greater the amount of hydroxyl group in ring B (e.g., (+)-catechin) of the structure, the greater the ability to donate hydrogen atoms;

(3) presence of unsaturation between C2 and C3 in conjugation with a 4-oxo group in the C-ring (e.g., quercetin), also increases the flavonoid's antioxidant capacity.

Through the detoxification of ROS, antioxidants, such as phenolic compounds, have the potential to prevent deleterious, and sometimes oncogenic, outcomes in healthy cells (Harris & DeNicola, 2020). Moreover, in these cells, antioxidant compounds may prevent apoptosis pathways by inhibiting JNK/AP1 activation and stimuli normal cell proliferation (Valko et al., 2006). On the other hand, in cancer cells, the antioxidant effect of phenolic compounds can decrease the ROS levels, impairing the pro-tumorigenic signaling, cell proliferation and tumor progression (Fig. 3). Lage et al. (2020) observed that phenolic compounds from *Prunus avium* exhibited breast anticancer activity by reduction of oxidative stress, Akt/mTOR, p38, and surviving regulation, suggesting antiproliferative and apoptotic mechanisms. In contrast, it is necessary caution, once this kind of therapy with antioxidants can interfere on cancer treatment that depends on ROS-induced cytotoxicity, like cisplatin, carboplatin and doxorubicin chemotherapeutics (Glasauer & Chandel, 2014).

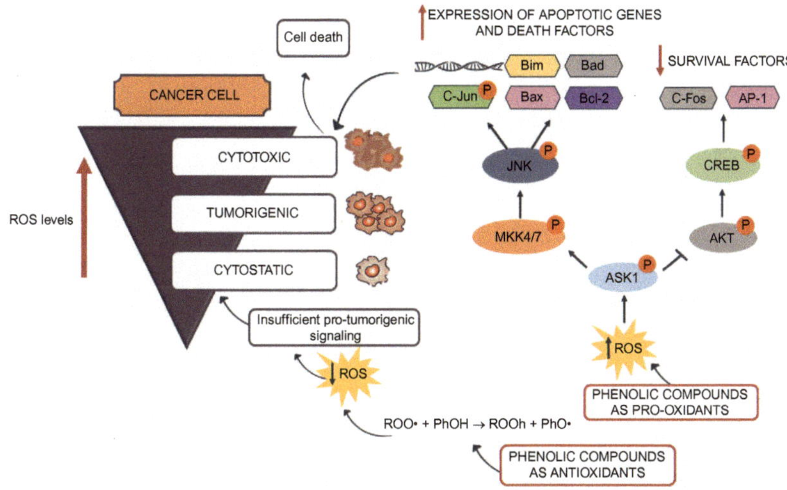

Fig. 3 Effects of phenolic compounds on ROS generation in cancer cells. Phenolic compounds can act as pro-oxidant agents, enhancing the ROS production into cytotoxic levels throughout apoptosis pathways and these compounds can also act as antioxidants, decreasing the ROS levels impairing the cell proliferation and metabolism. Optimal levels of ROS maintain the pro-tumorigenic signaling.

Thus, the antioxidant activity of phenolic compounds can prevent cell damage in normal cells and in cancer cells; they may reduce ROS levels, impairing the pro-tumorigenic signaling and consequently, the cell proliferation and tumor progression.

2.2 Phenolic compounds as pro-oxidant agents

Interestingly, the same structure features responsible to antioxidant activity of phenolic compounds, are also related to their pro-oxidant properties. In other words, the greater the antioxidant capacity of the phenolic compound, the greater its pro-oxidant potential. As mentioned above, the main factors that will determine whether the phenolic compound acts as an antioxidant or pro-oxidant are: pH, phenolic compound concentration, presence of oxygen and copper ions (Eghbaliferiz & Iranshahi, 2016).

The reduced transition metals are characterized by better solubility and stability at an acidic pH, decreasing their availability in pro-oxidative reactions. Thus, conditions like high pH, favor pro-oxidant activity of phenolic compounds (Olszowy, 2019).

Several studies displayed that pro-oxidant compounds enhance intracellular ROS levels to cytotoxic conditions in cancer cells (Khamphio, Barusrux, & Weerapreeyakul, 2016; Kumar & Mandal, 2019), but not in normal cells, showing a selective behavior (de Mejía, Song, Heck, & Ramírez-Mares, 2010; Fidelis et al., 2020). This effect can be explained by the higher concentration of copper ions in malignant cells when compared with healthy cells. In other words, cancer cells are more sensitive to the pro-oxidant activities of phenolic compounds, generating $OH^{\bullet}$ radicals by the following reactions:

$$PhOH + Cu(II) \rightarrow PhO^{\bullet} + Cu\ (I) + H^{+} \quad (1)$$

$$PhO^{\bullet} + O_2 \rightarrow PhO^{+} + O_2^{\bullet -} \quad (2)$$

$$O_2^{\bullet -} + 2H^{+} \rightarrow H_2O_2 + O_2 \quad (3)$$

$$Cu\ (I) + H_2O_2 \rightarrow Cu(II) + OH + {}^{\bullet}OH \quad (4)$$

The ${}^{\bullet}OH$ radicals produced by Fenton-type reaction eventually leads to DNA damage and stimuli the apoptosis pathways, inducing cell death. Thus, the pro-oxidant of natural antioxidants plays an important role in their selectivity cytotoxicity toward cancer cells, due their copper content (Eghbaliferiz & Iranshahi, 2016). Moreover, since exceedingly high levels of ROS can induce cell death and cell cycle arrest, cancer cells with an increased rate of copper ions and ROS, are potentially more vulnerable to damage by further ROS insult (Fig. 3) (Glasauer & Chandel, 2014). In addition, elevated levels of ROS can active the C-Jun N-terminal kinase (JNK), a member of mitogen-activated protein kinases (MAPKs), which plays vital roles in regulating cell processes such as cell apoptosis, proliferation, migration, survival, differentiation and inflammation through stimulating a number of nuclear and non-nuclear molecules (Xu & Hu, 2020). Enhanced ROS production active the apoptosis signal-regulating kinase (ASK) family that stimuli protein kinase kinases (MKK4 and MKK7) to phosphorylate JNK. Once activated, JNK induces expression of apoptotic genes and activates specific mitochondrial apoptosis-related proteins such as Bcl-2 family members Bcl-2, Bad, Bim, and Bax, which trigger the cell death (Shen & Liu, 2006; Xu & Hu, 2020). Furthermore, ASK1 inactivates AKT pathways that mediate cell survival and proliferation, decreasing the survival factors (Oh & Mouradian, 2018) and pro-tumorigenic signaling (Fig. 3).

Indeed, the antioxidant action of phenolic compounds may effectively exhibit cytoprotective effect on normal cells by inhibiting neoplastic transformation. In contrast, when the tumor is already installed, these compounds could exert cytostatic activity by acting as antioxidant and cytotoxic effect by acting as pro-oxidant (Fig. 3). Finally, phenolic compounds can produce positive effects in several cancers, but whether an antioxidant supplement would be helpful, harmful or neutral depends on the specific antioxidant, its dose, the chemotherapy drugs being used, and the type and stage of cancer (do Carmo et al., 2018). Thus, despite all evidences, it is still a challenge to determine an effective and security dose of phenolic compound to treat oxidative stress diseases-related, once the main of studies perform *in vitro* assays. In any case, we must invest in prevention before the disease sets in, and the phenolic compounds are effective for health promotion, once they protect cells against free radical damage as well as help in reducing the risk of many chronic diseases (Martín & Ramos, 2017; Singh, Singh, Kaur, & Singh, 2020).

3. Biomolecular assays

Thinking about the above evidences, which point out the mediating role of ROS in cell life, stress and death, and that their different concentrations play distinct cell behavior therein, there are many *in vitro* biomolecular assays to evaluate the effect of phenolic compounds on cell proliferation and ROS generation (Table 1). These methods are an important tool for screening foods, phytochemicals, and dietary supplements for potential biological activity.

3.1 *In vitro* methods of intracellular ROS measurement

Despite the lacks in the measurement of ROS, many types of methods are available, like the use of fluorescent and chemiluminescent probes, spectrophotometry and spectrometry methods, as well as chromatography and genetically encoded fluorescent protein-based assays (Dikalov & Harrison, 2014; Zhang, Menghong, & Zonghui, 2018). The mainly methods are described below.

3.1.1 Fluorescent methods

Fluorescent probes are often oxidant sensitive, and they are non-fluorescent before being oxidized by oxygen species. Most of these probes are oxidized by a one-electron free radical mechanism, yielding a probe radical intermediate and resulting in fluorescent products (Zhang et al., 2018).

3.1.1.1 Dichlorodihydro fluorescein diacetate (DCFH-DA)

Based on the oxidation of DCFH-DA, this probe is generally used for the direct measurement of intracellular redox states and this method is one of the most used. It may react with several RONS including hydroxyl radicals ($^{\bullet}OH$), compounds I and II formed from peroxidase or heme interaction

Table 1 Features of ROS detection methods.

Methods	Probes	Specificity	Applications and references
Fluorimetric methods	DCFH-DA	Hydrogen peroxide, hydroxyl radicals, hypochlorous acid and peroxynitrite	Normal and cancer cell lines—epicatechin and methylvescalagin from seed camu-camu (Carmo et al., 2019); apigenin and cyanidin from petals of *Centaurea cyanus L.* (Escher et al., 2018); delfinidin, cyanidin, gallic acid and catechins (Fidelis et al., 2020); human intestinal cell line—chlorogenic acid and 3,5-O-dicaffeoylquinic acid from edible artichoke heads (Garbetta et al., 2014); liver cancer cell line—*p*-Coumaric acid and ferulic acid from *Ziziphus jujuba* (Shan et al., 2019)
	Amplex red	Hydrogen peroxide	Cancer cell lines—Silk sericin from the cocoons of *Bombyx mori*, *Antheraea assamensis* and *Philosamia ricini* (Kumar & Mandal, 2019); lung bronchial epithelial cells and fetal hepatic cells—proanthocyanidins (Thilakarathna & Rupasinghe, 2019); lung carcinoma cells—resveratrol (Whitehouse et al., 2016)
	DHE	Mainly superoxide anion	Breast cancer cells—isoquercitrin and epicatechin from hawthorn fruit peels and fleshes (Li et al., 2013); prostate cancer cells—curcumin, resveratrol, and silibinin (Rodriguez-Garcia et al., 2017); human spermatozoa—resveratrol, genistein and THP (Aitken et al., 2016)

Spectrophotometry methods	NBT	Superoxide anion	Lung carcinoma cells—resveratrol (Whitehouse et al., 2016); liver and paw in mice—phenolic acids and flavonoids (Sdayria et al., 2018); pre-adipocytes of the 3 T3-L1 cell line—anthocyanins from berry fruit extract (Kowalska, Olejnik, Szwajgier, & Olkowicz, 2017)
	Boronate	Hydrogen peroxide and peroxynitrite	Human gastric cancer cell line—epigallocatechin-3-gallate from green tea (Ishii et al., 2008); epigallocatechin-3-gallate from green tea (Tanaka et al., 2011)
Chemiluminescence-derived methods	Luminol	Superoxide anion, hydroxyl radical, hydrogen peroxide, and eroxynitrite	Human neutrophils—gallic acid from ethanolic extract of *T. gardneriana* seeds (Lopes Neto et al., 2020); hydroxytyrosol from olive oil (O'Dowd et al., 2004); retinal pigment epithelium—chlorogenic acid and caffeic acid from yerba mate leaves (Tate, Marazita, Marquioni-Ramella, & Suburo, 2020)
	Lucigenin	Superoxide anion	Human neutrophils—hydroxytyrosol from olive oil (O'Dowd et al., 2004); PC12 cell line—epigallocatechin-3-gallate from green tea (Raza & John, 2005); Liver cells—phenolic acids, anthocyanins, flavonols, and flavan-3-ols from Cranberry peel (Zavodnik et al., 2019)
Fluorescent protein-based methods	HyPer, rxRFP, cpYFP, roGFP	Hydrogen peroxide and redox status changes	Yeast cells —cyanidin (Ruta, Oprea, Popa, & Farcasanu, 2020); epigallocatechin gallate from green tea (Maeta, Nomura, Takatsume, Izawa, & Inoue, 2007)

with H_2O_2, $^{\bullet}NO_2$ formed from the myeloperoxidase/H_2O_2/NO_2^- system, hypochlorous acid (HOCl), and reactive species formed from peroxynitrite (ONOO−/ONOOH) decomposition (Kalyanaraman et al., 2012). The assay employs the cell-permeable fluorogenic probe DCFH-DA, which diffuses into cells and is deacetylated by cellular esterases into the non-fluorescent DCFH. In the presence of ROS, DCFH is rapidly oxidized to highly fluorescent DCF (Dikalov & Harrison, 2014; Wolfe & Liu, 2007; Zhang et al., 2018). However, the intermediate radical, $DCF^{\bullet-}$, formed from the one-electron oxidation of DCFH, rapidly reacts with O_2 ($k=108\,M^{-1}\,s^{-1}$) to form superoxide anion ($O_2^{\bullet-}$) and its dismutation produces additional H_2O_2. These high levels of hydrogen peroxide can begin a redox-cycling mechanism overestimating the fluorescence signal intensity and consequently, the RONS levels. Moreover, Cytochrome c, a heme protein that is released from mitochondria to the cytosol during apoptosis, is capable of oxidizing DCFH directly or indirectly *via* peroxidase-type pathways, generating DCF. The increase in DCF fluorescence that occurs during apoptosis of cells loaded with DCFH-DA has frequently been associated with enhanced oxidant production (Kalyanaraman et al., 2012). Considering the "dual-face" behavior of phenolic compounds, they can increase or decrease the intracellular ROS levels, acting as pro-oxidant or antioxidant agents, respectively (Fig. 4). Unavoidable limitations of this method, such as auto-oxidation, photo-oxidation, photo-bleaching, photo-conversion, cytotoxicity and non-specificity may interfere with the production of fluorescence and thus the detection of ROS (Yazdani, 2015).

3.1.1.2 Amplex red

The Amplex Red/horseradish peroxidase (AR/HRP) assay for H_2O_2 is one of the most sensitive and simple approaches for H_2O_2 quantification, which is effected by measuring the highly-fluorescent resorufin formed from oxidation of AR by the oxidizing intermediates generated by reaction of HRP and H_2O_2 (Fig. 5) (Wang, Miller, Wang, & Waite, 2017). As limitations of this method, Zhang et al. (2018) lists the following aspects: (1) HRP can react with electron donors, such as NADPH and reduced glutathione, to generate H_2O_2, resulting in the production of additional resorufin; (2) Resorufin could be reduced to a colorless, non-fluorescent compound(s) by NADPH-CYP450 reductases, followed by the generation of superoxide or even H_2O_2, posing problems for identification of the latter; (3) This probe shows low affinity to the cell membrane, resulting in its inability to measure intracellular H_2O_2 precisely; due to the diffusing nature of H_2O_2, values measured in

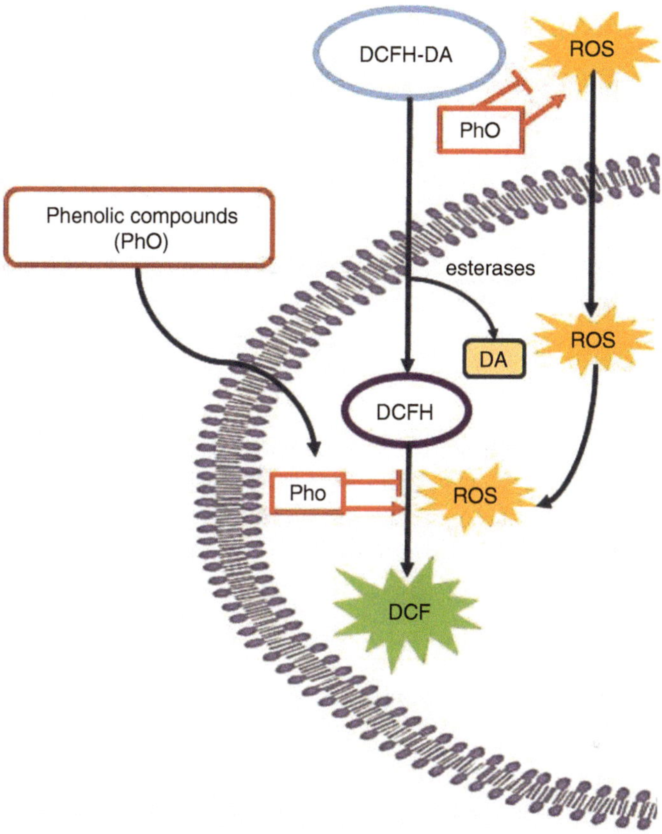

Fig. 4 Proposed principle of ROS measured by DCFH-DA fluorescent method. Phenolic compounds prevented or induce the oxidation of DCFH if they act as antioxidant or pro-oxidant agent, respectively.

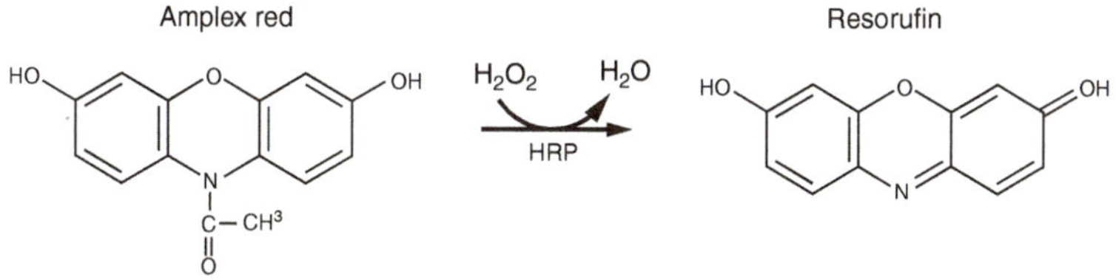

Fig. 5 Conversion of Amplex red into Resorufin.

extracellular buffers could serve as a reference for original sources in cells or tissues and (4) In various tissue samples like liver, kidney and cultured cells, Amplex Red could be transformed into resorufin by carboxylesterase without the requirement of H_2O_2, HRP or oxygen, implicating the necessity for a re-examination of similar assays.

3.1.1.3 Dihydroethidium (DHE)

The reaction of DHE is considered to be relatively specific for superoxide, with minimum reaction with hydrogen peroxide (Wilhelm, Vytasek, Adalova, & Ludek, 2009). DHE, the reduced form of ethidium bromide stain, is a probe that can mainly be oxidized by superoxide anions to form the specific red fluorescent 2-hydroxyethidium (2-OH-E+), or could be oxidized by other oxidants ($^{\bullet}OH$ or H_2O_2) to generate the non-specific red fluorescent ethidium (E^+). Both 2-OH-E^+ and E^+ can be determined indiscriminately by fluorescence-based methods. Thus, the probe DHE is an option for qualitative measurement of intracellular ROS (Dikalov & Harrison, 2014; Zhang et al., 2018). However, a freshly prepared DHE fluorescent probe has a light pink color, which darkens as it ages, eventually turning red due to auto-oxidation, which can interfere on sensitivity of assay through increased background fluorescence (Yazdani, 2015).

3.1.2 Spectrophotometry methods

The spectrophotometry methods work based on the reaction between radicals and redox substances, and the difference in absorbance at different wavelengths between the substrates and products makes it possible for the semi-quantification of free radicals (Zhang et al., 2018).

3.1.2.1 Nitroblue tetrazolium (NBT)

NBT is a yellow water-soluble nitro-substituted aromatic tetrazolium compound that reacts with cellular superoxide ions to form formazan derivative that can be monitored spectrophotometrically (Esfandiari, Sharma, Saleh, Thomas, & Agarwal, 2003).

NBT could be reduced by dehydrogenase and peroxidase, and it could also react with superoxide anion to generate water-insoluble diformazan (Zhang et al., 2018). Moreover, the cytoplasmic NADPH, which is produced by oxidation of glucose through the hexose monophosphate shunt, serves as an electron donor. The oxidase system available in the cytoplasm helps transfer electrons from NADPH to NBT and reduce NBT into formazan. Besides the method can lead to measurement errors due to the non-specific reduction of NBT by molecules other than superoxides (Javvaji et al., 2020) the NBT reaction indirectly reflects the intracellular ROS generation (Esfandiari et al., 2003).

3.1.2.2 Boronates

Very few small organic molecules stoichiometrically react with H_2O_2 to form a detectable intracellular fluorescent product (Kalyanaraman et al., 2012).

A palette of fluorescent and luminescent probes that react chemo selectively with hydrogen peroxide and peroxinitrite has been developed, utilizing a boronate oxidation trigger, forming a single stable phenolic product (Lin, Dickinson, & Chang, 2013). Red-fluorescent peroxyresorufin-1, green-fluorescent peroxyfluor-1, and blue-fluorescent peroxyxanthone-1 are a family of boronate-based probes, which fluoresce in the red, green, and blue ranges, respectively, upon exposure to peroxinitrite and H_2O_2 (Dikalov & Harrison, 2014). However, the rate constant for the reaction between boronates and H2O2 is relatively low ($0.1–1.0\,M^{-1}\,s^{-1}$). It is unlikely to compete with H_2O_2-detoxifying enzymes such as the peroxiredoxins, which react seven orders of magnitude faster with H_2O_2 (Kalyanaraman et al., 2012) (Fig. 6).

3.1.3 Chemiluminescence-derived methods

Chemiluminescence is defined as the emission of light when an electronically excited compound returns to the ground state after a chemical reaction. This property of chemiluminescence has been used in analytical chemistry to conduct sensitive assays without the need for expensive equipment. The most commonly used chemiluminescent compound in biological assays is luminol (Uy, Mcglashan, & Shaikh, 2011). Other chemiluminescent probes include lucigenin, MCLA, and coelenterazine (Dikalov & Harrison, 2014). These probes could react with superoxide anion to form a photon, which is captured by a photometer or scintillation counter without the requirement of an excitation light source (Zhang et al., 2018).

3.1.3.1 Luminol

Luminol detects global levels of ROS (superoxide anion, hydroxyl radical, hydrogen peroxide, and peroxynitrite) released from live cells, including spermatozoa (Uy et al., 2011). The luminol reaction cannot identify the

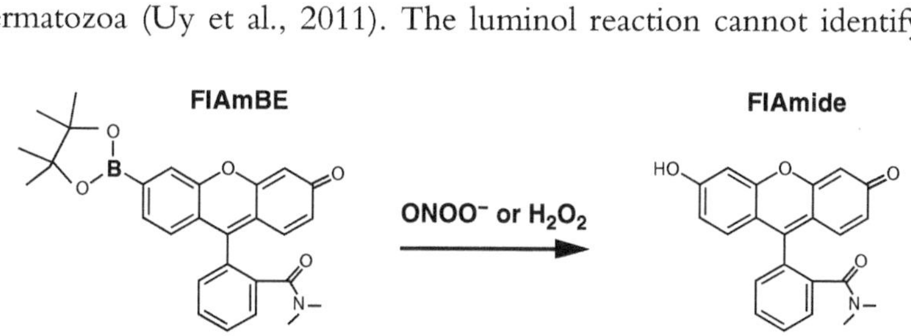

Fig. 6 Molecular structure and reactivity of boronate probe. *Source: Dikalov, S. I., & Harrison, D. G. (2014). Methods for detection of mitochondrial and cellular reactive oxygen species.* Antioxidants & Redox Signaling, *20(2), 372–382. https://doi.org/10.1089/ars.2012.4886.*

reactive species or provide the involved mechanisms of action. Thus, the assay has very little ability to discriminate between individual reactive oxygen species in biological systems (Winterbourn, 2014). There are several steps in the conversion of luminol to the last luminescent molecule, and the first oxidizing reaction involving its conversion to the intermediate radical ($LH^{\bullet}$) is not specific to superoxide. Oxidants such as $^{\bullet}OH$ and $CO_3^{\bullet -}$ seem to be more efficient in this reaction (Zhang et al., 2018).

3.1.3.2 Lucigenin

Lucigenin-chemiluminescence is a common method for measuring intracellular $O_2^{\bullet -}$ production by non-phagocytic cells due to its great cell membrane permeability, minimal cellular toxicity, high sensitivity and specificity of reaction with $O_2^{\bullet -}$, which makes lucigenin chemiluminescence particularly useful in the detection of low levels of $O_2^{\bullet -}$ (Fan & Li, 2014) produced by macrophages, neutrophils, as well as extracellular isolated enzymes (Zhang et al., 2018). The initial reaction to generate the lucigenin radical is carried out by cellular reductases so the assay is sensitive to changes in reductase activity (Winterbourn, 2014).

3.1.4 Fluorescent protein-based methods

Fluorescent protein-based redox probes are designed through the combination of fluorescent proteins (FPs) and prokaryotic redox-sensitive proteins (Zhang et al., 2018). Recent techniques have been performed for dynamic detection of ROS and redox cell status throughout the intracellular introduction of either plasmids or adenoviruses. HyPer, rxRFP, cpYFP, roGFP are examples of this kind of probe, which have been successfully targeted to various intracellular organelles, including the nucleus, the cytoplasm, peroxisomes, the mitochondrial intermembrane space, and the mitochondrial matrix (Dikalov & Harrison, 2014; Zhang et al., 2018). Despite its pioneering discovery of GFP, a certain limitation, such as pH sensitivity, chloride ion sensitivity, and poor photostability, were described (Kong et al., 2020). Hyper probe is also sensible to pH alterations (Oparka et al., 2016).

3.2 *In vitro* cell viability and antiproliferative assays

In vitro assays provide an initial stage for cancer phenolic compounds discovery approaches and a wide number of methods have been developed to evaluate antiproliferative properties. We have pointed out some commonly utilized *in vitro* techniques used to evaluate cell viability/proliferation in cancer phenolic compounds discovery (Table 2).

Table 2 More common antiproliferative methods.

Antiproliferative methods	Principle	References with several cell lines
Colorimetric assays	MTT	Normal and cancer cell lines—polyphenols from hydroalcoholic *Myrciaria dubia* (camu-camu) seed extract (Carmo et al., 2019); Dalton's lymphoma cells—*Ampelocissus latifolia* (Roxb.) Planch. aqueous extract (Chaudhuri & Ray, 2020); liver and colon cancer cell lines—*Centaurea cyanus* L. aqueous extract (Escher et al., 2018); colon cancer cell line—polyphenols from green, black tea and olive leaves (Jilani et al., 2020); liver and colon cancer cell lines—anthocyanins from red chicory (*Cichorium intybus*) aqueous extract (Migliorini et al., 2019); normal and cancer cell lines—Chinese keemun black tea grades (*Camellia sinensis* var. sisnensis) (Zhang et al., 2019)
	MTS	Breast cell line—synthetic resveratrol-curcumin hybrid (de Freitas Silva et al., 2018); neonatal rat ventricular cardiomyocytes (NRVCs)—resveratrol (Xu et al., 2019); chondrocytes—resveratrol (Gu et al., 2017); acute myeloid leucemia and peripheral blood mononuclear cells—*Azorella glabra* Wedd. extract (Lamorte et al., 2020)
	Trypan blue	Dalton's lymphoma cells—*Ampelocissus latifolia* (Roxb.) Planch. aqueous extract (Chaudhuri & Ray, 2020); colon cancer cell line—anthocyanins isolated from bilberries (*Vaccinium myrtillus*) and blackcurrant (*Ribes nigrum*) (Anwar et al., 2016); breast cancer cells and embryonic rat cardiac tissue-derived H9C2 cardiomyoblasts—curcumin (Jain & Rani, 2018); T-47D cell line—curcumin (Coker-Gurkan et al., 2019)

Continued

Table 2 More common antiproliferative methods.—cont'd

Antiproliferative methods	Principle	References with several cell lines
Clonogenic assay	Trypan blue or Crystal violet	Leukemia cell lines—polyphenols from etheric and methanolic extract of fruit of *Laserpitium Krapffii* Crantz (Bogucka-Kocka, Vorobets, Chrzaszcz, Pietrzak, & Szewczyk, 2019); human oral squamous cell carcinoma—earthworm coelomic fluid of *Eudrilus eugeniae*, *Eisenia foetida* and *Perionyx excavatus* (Augustine, Rao, Anbu, & Chidambara Murthy, 2019); human breast cancer cells—proanthocyanidins and resveratrol (Gao & Tollefsbol, 2018); colon cancer cell line—resveratrol (Pouyafar et al., 2019)
Real-time assay	Microelectrodes	Human soft-tissue sarcoma cell lines—curcumin and *Viscum album* extract (Harati, Behr, Daigeler, et al., 2017; Harati, Behr, Wallner, et al., 2017); normal rabbit kidney epithelial cells—*Taraxacum officinale* F.H. Wigg (dandelion) extract (Miłek et al., 2019); human soft-tissue sarcoma cell lines—epigallocatechin-3-gallte (Harati, Behr, Daigeler, et al., 2017; Harati, Behr, Wallner, et al., 2017); melanoma and colon cancer cell lines—*Ginkgo biloba* kernel extract (Feodorova et al., 2020)

3.2.1 Cell viability by colorimetric test

In most of the colorimetric viability/antiproliferative assays, a substrate is converted to a colored product by intracellular enzymes present in viable cells and the color developed is relatively proportional to viable cell number (Ediriweera, Tennekoon, & Samarakoon, 2019).

The MTT (3-(4,5-dimethylthiazol-2-yl)-2,5-diphenyltetrazolium bromide) and MTS (3-(4,5-dimethylthiazol-2-yl)-5-(3-carboxymethoxyphenyl)-2-(4-sulfophenyl)-2H-tetrazolium) assays are the most commonly used methods for antiproliferative screening. They are used to evaluate the mitochondrial functionality and indirectly determine the number of metabolic intact and alive cells, based on reduction of the tetrazolium ring of MTT/MTS by the succinate dehydrogenase enzyme in the mitochondrial respiratory chain, yielding purple formazan crystals (Ionta et al., 2015; Jilani, Cilla, Barberá, & Hamdi, 2020). However, many non-mitochondrial dehydrogenases or flavin oxidases can also reduce MTT. It is taken up by cells through endocytosis and that reduced MTT formazan accumulates in the endosomal/lysosomal compartment and is then transported to the cell surface through exocytosis (Liu, Peterson, Kimura, & Schubert, 1997). For MTT the formazan produced need to be solubilized for spectrophotometric analysis, while MTS do not require this step once its solubilization occurs in the cell medium itself (Ediriweera et al., 2019).

Some important parameters to assess the cell viability are used: IC_{50}, GI_{50} and LC_{50}. The IC_{50} is the concentration of the agent that inhibits growth by 50% at moment of treatment, where $(T/C) \times 100 = 50$, T = number of cells at treatment time; C = control cells at time t of treatment. The GI_{50} is the concentration of the agent that inhibits growth by 50%, taking into account the untreated cells and number of cells from start of treatment, where $([T - T0]/[C - T0]) \times 100 = 50$, T0 is the number of cells at time zero, T and C are the number of treated and control cells, respectively, at time t of treatment and $T > T0$. LC_{50} is the concentration of the agent that results in a net loss of 50% cells, relative to the number at the start of treatment, where $([T - T0]/T0) \times 100 = -50$; $T < T0$ (do Carmo et al., 2018).

Trypan blue (TB) assay is another very common assay for evaluating cytotoxicity in experimental investigations (Avelar-Freitas et al., 2014), used to distinguish the live and dead cells by differences in the color pattern (Chaudhuri & Ray, 2020). The intact cell membranes cannot recruit this dye into cells. In contrast, trypan blue can enter into non-viable cells (Ediriweera et al., 2019). Thus, dead cells absorb trypan blue into the cytoplasm because of loss of membrane selectivity, whereas live cells remain unstained. Then, the relative number of dead and live cells is obtained by optical microscopy by counting the number of stained (dead) and unstained (live) cells using a Neubauer chamber (Avelar-Freitas et al., 2014). Besides extensively used, this method it is considered a time-consuming assay and when used for a large number of samples it can provide low precision results. In addition, exposing live cells to TB for longer than 5 min can cause inappropriate uptake of the dye into viable cells and therefore result in overestimation of dead cells in a sample (Damiani, Solorio, Doyle, & Wallace, 2019).

3.2.2 Clonogenic assay

Clonogenic or colony formation assay is an *in vitro* cell survival method based on the ability of a single cell to grow into a colony, which is defined to consist of at least 50 cells. This technique can be used to determine the effectiveness of cytotoxic agents (Franken, Rodermond, Stap, Haveman, & Van Bree, 2006). Formed colonies are counted after staining with a suitable dye such as trypan blue and crystal violet. A microscope is usually used to visualize cell colonies and counting is normally carried out manually (Ediriweera et al., 2019). However, the clonogenic assay cannot measure impact of cell–cell communication on cell proliferation, once cells are seemed at low densities to form colonies. Moreover, this method is not appropriate when a substance declines growth without interfering on DNA synthesis and/or cell cycle progression. This test is not indicate for testing agents, which inhibit growth solely by causing cytoskeletal damage, or by inducing apoptosis (Sumantran, 2011).

3.2.3 Real-time assay for viable cells

A recently developed approach for assessing viable cell number in real time utilizes an engineered luciferase derived from a marine shrimp and a small molecule pro-substrate. The reagent components are added directly to cells in culture. Viable cells with an active metabolism reduce the pro-substrate to create a substrate for luciferase that generates light (Fig. 7). Dead cells absent metabolic activity do not generate luciferase substrate and thus do not produce a luminescent signal. The reagent is well tolerated by cells and is stable in complete cell culture medium at 37 °C for at least 72 h, which allows measurements from the same sample for days without replacing the pro-substrate (Sittampalam et al., 2020). Another system of real-time assay utilizes a series of micro wells, the bottoms of which are covered at 80%, with microelectrodes that measure the cell viability. This method monitories cell proliferation and morphology, using a dimensionless unit called the cell

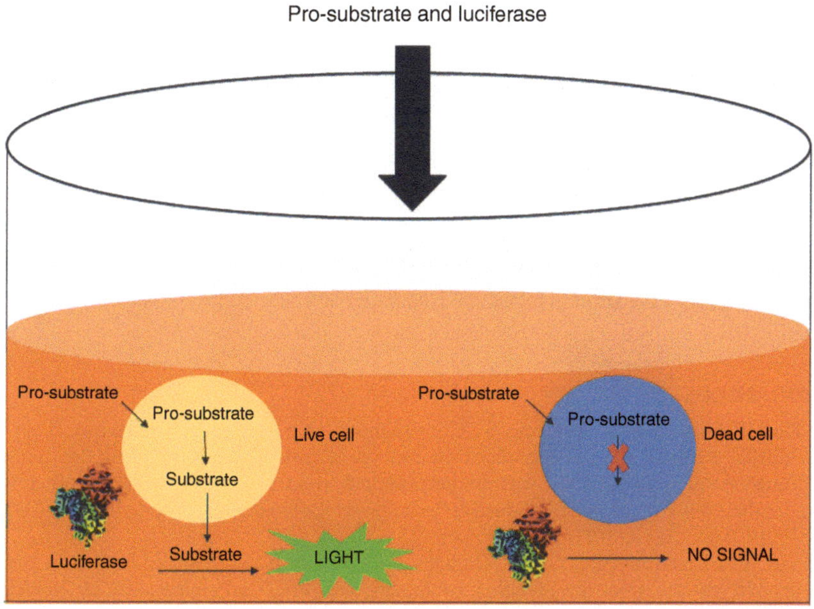

Fig. 7 Luminescence real-time assay. *Adapted from Sittampalam, G. S., Grossman, A., Brimacombe, K., Arkin, M., Auld, D., Austin, C. P., … James, L. (2020). Assay guidance manual.*

index that is based on the impedance variations caused by cells interacting with the microelectrodes (Miłek, Marcinčáková, & Legáth, 2019). The main disadvantage of real-time system is the cost of experiments. *E*-plates are expensive, single use, and disposable. However, some authors propose that E-plates can be regenerated and reused several times without significantly affecting experimental results (Stefanowicz-Hajduk & Ochocka, 2020).

Besides these antiproliferative methods, other *in vitro* assays are commonly used in discovering of compounds against cancer, in order to clarify the mechanisms and pathways involved on their cytotoxic potential, such as assays for apoptosis, cell migration and invasion, angiogenesis, cell cycle, cellular senescence and techniques to detect gene mutations and chromosomal aberrations.

4. Conclusions and perspectives

In this chapter, we pointed out the antiproliferative potential of phenolic compounds due their antioxidant and pro-oxidant activities. It is known that these compounds interfere numerous vital pathways and targets associated with antiproliferative effects, although, there is still few clinical trials with humans, testing their role in inhibiting tumor growth. The selection of the appropriate *in vitro* cell viability and ROS assays will depend largely on the availability in the laboratory where the study is to be performed, test compounds, detection mechanism, available resources, specificity, and sensitivity. Thus, *in vitro* methods may revealed their usefulness as an initial screening in understanding the mechanisms of potential antiproliferative compounds for the development of innovative food and chemical structures to prevent many diseases and offer multitude health benefits.

References

Aitken, R. J., Muscio, L., Whiting, S., Connaughton, H. S., Fraser, B. A., Nixon, B., et al. (2016). Analysis of the effects of polyphenols on human spermatozoa reveals unexpected impacts on mitochondrial membrane potential, oxidative stress and DNA integrity; implications for assisted reproductive technology. *Biochemical Pharmacology*, *121*, 78–96. https://doi.org/10.1016/j.bcp.2016.09.015.

Anwar, S., Fratantonio, D., Ferrari, D., Saija, A., Cimino, F., & Speciale, A. (2016). Berry anthocyanins reduce proliferation of human colorectal carcinoma cells by inducing caspase-3 activation and p21 upregulation. *Molecular Medicine Reports*, *14*(2), 1397–1403. https://doi.org/10.3892/mmr.2016.5397.

Augustine, D., Rao, R. S., Anbu, J., & Chidambara Murthy, K. N. (2019). In vitro cytotoxic and apoptotic induction effect of earthworm coelomic fluid of Eudrilus eugeniae, Eisenia foetida, and Perionyx excavatus on human oral squamous cell carcinoma-9 cell line. *Toxicology Reports*, *6*, 347–357. https://doi.org/10.1016/j.toxrep.2019.04.005.

Avelar-Freitas, B., Almeida, V., Pinto, M., Mourão, F., Massensini, A., Martins-Filho, O., et al. (2014). Trypan blue exclusion assay by flow cytometry. *Brazilian Journal of Medical and Biological Research*, *47*, 307–315.

Bogucka-Kocka, A., Vorobets, N., Chrzaszcz, M., Pietrzak, W., & Szewczyk, K. (2019). Polyphenol composition of extracts of the fruits of *Laserpitium krapffii* Crantz and their antioxidant and cytotoxic activity. *Antioxidants*, *8*(9), 363.

Carmo, M. A., Do, V., Fidelis, M., Pressete, C. G., Marques, M. J., Castro-Gamero, A. M., et al. (2019). Hydroalcoholic Myrciaria dubia (camu-camu) seed extracts prevent chromosome damage and act as antioxidant and cytotoxic agents. *Food Research International*, *125*(June), 108551. https://doi.org/10.1016/j.foodres.2019.108551.

Chaudhuri, A., & Ray, S. (2020). Cytotoxic and antiproliferative activities of aqueous extract from aerial parts of Ampelocissus latifolia (Roxb.) Planch. on Dalton's lymphoma cells. *Journal of King Saud University—Science*, *32*(3), 1978–1985. https://doi.org/10.1016/j.jksus.2020.01.046.

Chen, Z., Tian, R., She, Z., Cai, J., & Li, H. (2020). Role of oxidative stress in the pathogenesis of nonalcoholic fatty liver disease. *Free Radical Biology and Medicine*, *152*(February), 116–141. https://doi.org/10.1016/j.freeradbiomed.2020.02.025.

Coker-Gurkan, A., Bulut, D., Genc, R., Arisan, E. D., Obakan-Yerlikaya, P., & Palavan-Unsal, N. (2019). Curcumin prevented human autocrine growth hormone (GH) signaling mediated NF-κB activation and miR-183-96-182 cluster stimulated epithelial mesenchymal transition in T47D breast cancer cells. *Molecular Biology Reports*, *46*(1), 355–369. https://doi.org/10.1007/s11033-018-4479-y.

Damiani, E., Solorio, J. A., Doyle, A. P., & Wallace, H. M. (2019). How reliable are in vitro IC50 values? Values vary with cytotoxicity assays in human glioblastoma cells. *Toxicology Letters*, *302*, 28–34. https://doi.org/10.1016/j.toxlet.2018.12.004.

de Freitas Silva, M., Coelho, L. F., Guirelli, I. M., Pereira, R. M., Ferreira-Silva, G.Á., Graravelli, G. Y., et al. (2018). Synthetic resveratrol-curcumin hybrid derivative inhibits mitosis progression in estrogen positive MCF-7 breast cancer cells. *Toxicology In Vitro, 50*, 75–85. https://doi.org/10.1016/j.tiv.2018.02.020.

de Mejía, E. G., Song, Y. S., Heck, C. I., & Ramírez-Mares, M. (2010). Yerba mate tea (Ilex paraguariensis): Phenolics, antioxidant capacity and in vitro inhibition of colon cancer cell proliferation. *Journal of Functional Foods, 2*(1), 23–34. https://doi.org/10.1016/j.jff.2009.12.003.

Dikalov, S. I., & Harrison, D. G. (2014). Methods for detection of mitochondrial and cellular reactive oxygen species. *Antioxidants & Redox Signaling, 20*(2), 372–382. https://doi.org/10.1089/ars.2012.4886.

do Carmo, M. A. V., Pressete, C. G., Marques, M. J., Granato, D., & Azevedo, L. (2018). Polyphenols as potential antiproliferative agents: Scientific trends. *Current Opinion in Food Science, 24*, 26–35. https://doi.org/10.1016/J.COFS.2018.10.013.

Drzeżdżon, J., Jacewicz, D., & Chmurzyński, L. (2018). The impact of environmental contamination on the generation of reactive oxygen and nitrogen species—Consequences for plants and humans. *Environment International, 119*, 133–151. https://doi.org/10.1016/j.envint.2018.06.019.

Ediriweera, M. K., Tennekoon, K. H., & Samarakoon, S. R. (2019). In vitro assays and techniques utilized in anticancer drug discovery. *Journal of Applied Toxicology, 39*(1), 38–71. https://doi.org/10.1002/jat.3658.

Eghbaliferiz, S., & Iranshahi, M. (2016). Prooxidant activity of polyphenols, flavonoids, anthocyanins and carotenoids: Updated review of mechanisms and catalyzing metals. *Phytotherapy Research, 1391*(April), 1379–1391. https://doi.org/10.1002/ptr.5643.

Escher, G. B., Santos, J. S., Rosso, N. D., Marques, M. B., Azevedo, L., do Carmo, M. A. V., et al. (2018). Chemical study, antioxidant, anti-hypertensive, and cytotoxic/cytoprotective activities of Centaurea cyanus L. petals aqueous extract. *Food and Chemical Toxicology, 118*(April), 439–453. https://doi.org/10.1016/j.fct.2018.05.046.

Esfandiari, N., Sharma, R. K., Saleh, R. A., Thomas, A. J., Jr., & Agarwal, A. (2003). Utility of the nitroblue tetrazolium reduction test for seminal leukocytes and spermatozoa. *Journal of Andrology, 24*(6), 862–870.

Fan, L. M., & Li, J.-M. (2014). Evaluation of methods of detecting cell reactive oxygen species production for drug screening and cell cycle studies. *Journal of Pharmacological and Toxicological Methods, 70*(1), 40–47. https://doi.org/10.1016/j.vascn.2014.03.173.

Feodorova, Y., Tomova, T., Minchev, D., Turiyski, V., Draganov, M., & Argirova, M. (2020). Cytotoxic effect of Ginkgo biloba kernel extract on HCT116 and A2058 cancer cell lines. *Heliyon, 6*(9), e04941. https://doi.org/10.1016/j.heliyon.2020.e04941.

Fidelis, M., Vieira do Carmo, M. A., Azevedo, L., Cruz, T. M., Marques, M. B., Myoda, T., et al. (2020). Response surface optimization of phenolic compounds from jabuticaba (*Myrciaria cauliflora* [Mart.] O.Berg) seeds: Antioxidant, antimicrobial, antihyperglycemic, antihypertensive and cytotoxic assessments. *Food and Chemical Toxicology, 142*, 111439. https://doi.org/10.1016/j.fct.2020.111439.

Franken, N. A. P., Rodermond, H. M., Stap, J., Haveman, J., & Van Bree, C. (2006). Clonogenic assay of cells in vitro. *Nature Protocols, 1*(5), 2315–2319. https://doi.org/10.1038/nprot.2006.339.

Gao, Y., & Tollefsbol, T. O. (2018). Combinational proanthocyanidins and resveratrol synergistically inhibit human breast cancer cells and impact epigenetic–mediating machinery. *International Journal of Molecular Sciences, 19*(8), 1–18. https://doi.org/10.3390/ijms19082204.

Garbetta, A., Capotorto, I., Cardinali, A., D'Antuono, I., Linsalata, V., Pizzi, F., et al. (2014). Antioxidant activity induced by main polyphenols present in edible artichoke heads: Influence of in vitro gastro-intestinal digestion. *Journal of Functional Foods, 10*, 456–464. https://doi.org/10.1016/j.jff.2014.07.019.

Ge, L., Li, S.-P., & Lisak, G. (2020). Advanced sensing technologies of phenolic compounds for pharmaceutical and biomedical analysis. *Journal of Pharmaceutical and Biomedical Analysis*, *179*, 112913. https://doi.org/10.1016/j.jpba.2019.112913.

Glasauer, A., & Chandel, N. S. (2014). Targeting antioxidants for cancer therapy. *Biochemical Pharmacology*, *92*(1), 90–101. https://doi.org/10.1016/j.bcp.2014.07.017.

Griendling, K. K., Touyz, R. M., Zweier, J. L., Dikalov, S., Chilian, W., Chen, Y. R., et al. (2016). Measurement of reactive oxygen species, reactive nitrogen species, and redox-dependent signaling in the cardiovascular system: A scientific statement from the American Heart Association. *Circulation Research*, *119*(5), e39–e75. https://doi.org/10.1161/RES.0000000000000110.

Gu, H., Jiao, Y., Yu, X., Li, X., Wang, W., Ding, L., et al. (2017). Resveratrol inhibits the IL-1β-induced expression ofMMP-13 and IL-6 in human articular chondrocytes viaTLR4/MyD88-dependent and-independent signaling cascades. *International Journal of Molecular Medicine*, *39*(3), 734–740. https://doi.org/10.3892/ijmm.2017.2885.

Gupta, P. B., Pastushenko, I., Skibinski, A., & Blanpain, C. (2019). Phenotypic plasticity: driver of cancer initiation, progression, and therapy resistance. *Stem Cell*, *24*(1), 65–78. https://doi.org/10.1016/j.stem.2018.11.011.

Harati, K., Behr, B., Daigeler, A., Hirsch, T., Jacobsen, F., Renner, M., et al. (2017). Curcumin and Viscum album extract decrease proliferation and cell viability of soft-tissue sarcoma cells: An in vitro analysis of eight cell lines using real-time monitoring and colorimetric assays. *Nutrition and Cancer*, *69*(2), 340–351. https://doi.org/10.1080/01635581.2017.1263349.

Harati, K., Behr, B., Wallner, C., Daigeler, A., Hirsch, T., Jacobsen, F., et al. (2017). Anti-proliferative activity of epigallocatechin-3-gallate and silibinin on soft tissue sarcoma cells. *Molecular Medicine Reports*, *15*(1), 103–110. https://doi.org/10.3892/mmr.2016.5969.

Harris, I. S., & DeNicola, G. M. (2020). The complex interplay between antioxidants and ROS in cancer. *Trends in Cell Biology*, *30*(6), 440–451. https://doi.org/10.1016/j.tcb.2020.03.002.

Hider, R. C., Liu, Z. D., & Khodr, H. H. (2001). Metal chelation of polyphenols. *Methods in Enzimology*, *335*, 190–203. https://doi.org/10.1016/S0076-6879(01)35243-6.

Hu, K., Liu, Y.-L., Oleinick, A., Mirkin, M. V., Huang, W.-H., & Amatore, C. (2020). Nanoelectrodes for intracellular measurements of reactive oxygen and nitrogen species in single living cells. *Current Opinion in Electrochemistry*, *22*, 44–50. https://doi.org/10.1016/j.coelec.2020.04.003.

Ionta, M., Ferreira-Silva, G. A., Niero, E. L., Costa, É. D. M., Martens, A. A., Rosa, W., et al. (2015). 7-Epiclusianone, a benzophenone extracted from Garcinia brasiliensis (Clusiaceae), induces cell cycle arrest in G1/S transition in A549 cells. *Molecules*, *20*(7), 12804–12816. https://doi.org/10.3390/molecules200712804.

Ishii, T., Mori, T., Tanaka, T., Mizuno, D., Yamaji, R., Kumazawa, S., et al. (2008). Covalent modification of proteins by green tea polyphenol (−)-epigallocatechin-3-gallate through autoxidation. *Free Radical Biology and Medicine*, *45*(10), 1384–1394. https://doi.org/10.1016/j.freeradbiomed.2008.07.023.

Jain, A., & Rani, V. (2018). Assessment of herb-drug synergy to combat doxorubicin induced cardiotoxicity. *Life Sciences*, *205*(December 2017), 97–106. https://doi.org/10.1016/j.lfs.2018.05.021.

Javvaji, P. K., Dhali, A., Francis, J. R., Kolte, A. P., Mech, A., Roy, S. C., et al. (2020). An efficient nitroblue tetrazolium staining and bright-field microscopy based method for detecting and quantifying intracellular reactive oxygen species in oocytes, cumulus cells and embryos. *Frontiers in Cell and Developmental Biology*, *8*(August), 1–12. https://doi.org/10.3389/fcell.2020.00764.

Jilani, H., Cilla, A., Barberá, R., & Hamdi, M. (2020). Antiproliferative activity of green, black tea and olive leaves polyphenols subjected to biosorption and in vitro gastrointestinal digestion in Caco-2 cells. *Food Research International, 136*(May), 109317. https://doi.org/10.1016/j.foodres.2020.109317.

Kalyanaraman, B., Darley-Usmar, V., Davies, K. J. A., Dennery, P. A., Forman, H. J., Grisham, M. B., et al. (2012). Measuring reactive oxygen and nitrogen species with fluorescent probes: Challenges and limitations. *Free Radical Biology and Medicine, 52*(1), 1–6. https://doi.org/10.1016/j.freeradbiomed.2011.09.030.

Khamphio, M., Barusrux, S., & Weerapreeyakul, N. (2016). Sesamol induces mitochondrial apoptosis pathway in HCT116 human colon cancer cells via pro-oxidant effect. *Life Sciences, 158*, 46–56. https://doi.org/10.1016/j.lfs.2016.06.017.

Kong, J., Wang, Y., Qi, W., Huang, M., Su, R., & He, Z. (2020). Green fluorescent protein inspired fluorophores. *Advances in Colloid and Interface Science, 285*, 102286. https://doi.org/10.1016/j.cis.2020.102286.

Kowalska, K., Olejnik, A., Szwajgier, D., & Olkowicz, M. (2017). Inhibitory activity of chokeberry, bilberry, raspberry and cranberry polyphenol-rich extract towards adipogenesis and oxidative stress in differentiated 3T3-L1 adipose cells. *PLoS One, 12*(11), 1–15. https://doi.org/10.1371/journal.pone.0188583.

Kumar, J. P., & Mandal, B. B. (2019). Silk sericin induced pro-oxidative stress leads to apoptosis in human cancer cells. *Food and Chemical Toxicology, 123*, 275–287. https://doi.org/10.1016/j.fct.2018.10.063.

Lage, N. N., Layosa, M. A. A., Arbizu, S., Chew, B. P., Pedrosa, M. L., Mertens-Talcott, S., et al. (2020). Dark sweet cherry (Prunus avium) phenolics enriched in anthocyanins exhibit enhanced activity against the most aggressive breast cancer subtypes without toxicity to normal breast cells. *Journal of Functional Foods, 64*, 103710. https://doi.org/10.1016/j.jff.2019.103710.

Lamorte, D., Faraone, I., Laurenzana, I., Trino, S., Russo, D., Rai, D. K., et al. (2020). Advances in Azorella glabra Wedd. Extract research: In vitro antioxidant activity, antiproliferative effects on acute myeloid leukemia cells and bioactive compound characterization. *Molecules (Basel, Switzerland), 25*(21), 4890. https://doi.org/10.3390/molecules25214890.

Li, T., Zhu, J., Guo, L., Shi, X., Liu, Y., & Yang, X. (2013). Differential effects of polyphenols-enriched extracts from hawthorn fruit peels and fleshes on cell cycle and apoptosis in human MCF-7 breast carcinoma cells. *Food Chemistry, 141*(2), 1008–1018. https://doi.org/10.1016/j.foodchem.2013.04.050.

Lin, V. S., Dickinson, B. C., & Chang, C. J. (2013). Chapter two—Boronate-based fluorescent probes: Imaging hydrogen peroxide in living systems. In E. Cadenas, & L. Packer (Eds.), *Hydrogen peroxide and cell signaling, Part A: Vol. 526. Methods in Enzymology* (pp. 19–43). https://doi.org/10.1016/B978-0-12-405883-5.00002-8.

Liu, Y., Peterson, D. A., Kimura, H., & Schubert, D. (1997). Mechanism of cellular 3-(4,5-dimethylthiazol-2-yl)-2,5-diphenyltetrazolium bromide (MTT) reduction. *Journal of Neurochemistry, 69*(2), 581.

Lopes Neto, J. J., Silva de Almeida, T., Almeida Filho, L. C. P., Rocha, T. M., Nogara, P. A., Nogara, K. F., et al. (2020). Triplaris gardneriana seeds extract exhibits in vitro anti-inflammatory properties in human neutrophils after oxidative treatment. *Journal of Ethnopharmacology, 250*, 112474. https://doi.org/10.1016/j.jep.2019.112474.

Maeta, K., Nomura, W., Takatsume, Y., Izawa, S., & Inoue, Y. (2007). Green tea polyphenols function as prooxidants to activate oxidative-stress-responsive transcription factors in yeasts. *Applied and Environmental Microbiology, 73*(2), 572–580. https://doi.org/10.1128/AEM.01963-06.

Martín, M.Á., & Ramos, S. (2017). Health beneficial effects of cocoa phenolic compounds: A mini-review. *Current Opinion in Food Science, 14*, 20–25. https://doi.org/10.1016/j.cofs.2016.12.002.

Migliorini, A. A., Piroski, C. S., Daniel, T. G., Cruz, T. M., Escher, G. B., do Carmo, M. A. V., et al. (2019). Red chicory (Cichorium intybus) extract rich in anthocyanins: Chemical stability, antioxidant activity, and antiproliferative activity in vitro. *Journal of Food Science, 84*(5), 990–1001. https://doi.org/10.1111/1750-3841.14506.

Miłek, M., Marcinčáková, D., & Legáth, J. (2019). Polyphenols content, antioxidant activity, and cytotoxicity assessment of taraxacum officinale extracts prepared through the micelle-mediated extraction method. *Molecules, 24*(6), 1–14. https://doi.org/10.3390/molecules24061025.

Morry, J., Ngamcherdtrakul, W., & Yantasee, W. (2017). Oxidative stress in cancer and fibrosis: Opportunity for therapeutic intervention with antioxidant compounds, enzymes, and nanoparticles. *Redox Biology, 11*(December 2016), 240–253. https://doi.org/10.1016/j.redox.2016.12.011.

O'Dowd, Y., Driss, F., Dang, P. M. C., Elbim, C., Gougerot-Pocidalo, M. A., Pasquier, C., et al. (2004). Antioxidant effect of hydroxytyrosol, a polyphenol from olive oil: Scavenging of hydrogen peroxide but not superoxide anion produced by human neutrophils. *Biochemical Pharmacology, 68*(10), 2003–2008. https://doi.org/10.1016/j.bcp.2004.06.023.

Oh, S. E., & Mouradian, M. M. (2018). Cytoprotective mechanisms of DJ-1 against oxidative stress through modulating ERK1/2 and ASK1 signal transduction. *Redox Biology, 14*, 211–217. https://doi.org/10.1016/j.redox.2017.09.008.

Olszowy, M. (2019). What is responsible for antioxidant properties of polyphenolic compounds from plants? *Plant Physiology and Biochemistry, 144*, 135–143. https://doi.org/10.1016/j.plaphy.2019.09.039.

Oparka, M., Walczak, J., Malinska, D., van Oppen, L. M. P. E., Szczepanowska, J., Koopman, W. J. H., et al. (2016). Quantifying ROS levels using CM-H2DCFDA and HyPer. *Methods, 109*, 3–11. https://doi.org/10.1016/j.ymeth.2016.06.008.

Pouyafar, A., Rezabakhsh, A., Rahbarghazi, R., Heydarabad, M. Z., Shokrollahi, E., Sokullu, E., et al. (2019). Treatment of cancer stem cells from human colon adenocarcinoma cell line HT-29 with resveratrol and sulindac induced mesenchymal-endothelial transition rate. *Cell and Tissue Research, 376*, 377–388. https://doi.org/10.1007/s00441-019-02998-9.

Raza, H., & John, A. (2005). Green tea polyphenol epigallocatechin-3-gallate differentially modulates oxidative stress in PC12 cell compartments. *Toxicology and Applied Pharmacology, 207*(3), 212–220. https://doi.org/10.1016/j.taap.2005.01.004.

Rodriguez-Garcia, A., Hevia, D., Mayo, J. C., Gonzalez-Menendez, P., Coppo, L., Lu, J., et al. (2017). Thioredoxin 1 modulates apoptosis induced by bioactive compounds in prostate cancer cells. *Redox Biology, 12*, 634–647. https://doi.org/10.1016/j.redox.2017.03.025.

Ruta, L. L., Oprea, E., Popa, C. V., & Farcasanu, I. C. (2020). Saccharomyces cerevisiae cells lacking transcription factors Skn7 or Yap1 exhibit different susceptibility to cyanidin. *Heliyon, 6*(10), e05352. https://doi.org/10.1016/j.heliyon.2020.e05352.

Santos, J. S., Alvarenga Brizola, V. R., & Granato, D. (2017). High-throughput assay comparison and standardization for metal chelating capacity screening: A proposal and application. *Food Chemistry, 214*, 515–522. https://doi.org/10.1016/J.FOODCHEM.2016.07.091.

Sdayria, J., Rjeibi, I., Feriani, A., Ncib, S., Bouguerra, W., Hfaiedh, N., et al. (2018). Chemical composition and antioxidant, analgesic, and anti-inflammatory effects of methanolic extract of Euphorbia retusa in mice. *Pain Research and Management, 2018*, 4838413. https://doi.org/10.1155/2018/4838413.

Shan, S., Xie, Y., Zhao, H., Niu, J., Zhang, S., Zhang, X., et al. (2019). Bound polyphenol extracted from jujube pulp triggers mitochondria-mediated apoptosis and cell cycle arrest of HepG2 cell in vitro and in vivo. *Journal of Functional Foods*, *53*(December 2018), 187–196. https://doi.org/10.1016/j.jff.2018.12.017.

Shen, H.-M., & Liu, Z. (2006). JNK signaling pathway is a key modulator in cell death mediated by reactive oxygen and nitrogen species. *Free Radical Biology and Medicine*, *40*(6), 928–939. https://doi.org/10.1016/j.freeradbiomed.2005.10.056.

Singh, B., Singh, J. P., Kaur, A., & Singh, N. (2020). Phenolic composition, antioxidant potential and health benefits of citrus peel. *Food Research International*, *132*, 109114. https://doi.org/10.1016/j.foodres.2020.109114.

Sittampalam, G. S., Grossman, A., Brimacombe, K., Arkin, M., Auld, D., Austin, C. P., et al. (2020). *Assay guidance manual*. National Center for Advancing Translational Sciences.

Stefanowicz-Hajduk, J., & Ochocka, J. R. (2020). Real-time cell analysis system in cytotoxicity applications: Usefulness and comparison with tetrazolium salt assays. *Toxicology Reports*, 7(February), 335–344. https://doi.org/10.1016/j.toxrep.2020.02.002.

Sumantran, V. N. (2011). Cellular chemosensitivity assays: An overview. *Methods in Molecular Biology*, *731*, 219–236. https://doi.org/10.1007/978-1-61779-80-5_19.

Tanaka, T., Ishii, T., Mizuno, D., Mori, T., Yamaji, R., Nakamura, Y., et al. (2011). (−)-Epigallocatechin-3-gallate suppresses growth of AZ521 human gastric cancer cells by targeting the DEAD-box RNA helicase p68. *Free Radical Biology and Medicine*, *50*(10), 1324–1335. https://doi.org/10.1016/j.freeradbiomed.2011.01.024.

Tate, P. S., Marazita, M. C., Marquioni-Ramella, M. D., & Suburo, A. M. (2020). Ilex paraguariensis extracts and its polyphenols prevent oxidative damage and senescence of human retinal pigment epithelium cells. *Journal of Functional Foods*, *67*(February), 103833. https://doi.org/10.1016/j.jff.2020.103833.

Thilakarathna, W. P. D. W., & Rupasinghe, H. P. V. (2019). Microbial metabolites of proanthocyanidins reduce chemical carcinogen-induced DNA damage in human lung epithelial and fetal hepatic cells in vitro. *Food and Chemical Toxicology*, *125*(January), 479–493. https://doi.org/10.1016/j.fct.2019.02.010.

Uy, B., Mcglashan, S. R., & Shaikh, S. B. (2011). Measurement of reactive oxygen species in the culture media using acridan lumigen PS-3 assay. *Journal of Biomolecular Techniques*, 95–107.

Valko, M., Rhodes, C. J., Moncol, J., Izakovic, M., & Mazur, M. (2006). Free radicals, metals and antioxidants in oxidative stress-induced cancer. *Chemico-Biological Interactions*, *160*, 1–40. https://doi.org/10.1016/j.cbi.2005.12.009.

Wang, N., Miller, C. J., Wang, P., & Waite, T. D. (2017). Quantitative determination of trace hydrogen peroxide in the presence of sulfide using the Amplex red/horseradish peroxidase assay. *Analytica Chimica Acta*, *963*, 61–67. https://doi.org/10.1016/j.aca.2017.02.033.

Whitehouse, S., Chen, P. L., Greenshields, A. L., Nightingale, M., Hoskin, D. W., & Bedard, K. (2016). Resveratrol, piperine and apigenin differ in their NADPH-oxidase inhibitory and reactive oxygen species-scavenging properties. *Phytomedicine*, *23*(12), 1494–1503. https://doi.org/10.1016/j.phymed.2016.08.011.

Wilhelm, J., Vytasek, R., Adalova, I., & Ludek, V. (2009). Evaluation of different methods detecting intracellular generation of free radicals. *Molecular and Cellular Biochemistry*, *328*, 167–176. https://doi.org/10.1007/s11010-009-0086-5.

Winterbourn, C. C. (2014). The challenges of using fluorescent probes to detect and quantify specific reactive oxygen species in living cells. *Biochimica et Biophysica Acta - General Subjects*, *1840*(2), 730–738. https://doi.org/10.1016/j.bbagen.2013.05.004.

Wolfe, K. L., & Liu, R. H. (2007). Cellular antioxidant activity (CAA) assay for assessing antioxidants, foods, and dietary supplements. *Journal of Agricultural and Food Chemistry*, *55*(22), 8896–8907. https://doi.org/10.1021/jf0715166.

Xu, H., Cheng, J., Wang, X., Liu, H., Wang, S., Wu, J., et al. (2019). Resveratrol pretreatment alleviates myocardial ischemia/reperfusion injury by inhibiting STIM1-mediated intracellular calcium accumulation. *Journal of Physiology and Biochemistry*, *75*(4), 607–618. https://doi.org/10.1007/s13105-019-00704-5.

Xu, R., & Hu, J. (2020). The role of JNK in prostate cancer progression and therapeutic strategies. *Biomedicine & Pharmacotherapy*, *121*, 109679. https://doi.org/10.1016/j.biopha.2019.109679.

Yazdani, M. (2015). Concerns in the application of fluorescent probes DCDHF-DA, DHR 123 and DHE to measure reactive oxygen species in vitro. *Toxicology In Vitro*, *30*(1), 578–582. https://doi.org/10.1016/j.tiv.2015.08.010.

Zavodnik, I., Buko, V., Lukivskaya, O., Lapshina, E., Ilyich, T., Belonovskaya, E., et al. (2019). Cranberry (Vaccinium macrocarpon) peel polyphenol-rich extract attenuates rat liver mitochondria impairments in alcoholic steatohepatitis in vivo and after oxidative treatment in vitro. *Journal of Functional Foods*, *57*(April), 83–94. https://doi.org/10.1016/j.jff.2019.04.001.

Zhang, Y., Menghong, D., & Zonghui, Y. (2018). Methods for the detection of reactive oxygen. *Analytical Methods*, *10*, 4625–4638. https://doi.org/10.1039/c8ay01339j.

Zhang, L., Santos, J. S., Cruz, T. M., Marques, M. B., do Carmo, M. A. V., Azevedo, L., et al. (2019). Multivariate effects of Chinese keemun black tea grades (*Camellia sinensis var. sinensis*) on the phenolic composition, antioxidant, antihemolytic and cytotoxic/cytoprotection activities. *Food Research International*, *125*, 108516. https://doi.org/10.1016/j.foodres.2019.108516.

CHAPTER EIGHT

Protein-polyphenol conjugates: Preparation, functional properties, bioactivities and applications in foods and nutraceuticals

Soottawat Benjakul*, Avtar Singh, Lalita Chotphruethipong, and Ajay Mittal
International Center of Excellence in Seafood Science and Innovation, Faculty of Agro-Industry, Prince of Songkla University, Hat Yai, Songkhla, Thailand
*Corresponding author: e-mail address: soottawat.b@psu.ac.th

Contents

Advances in Food and Nutrition Research, Volume 98
ISSN 1043-4526
https://doi.org/10.1016/bs.afnr.2021.02.011

Abstract

Protein is a crucial nutritional ingredient in the daily human diet. Polyphenols (PPNs) are the abundant phytochemicals in plants, which are associated with health promotion as well as affect functionality in food systems. Both ingredients possess different types of functionalities (crosslinking, gelling, emulsifying, film-forming, etc.) and bioactivities (antioxidant, antimicrobial, anti-inflammatory, etc.). In the past decade, various methods have been implemented to enhance the functionalities and bioactivities of foods. Conjugation or grafting methods has been introduced widely. Conjugations of PPNs with proteins through various methods have been performed for the synthesis of the protein-polyphenol conjugate. Those potential grafting methods are alkaline associated, free-radical mediated, enzyme catalyzed, and chemical coupling methods. Several factors such as reaction conditions, type of proteins, and PPNs also influenced the conjugation efficiency. Various technologies, e.g., mass spectroscopy, fluorescence spectroscopy, UV spectroscopy, Fourier transform infrared spectroscopy, circular dichroism, and sodium dodecyl sulfate polyacrylamide gel electrophoresis have been used to elucidate conjugation and structural alternation of proteins and some properties of resulting conjugates. The prepared protein-PPN conjugates have been documented to enhance the bioactivities and functional properties of an initial protein. Moreover, conjugates have been employed in emulsions or as nanoparticles for nutraceutical delivery. Edible-films for food packaging and hydrogels for controlled drug release have been developed using protein-PPN conjugates. This chapter focuses on the methodologies and characteristics of protein-PPN conjugates and their applications in various food systems and nutraceutical field.

1. Introduction

Grafting or conjugation is gaining interest in various foods, particularly in health sectors as it is known to improve various functional and structural properties as well as bioactivities of compounds of interest (Quan, Benjakul, Sae-leaw, Balange, & Maqsood, 2019). Several biomolecules such as proteins, chitosan, chitooligosaccharides, etc. have been modified via grafting or conjugation with polyphenols (PPNs) such as gallic acid, catechin, epigallocatechin gallate (EGCG), tannic acid etc. (Cho, Kim, Ahn, & Je, 2011; Feng, Cai, Wang, Li, & Liu, 2018; Gu et al., 2017; Quan et al., 2019; Singh, Benjakul, Huda, Xu, & Wu, 2020). Protein is a major nutritional and functional component of several foods. Protein structure and its functionalities are mainly dependent on amino acid compositions and sequences (Quan et al., 2019). Native forms of proteins possessed the impaired functional properties such as solubility, gelling, emulsifying, and foaming properties. Therefore, modification of various functional groups of proteins is performed to improve those functional properties.

PPNs are natural phytochemicals containing aromatic-ring, occupied with one or more OH-groups. PPNs having antioxidant activity exhibit outstanding potential to prevent or retard the oxidative damages caused by reactive oxygen species (ROS) or free-radicals (Liu, Sun, Yang, Yuan, & Gao, 2015; Quan et al., 2019). The intake of foods rich in PPNs has been known to curb oxidative stress related to health problems or diseases, e.g., heart-related disease, diabetes, and cancer. This is well related to their excellent antioxidant, antimicrobial, anti-inflammatory, anticancer activities, etc. (Xiao & Kai, 2012). Several biomolecules, including proteins or peptides, have been conjugated to augment their various functional properties and bioactivities (Quan et al., 2019; Zhao & Sun, 2017).

Therefore, this book chapter will mainly cover the information on techniques used for conjugation of proteins or peptides with PPNs. Functional properties and bioactivities of the protein-PPN conjugates or peptides-PPN conjugates and their uses are revisited (Fig. 1).

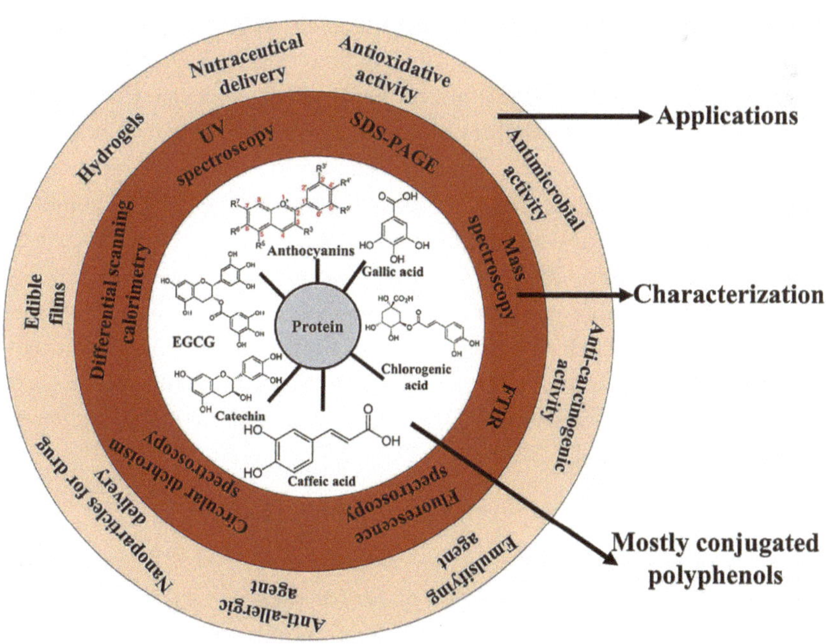

Fig. 1 Overview of protein-polyphenol conjugate.

2. Role of proteins and PPNs in foods and health

2.1 Proteins and peptides

The nutritional and functional properties of food proteins are affected by their sources, extraction/purification methods, modification during processing, and their interactions with other components such as carbohydrates, metals, lipids, etc. (Loveday, 2019). Proteins also contribute to texture, color, flavor, and other properties of foods (Vaclavik & Christian, 2008). Proteins have been utilized widely as thickening, binding, gelling, emulsifying, and foaming agents (Quan et al., 2019; Singh & Benjakul, 2017b; Singh, Benjakul, & Karnjanapratum, 2019; Vaclavik & Christian, 2008). The functional properties of a protein are governed by amino acid composition and sequences, which determine conformational characteristics of the proteins (Singh, Benjakul, & Kishimura, 2017; Vaclavik & Christian, 2008). Moreover, chemical composition and environmental factor affect solubility, gelling, viscoelasticity, and foaming properties of food proteins (Singh & Benjakul, 2019). Therefore, modification of proteins has been made via partial or complete hydrolysis to alter or amend the native conformation, in which their functional properties, as well as various bioactivities, can be improved (Guo, Zhang, Ma, & Tian, 2013; Singh & Benjakul, 2019; Zeng et al., 2013). Hydrolysis results in the liberation of bioactive peptides containing 2–20 amino acids having enhanced antioxidant, antimicrobial, anticancer activities etc. than the initial protein (Chakrabarti, Guha, & Majumder, 2018; Thiansilakul, Benjakul, & Shahidi, 2007). Protein hydrolysates or peptides commonly possess an improved solubility; however, they have lower foaming, emulsifying, and gelling properties (Singh & Benjakul, 2019; Wouters, Rombouts, Fierens, Brijs, & Delcour, 2016). Moreover, protein hydrolysates are known to possess higher bioactivities such as wound healing, skin nourishment, antioxidant, anti-inflammatory, anticancer activities, etc. than their native form (Chotphruethipong, Sukketsiri, Battino, & Benjakul, 2021; Tkaczewska et al., 2020; Wen et al., 2020).

2.2 Polyphenols (PPNs)

In general, PPNs are synthesized by plants for their various defense mechanisms and they serve as bioactive compounds regularly applied in therapeutics, cosmetics, and food industries (González-Minero, Bravo-Díaz, & Ayala-Gómez, 2020; Jimenez-Lopez et al., 2020). PPNs serve as an alternative to synthetic compounds, that are harmful to human as well as the environment. Plant PPNs consist of simple flavonoids, tannins, phenols, phenolic acids, coumarins, lignans, and lignin. Through the shikimate pathway from L-phenylalanine and L-tyrosine, they are synthesized and contain one or more OH-groups attached to aromatic-ring (Chirinos et al., 2009; Kumar & Goel, 2019). Their origin, extraction, characterization, biosynthesis as well as applications have been thoroughly reviewed by Kumar and Goel (2019). PPNs can affect flavor, astringency, texture, color, sensory qualities, extend shelf-life as well as augment antioxidant properties in various foods (Chotphruethipong & Benjakul, 2019; Nilsuwan, Guerrero, de la Caba, Benjakul, & Prodpran, 2020; Sae-leaw & Benjakul, 2019; Shiekh & Benjakul, 2020a, 2020b). Apart from food applications, PPNs such as hydroxybenzoic and hydroxycinnamic acids and their derivatives have been used for cancer treatment (Huang, Cai, & Zhang, 2009; Kumar et al., 2019; Kumar & Goel, 2019; Kumar, Goel, Yadav, & Pruthi, 2017). Additionally, diabetes associated with oxidative stress has been known to be controlled by the PPNs. Antioxidant activity of PPNs could reduce the production of reactive oxygen species (ROS) or inhibiting enzymes responsible for ROS production (Furukawa et al., 2017; Wu, Ding, Tanaka, & Zhang, 2014). Moreover, PPNs have the excellent efficiency in reducing insulin content and inhibiting α-glucosidase or α-amylase, the major factors to enhance diabetes (Gandhi et al., 2014; Hanhineva et al., 2010; Jung, Lee, Park, Jeon, & Choi, 2006; Prabhakar & Doble, 2011; Vinayagam, Jayachandran, & Xu, 2016). Therefore, considering the advantage of each component and their limitation, proteins and PPNs have been combined via grafting or conjugation to gain the maximal benefits.

3. Principle and methods of conjugation/grafting between proteins and PPNs

Protein and PPNs conjugation are based on non-covalent (hydrophobic, hydrogen or ionic bonding, etc.) or covalent bonding (Quan et al., 2019). These non-covalent and covalent bondings between proteins and PPNs have been introduced with help of non-enzymatic or chemical (alkaline reaction, free-radical grafting or chemical coupling) or enzymatic (polyphenoloxidase, tyrosinase, laccase, etc.) reactions (Gu, Peng, et al., 2017; Liu et al., 2015; Quan et al., 2019) (Table 1).

Table 1 Methods of preparation, characterization and applications of protein-polyphenol conjugates.

Proteins	Polyphenols	Conjugation methods	Structural characterization techniques	Functional properties/ applications/bioactivities	References
Egg white protein	Tea polyphenols	Ultrasound generated free radical and alkaline	FTIR	Improved antioxidant activity	Jing et al. (2020)
	Catechin and chlorogenic acid	Free radical	Liquid chromatography (LC)-MS/MS analysis	Improved emulsifying stability	Sun et al. (2020)
	Catechin and EGCG	Alkaline	Surface potential, fluorescence, and circular dichroism (CD) spectroscopy	Improved emulsifying stability	Sun et al. (2020)
Selenium-biofortified soybean protein	Gallic acid and EGCG	Alkaline	FTIR, UV, fluorescence, and CD spectroscopy	–	Zhang, He, and Hou (2020)
Soy protein isolate	Tea polyphenols	Alkaline	SDS-PAGE	–	Djuardi, Yuliana, Ogawa, Akazawa, and Suhartono (2020)
	EGCG	Alkaline	Mass Spectroscopy	–	
Duck albumen hydrolysate	EGCG	Free radical	FTIR	Extended shelf-life of fish tofu at refrigerated temperature	Quan and Benjakul (2019); Quan, Benjakul, and Hozzein (2020)

β-lactoglobulin	EGCG and chlorogenic acid	Free radical	SDS-PAGE, MALDI-TOF-MS, FTIR, fluorescence, and CD spectroscopy	–	Wu et al. (2018)
	Caffeic acid	Carbodiimide reaction	MALDI-TOF-MS, FTIR, fluorescence, and CD spectroscopy	–	Abd El-Maksoud et al. (2018)
Zein	EGCG	Alkaline	RAMAN spectrum, FTIR, TEM	–	Liu et al. (2018)
Ovalbumin	Catechin	Free radical	SDS-PAGE, MALDI-TOF-MS, FTIR, fluorescence, and CD spectroscopy, DSC	Improved oxidative stability of fish oil emulsion	Feng et al. (2018)
Bovine serum albumin	Caffeic acid	Free radical	SDS-PAGE, CD	–	Fan, Liu, Gao, Zhang, and Yi (2018a)
Cuttlefish skin gelatin	Gallic acid, catechin, tannic acid, ferulic acid	Enzymatic (Laccase)	SDS-PAGE, SEM	Enhanced gelling properties	Temdee and Benjakul (2015)
Gelatin	Catechin	Laccase	UV	Improved antioxidant activity	Chung, Kurisawa, Uyama, and Kobayashi (2003)
Bombyx mori silk fibroins	Catechin	Tyrosinase	^{1}H NMR	Improved antioxidant activity and better durability of silk fibroin membrane	Qi et al. (2016)

Continued

Table 1 Methods of preparation, characterization and applications of protein-polyphenol conjugates.—cont'd

Proteins	Polyphenols	Conjugation methods	Structural characterization techniques	Functional properties/ applications/bioactivities	References
Poly(3-lysine)	Catechin	Laccase		Inhibition against disease-related enzymes (collagenase, hyaluronidase, and xanthine oxidase)	Ihara et al. (2004)
Milk whey protein	Coffee specific phenolics	Alkaline and polyphenol oxidase	DSC, SDS-PAGE, MALDI-TOF-MS, UV, CD spectroscopy	Emulsions for lutein ester delivery with enhanced antioxidative properties	Ali, Homann, Khalil, Kruse, and Rawel (2013)
Porcine bone protein hydrolysates	Rutin	Alkaline	Surface hydrophobicity, fluorescence, and CD spectroscopy	Improved emulsifying property	Liu, Han, Zhang, Liu, and Kong (2019)
Chitosan/gelatin	Plant polyphenols	Laccase	–	Inhibit myeloperoxidase and collagenase activities	Rocasalbas et al. (2013)
Gelatin	Gallic acid	Free radical	–	Enhanced anticancer activity, used for chemoprevention of Alzheimer's disease	Cirillo et al. (2010)
Gelatin	EGCG	Chemical synthesis using 4-(4,6-dimethoxy-1,3,5-triazin-2-yl)-4-morpholinium chloride (DMT-MM)	FTIR, SEM	Osteogenic capability	Honda et al. (2015)

Chitosan/gelatin	Gallic acid	EDC/HOBt	–	Biomaterials in food packaging	Rui et al. (2017)
Gelatin	Tannic acid	Alkaline	–	Used as encapsulated agent with flaxseed mucilage	Mohseni and Goli (2019)
Gelatin	Chlorogenic acid	NHS/EDC	^{1}H NMR, FTIR	Improved antioxidant activity, and antibacterial activity against *Escherichia coli, Pseudomonas aeruginosa, Listeria monocytogenes* and *Staphylococcus aureus*	Fu et al. (2017)
Ovalbumin	EGCG	Alkaline	SDS-PAGE, MALDI-TOF-MS, LC-MS/MS, FTIR, DSC, Fluorescence and CD spectroscopy	Decreased IgE-binding activities, increased digestibility	He et al. (2019)
Whey protein isolate	EGCG	Free radical	SDS-PAGE, CD, FTIR, ESI-MS	Protection of fish oil against deterioration	Fan, Liu, Gao, Zhang, and Yi (2018b)
Gelatin	Tannic acid	Alkaline		Stabilized menhaden oil-in-water emulsion	

3.1 Chemical methods

3.1.1 Free-radical associated grafting

Free-radical associated grafting has been commonly used in the conjugation of PPNs with various biomolecules such as chitosan (Mittal, Singh, Aluko, & Benjakul, 2020), chitooligosaccharide (Singh et al., 2020), polysaccharides (Liu, Wang, Yong, Kan, & Jin, 2018) along with proteins or peptides (Gu, Peng, et al., 2017; Quan & Benjakul, 2019). This method with low toxicity is economical and eco-friendly, thus expanding its applications in the food and biomedical industries. Practically, this reaction is feasible at room temperature, thus lowering deterioration or degradation of PPNs or proteins. The possible mechanism of free-radical induced conjugation is based on the redox reaction of grafting chemicals (Fig. 2A). Grafting chemicals such as ascorbic acid undergo oxidation to generate ascorbate ion, which further donates pair of H-ion or electron in the presence of H_2O_2 and ultimately ascorbate is converted into resonance stabilized tricarbonyl ascorbate free-radicals through ascorbate radicals (Gu, Peng, et al., 2017; Liu et al., 2019; Quan et al., 2019). Subsequently, the previously produced H-ion or electron reacts with H_2O_2 to generate OH-radicals. Afterward, hydrogen is abstracted from functional groups (NH_2 and -OH) of protein via OH-radical to generate protein-macroradicals. Ultimately, PPN molecules present in the reaction mixture are covalently bound with macro-radicals. PPNs are known to be the superior H-donor, which can form H-bonds with the C=O group in protein chain (Buitimea-Cantúa, Gutiérrez-Uribe, & Serna-Saldivar, 2018). Several protein-PPN conjugates such as gelatin-gallic acid/catechin, β-lactoglobulin-catechin, α-lactalbumin-catechin, lactoferrin-chlorogenic acid/EGCG/gallic acid, ovalbumin-catechin, etc. have been prepared using this grafting method (Feng et al., 2018; Liu et al., 2015; Spizzirri et al., 2009; Yi, Fan, Zhang, & Zhao, 2016; Yi, Zhang, Liang, Zhong, & Ma, 2015). Via interaction between protein and PPN, the exact binding sites used in free-radical grafting is still unclear. Spizzirri et al. (2009) hypothesized that the binding sites of catechin are H-2′/H-5′ at ring B and H-6/H-8 at ring A, whereas heteroatom-centered groups of gelatin serve as the binding sites. Protein-macroradical preferably attacks at OH-group of ortho- and para-positions in the benzene ring of PPNs (Kobayashi & Higashimura, 2003; Uyama, Maruichi, Tonami, & Kobayashi, 2002). Positions and types of amino acids in the protein chain also influence grafting efficacy. You, Luo, and Wu (2014) also demonstrated that lysine (residue 327) and glutamic acid (residue 186) of ovo-transferrin-bond covalently with catechin to form the conjugate.

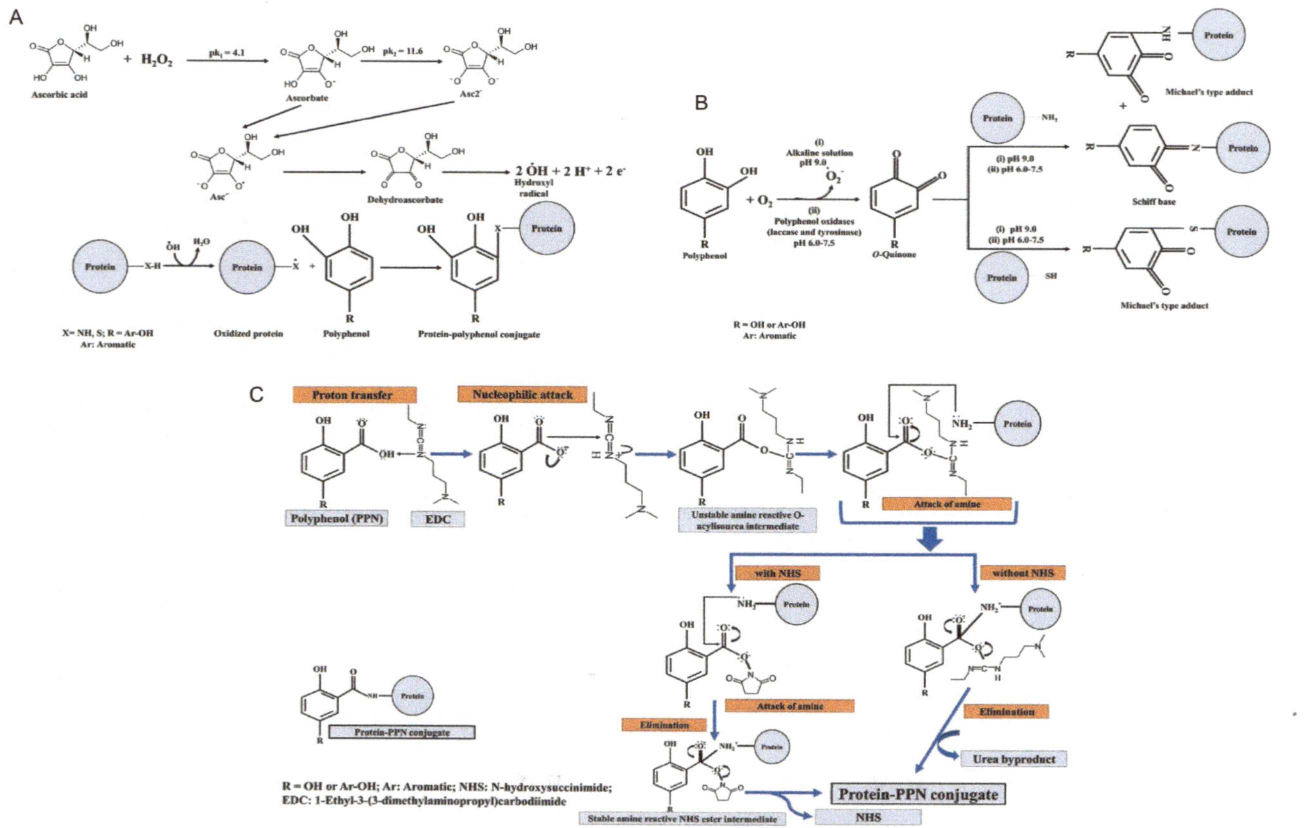

Fig. 2 Covalent conjugation of proteins and polyphenols via free radical grafting (A), alkaline (i) and enzymatic (ii) (B) and chemical coupling methods (C).

3.1.2 Alkali reaction

In this method, firstly, PPNs are converted into their respective quinone or semiquinone radical intermediates with the aid of alkali (pH 9) along with atmospheric oxygen. Those intermediates further react with nucleophiles, e.g., lysine, cysteine, and tryptophan of proteins (Liu, Yong, et al., 2019). The reaction mechanism for the covalent interactions between PPNs and proteins is illustrated in Fig. 2B (i). Briefly, the diphenol of PPN is oxidized to an ortho-quinone, which further forms dimer or polymer. Subsequently, it reacts with NH- or SH-groups of protein. Moreover, covalent C–N or C–S bonds with the phenolic ring are formed, resulting in the formation of hydroquinone (Cao & Xiong, 2015; Jongberg, Gislason, Lund, Skibsted, & Waterhouse, 2011). Further oxidation of hydroquinone resulted in the formation of quinone, in which protein-protein crosslinks can be generated. Generally, radical intermediate formation solely depends on the configuration and type of PPNs. The compounds containing catechol such as quercetin, caffeic acid, gallic acid, and myricetin undergo the formation of quinone radical intermediates (Rawel, Czajka, Rohn, & Kroll, 2002). On the other hand, semiquinone radical intermediates are generated from sinapic, coumaric, and ferulic acids containing monophenols (Cilliers & Singleton, 1991; Pham, Wang, Zisu, & Adhikari, 2019; Rawel, Czajka, et al., 2002; Strauss & Gibson, 2004). Wei, Yang, Fan, Yuan, and Gao (2015) proposed mechanism for the formation of protein-EGCG conjugates via Schiff-bases (C=N) and Michael-type (C-NH) adducts based on the interaction of NH- and SH-groups of protein, respectively. Moreover, EGCG has been documented to act as a bridge between β-lactoglobulin, which resulted in the dimerization of β-lactoglobulin (Tao et al., 2019). Various types of flavonoids possessed varying conjugation efficiency based on the position of OH-group in B and C rings. Rawel, Ranters, Rohn, and Kroll (2004) reported that ortho-OH groups in catechol moiety on ring B of flavonoids are responsible for covalent bonding. On the other hand, catechol moieties on ring A have much lower reactivity.

3.1.3 Chemical coupling method

In this method, a crosslinker, such as glutaraldehyde (GLA), has been used for conjugation or protein crosslinking. It reacts with nucleophilic side chains of proteins, such as amine, thiol, phenol and imidazole (Liu, Yong, et al., 2019). Additionally, 4-(4,6-dimethoxy-1,3,5-triazin-2-yl)-4-methyl-morpholinium chloride, *N*-methylmorpholine, 1-ethyl-3-(3-dimethylaminopropyl) carbodiimide hydrochloride (EDC) and *N*-hydroxysuccinimide (NHS) have been used as chemical coupling agents (Fu et al., 2017; Honda et al., 2015; Wu, Zhuo, He, Liao, & Shi, 2009). The reaction mechanism of coupling methods using EDC/NHS is explained in Fig. 2C. Firstly, EDC couples carboxyl or phosphate groups to primary amines, which further reacts with the carboxyl group of phenolic acids, in which an amine-reactive O-acylisourea intermediate is formed. This intermediate is quite unstable without the NHS. It might convert back into phenolic acids. However, NHS can stabilize the amine-reactive intermediate by converting it to an amine-reactive sulfo-NHS ester, which can further react with the primary amine of proteins. Therefore, the protein-PPN conjugates can be produced continuously (Conde et al., 2014). Moreover, in the presence of protein, the ortho-acylisourea intermediate may also react with a primary amine of protein, resulting in the formation of protein-PPN conjugate without the involvement with NHS agent (Conde et al., 2014). Despite high grafting efficiency for PPN–protein conjugate synthesis, a large amount of cross-linking reagents is required. Those chemicals are expensive and toxic. The aforementioned drawbacks limit the application of the method (Table 1).

3.2 Enzymatic method

As compared to the chemical grafting methods, enzymatic method for the synthesis of protein-PPN conjugates is still limited. Polyphenol oxidase (PPO), laccase, and tyrosinase are the most commonly used enzymes, which can transform phenolic compounds into electrophilic quinones (Chung et al., 2003). These quinones undergo reaction with the nucleophilic NH-groups of protein in a similar way to the alkali method (Fig. 2B, ii). The covalent bonds formed between PPNs and proteins are Schiff-bases and Michael-type adducts (Qi et al., 2016). Qi et al. (2016) found the decrease in free amine groups in the reaction mixture of catechin, tyrosinase, and ε-polylysine, reflecting the efficient grafting associated with oxidation of catechin. Moreover, insoluble matters were generated after 8 h of reaction, mainly caused by the grafting of catechin onto ε-polylysine. However, the drawback associated with the enzymatic grafting is the lower availability of amino acids such as ε-amino groups of lysine and thiol group of cysteine. The limited studies, especially with catechin, have been documented for conjugation with some proteins via enzymatic procedure (Chung et al., 2003; Ihara et al., 2004; Qi et al., 2016) (Table 1).

Although various proteins have been modified using grafting methods, only a few reports have been available on the protein hydrolysates or peptides. Chen, Jiang, Chen, Liu, and Kong (2019) prepared conjugates between porcine plasma protein hydrolysates (PPPH) and oxidized tannic acid or chlorogenic acid with the help of the alkali method. Similarly, porcine bone protein hydrolysates (PBPH) were conjugated with rutin using alkali method as confirmed by the decreasing amount of free amino acids, sulfhydryl (SH) and tyrosine groups in PBPH. The duck albumin hydrolysate was grafted with EGCG at different concentrations (2–5%, w/w) using free-radical grafting methods (Feng et al., 2018). This conjugation was also confirmed by higher surface hydrophobicity of hydrolysates, which was mainly governed by EGCG. Total SH group and carbonyl contents became lowered as EGCG, especially at greater levels, was used for grafting (Quan & Benjakul, 2019). Similarly, rice protein hydrolysate-chlorogenic acid conjugate was prepared using alkali method (Pan et al., 2019). Generally, amino acids in protein hydrolysate are more exposed and readily available for the interactions with PPNs (Quan & Benjakul, 2019).

3.3 Factors affecting conjugation efficiency

3.3.1 Reaction conditions

Reaction conditions such as temperature and pH have a marked effect on conformational changes of both proteins and PPNs. Heating more likely causes unfolding of protein molecules, thus exposing buried hydrophobic groups, which might induce hydrophobic-hydrophobic interaction or H-bonding with PPNs (Quan et al., 2019; Wang, Wang, Xue, & Li, 2019). In addition to hydrophobic-hydrophobic interactions, heating could induce covalent interaction between proteins and PPNs. When the temperature of the reaction medium was augmented from 10 to 45 °C, higher interaction of caffeic acid and helianthinin (protein from sunflower seeds) was observed, more likely because of the exposure of buried lysine, tryptophan, and tyrosine residues for ionic or H-bonding. Nevertheless, when the temperature became higher (40–90 °C), the poorer affinity between bovine serum albumin (BSA) and PPNs via hydrophobic-hydrophobic interaction was obtained (Rawel, Meidtner, & Kroll, 2005). High aggregation of proteins or excessive oxidation of PPNs led to lower interaction between both compounds (Ozdal, Capanoglu, & Altay, 2013; Quan et al., 2019). Under the controlled temperature, the oxidation of PPNs still occurs and quinone derivatives are continuously formed.

In addition to temperature, pH also has a profound effect on the protein-PPN conjugations. Depending on the pH (4.0–10.0), conjugation occurred via either covalent or non-covalent bonding (Wang et al., 2014).

In general, pH also influences the conformation of the protein, which can expose (low pH < 7.0) the binding sites for interaction with PPN, especially via ionic interaction (Ozdal et al., 2013; Quan et al., 2019). Numerous globular proteins, e.g., BSA had a greater affinity for binding with tannic acid at pH 4.9 than at pH 7.8. Le Bourvellec and Renard (2012) reviewed all the possible factors such as temperature, pH, and concentrations influencing conjugation between protein and PPNs. It was suggested that both temperature and pH had the marked impacts on both protein conformation or PPN structure. This inevitably determines covalent or non-covalent interaction, hydrophobic-hydrophobic or ionic interactions of protein-PPN conjugates (Le Bourvellec & Renard, 2012).

3.3.2 Type of proteins and PPN

Amino acid compositions and sequence of proteins also influence various functional and structural properties such as hydrophobicity, isoelectric point, solubility, etc., which indirectly determine the conjugation efficiency with PPNs (Ozdal et al., 2013; Quan et al., 2019). Proteins with unfolded conformation have a higher affinity to PPNs than those with tightly compact counterpart. This is associated with unavailability of amino acid residues, especially hydrophobic amino acids for the interaction with a PPN of the latter (Yildirim-Elikoglu & Erdem, 2018). Among various amino acids, proline, particularly on the surface, is known to be the primary site for binding of several PPNs (Bohin, Vincken, van der Hijden, & Gruppen, 2012). Milk proteins, especially caseins rich in proline, have a high tendency to react with PPNs. The BSA and human serum albumin had greater affinity binding capacity than skimmed milk protein and β-lactoglobulin. Although BSA has globular and compact conformation, the presence of Trp 134 residue on the protein surface and Try 213 having hydrophobicity enhanced binding efficiency of BSA, especially when conjugated with epicatechin and catechin (Yildirim-Elikoglu & Erdem, 2018).

PPNs are diverse compounds with very low molecular weight (MW) to high MW. Four major groups including phenolic acids, flavonoids, lignans, and hydroxycinnamic acids have been classified (Czubinski & Dwiecki, 2017; Quan et al., 2019). For each group, subclasses are categorized according to the degree of the oxidation of the heterocyclic ring, which further influences the interactions toward proteins (Ozdal et al., 2013). Dubeau, Samson, and Tajmir-Riahi (2010) reported that the larger PPNs, e.g., thearubigin and theaflavin from black tea preferably interact with milk proteins than flavanol (catechin) monomer due to their higher binding sites for interaction with proteins. PPNs with varying MWs possessed different affinity towards methylation, hydroxylation, and glycosylation (Martini, Bonechi, & Rossi, 2008; Xiao et al., 2011; Xiao & Kai, 2012). Martini et al. (2008) observed the strong interactions between quercetin aglycone (glycosylated form) and BSA. Xiao et al. (2011) reported that methylation of PPNs lowered conjugation efficiency with proteins, while the hydroxylation of the A and B rings enhanced the binding efficiency of flavonoids. On the other hand, hydrogenation of PPNs depleted binding affinity. Xiao et al. (2011) found the decreased binding affinity of PPNs towards milk proteins after being hydrogenated with C2=C3 double bond of flavonoids.

4. Molecular characteristics

To elucidate molecular characteristics of conjugates, a wide range of analytical techniques involving mass spectrometry, size exclusion chromatography, electrophoresis, Fourier-transform infrared spectroscopy (FTIR), circular dichroism spectroscopy (CD), electrophoresis, ultraviolet-visible spectroscopy, and differential scanning calorimetry (DSC) have been used (Table 1) (Fan et al., 2018b; Liu, Ma, Gao, & McClements, 2017; Ozdal, Yalcinkaya, Toydemir, & Capanoglu, 2019; Quan & Benjakul, 2019).

4.1 Mass spectrometry (MS)

MS can be used to identify the conjugate between proteins and PPNs via separation of the charged particles based on their mass (the mass/charge ratio, m/z) (Ozdal et al., 2019). In general, matrix-assisted laser desorption/ionization-time-of-flight-mass mass spectrometry (MALDI-TOF-MS) and electrospray ionization mass spectrometry (ESI-MS) have been applied for characterization of conjugates (Liu, Ma, Gao, & McClements, 2017). Wu et al. (2018) used MALDI-TOF-MS for determining MW of β-lactoglobulin-EGCG conjugate after covalent conjugation under heat treatment. Also, change in MW of Ara h1 peanut protein-EGCG/chlorogenic acid conjugates prepared by alkaline treatment can be detected using MALDI-TOF-MS (He et al., 2020). Covalent binding of ovalbumin with PPNs (catechin, epigallocatechin or EGCG) produced the conjugate through free-radical grafting method as confirmed by MALDI-TOF-MS (Feng et al., 2018). Apart from MALDI-TOF-MS, some studies used ESI-MS to identify covalent binding of conjugates (Sun, Jing, Mu, et al., 2020; Yi et al., 2016). Dias, Perez-Gregorio, Mateus, and De Freitas (2016) identified the grafting between wheat derived-peptide and tannin (procyanidin B3) using ESI-MS. Moreover, binding sites of egg white proteins conjugated with catechin/caffeic acid prepared under alkaline condition could be detected using ESI-MS (Sun, Jing, Mu, et al., 2020). Yi et al. (2016) used ESI-MS to confirm covalent bonding and evaluate the change in MW of α-lactalbumin grafted with catechin via radical grafting method, in which the augmented MW was attained after conjugation with catechin.

4.2 Size-exclusion chromatography (SEC)

SEC is a separation technique based on the molecular size of the components (Chirayil, Abraham, Mishra, George, & Thomas, 2017). PPNs can interact with the gel filtration material and the dilution by eluents can occur (Bartolome, Estrella, & Hernandez, 2000). This is the disadvantage of the method. The molecular properties of soy protein isolate (SPI) conjugated with tea PPN extracts prepared using alkaline treatment have been characterized by SEC (Djuardi et al., 2020). The result showed that MW of protein increased when conjugated with tea PPN extracts (Djuardi et al., 2020). Conjugation between zein and EGCG/caffeic acid under alkaline treatment increased MW of the resulting conjugates via covalent binding between them or promoting aggregation or crosslinking of proteins (Liu, Ma, McClements, & Gao, 2017). Yi et al. (2015) demonstrated that the increase in MW of β-Lactoglobulin was obvious via conjugation with catechin. A similar result was documented by Quan and Benjakul (2019) who reported that MW of duck albumen hydrolysates-EGCG conjugate was higher than that of unconjugated protein using the radical grafting method. Modification with PPN either by covalent or by non-covalent bonding likely enlarges MW of protein, thus influencing physiochemical and functional properties of the resulting conjugates.

4.3 Sodium dodecyl sulfate-polyacrylamide gel electrophoresis (SDS-PAGE)

SDS-PAGE is one of the techniques used for separating and purification of proteins, according to their size or molecular mass (Dunn, 2014). Numerous studies used this technique to confirm the formation of covalent conjugates (Jiang, Zhang, Zhao, & Liu, 2018; von Staszewski et al., 2012; Wu et al., 2013). Liu, Ma, McClements, and Gao (2017) found the increased MW of zein protein after conjugation with EGCG/quercetagetin under alkaline treatment, mainly mediated by covalent interaction.

Changes in MW of SPI were found when tea PPN extracts were incorporated under alkaline condition (Djuardi et al., 2020). Proteins having the MW of 22, 37 kDa or higher MW (50–76 kDa) were found after conjugation (Djuardi et al., 2020). Furthermore, the decreases in protein with MW of 28 kDa or the lower MW proteins (16–21 kDa) were observed (Djuardi et al., 2020). The finding was in line with the electrophoretic results of an ovotransferrin-catechin conjugate, lactoferrin-caffeic acid conjugate, whey protein-EGCG conjugate, and peanut protein isolate-CA-dextran conjugate, indicating that the grafting between protein and PPNs affected MW of produced conjugates (Liu et al., 2020; Liu, Ma, McClements, & Gao, 2016; You et al., 2014).

4.4 Fourier transform infrared spectroscopy (FTIR)

FTIR spectroscopy has been widely used for the identification of functional groups in organic and inorganic compounds via measuring their absorption of infrared radiation (Smith, 2011). It can be used for examining the conformational changes of protein and hydrophobic-hydrophobic interactions in the protein-PPN complex. In general, amide I band of FTIR spectrum can indicate the secondary structure of proteins (Feng et al., 2018). Thus, the changes in both the position and intensity of amide I peak after grafting between proteins and PPNs were more likely related to alterations in the secondary conformation (Feng et al., 2018). Peaks of amide I can be classified into α-helix (1658–1650 cm^{-1}), β-sheet (1640–1615 cm^{-1}), β-turn (1700–1660 cm^{-1}) and random coil (1650–1640 cm^{-1}) (Liu, Yong, et al., 2019). Moreover, amide II (N–H bending and C–N stretching, 1480–1575 cm^{-1}), amide III (C–N stretching and N–H bending, 1229–1301 cm^{-1}), amide A (N–H stretching, 3300 cm^{-1}) and amide B (C–H stretching and NH_3^+, 3000–2900 cm^{-1}) are used to determine the structure of protein-PPN conjugates (Liu, Yong, et al., 2019; Quan et al., 2019). An increase in wavenumber of amide I and a decrease in wavenumber of amide II was found after whey protein isolate was conjugated with EGCG via the C=O group of EGCG (Fan et al., 2018a, 2018b). Wu et al. (2018) revealed that covalent conjugate of EGCG/caffeic acid changed the structure of β-lactoglobulin as evidenced by the shifts of amide I and amide II peaks to lower wavenumber. Slight differences in FTIR spectra of zein-PPNs (EGCG/quercetagetin) conjugates were noted as compared to native zein, indicating the interactions between protein and PPNs (Liu, Ma, McClements, & Gao, 2017). These differences mainly took place at the amide A and amide II bands, and may represent non-covalent interactions between PPNs and zein via H-bonding or hydrophobic-hydrophobic interactions (Liu, Ma, McClements, & Gao, 2017). Modification of duck egg albumen hydrolysates (DAH) using EGCG resulted in some changes in FTIR spectra, indicating the interaction between DAH and EGCG via several bondings (Quan & Benjakul, 2019). With grafting using EGCG, the shift to lower wavelength of amide A peak (from 3292 to 3290 cm^{-1}) in DAH-EGCG conjugate was noted. Moreover, amide B peak (3083 cm^{-1}) could be detected in DAH-EGCG conjugate.

4.5 Circular dichroism (CD) spectroscopy

CD spectroscopy is an excellent technique used for determination of the secondary structures of proteins after conjugation with PPNs (Liu et al., 2015; Liu, Yong, et al., 2019; Yan et al., 2020). Commonly, the chromophore of the peptide bond in proteins is detected at wavelengths of 190–250 nm (Büyükköroğlu, Dora, Özdemir, & Hızel, 2018). The specific CD spectra of aromatic amino acids and disulfide bonds are detected at wavelengths of 250–350 nm (Büyükköroğlu et al., 2018). The α-helical conformation has negative bands at 208 and 222 nm (Liu, Yong, et al., 2019). The β-sheet conformation shows the negative band at 216 nm and a positive band at the range of 195–200 nm (Liu, Yong, et al., 2019). Liu et al. (2015) documented that the grafting between lactoferrin and PPNs (EGCG, caffeic acid or gallic acid) using free-radical method led to the increased fraction of α-helix with a coincidental decrease in the random coil, indicating the destruction effect on lactoferrin. Additionally, grafting of caffeic acid with BSA caused the decreased content of α-helix, while the increased contents of other structures were attained (Rawel, Rohn, Kruse, & Kroll, 2002). Yi et al. (2015) found that negative band of β-lactoglobulin shifted from 216 to 206 nm after grafting with catechin using the free-radical method, suggesting the decreased β-sheet content in β-lactoglobulin. Conversely, the disordered structure of β-lactoglobulin was enhanced when catechin was grafted, mainly due to the result of the exposure of interior amino acids (Yi et al., 2016). Moreover, the decrease in the α-helix fraction was also found, while β-sheet fraction was increased in α-lactalbumin after grafting with catechin via the free-radical method (Yi et al., 2016). Recently, Yan et al. (2020) reported the decreased β-strand content after soybean protein isolate was grafted with EGCG using an alkaline treatment, whereas the α-helix and random coil contents increased. Exposure of hydrophobic groups inside the protein as induced by EGCG incorporated brings about alternation of protein conformation. Liu, Han, et al. (2019) pointed out that the amounts of β-sheet and β-turn decreased, while the amounts of α-helix and random coil increased after porcine bone protein hydrolysate underwent interaction covalently with rutin. In addition, Damodaran, Parkin, and Fennema (2007) found that α-helix and random coil were more flexible than the β-sheet. Covalent bonding resulted in a more flexible secondary conformation of soybean protein isolate via conjugation. Conformational changes of the conjugates are therefore associated with the different types of protein and PPNs grafted as well as the methods used for preparing the conjugates.

4.6 Ultraviolet-visible (UV) spectroscopy

UV-spectroscopy is one of the most simplified methods for examining protein-PPN conjugates (Haupt & Mosbach, 2000). PPNs significantly differ from proteins in terms of UV absorption (Liu, Xing, Fu, Zhou, & Zhang, 2016). Phenolic acids have UV absorption at 270–280nm and 305–330nm, while flavonoids possess the absorption of UV in 270–280nm and 310–350nm (Liu, Han, et al., 2019). Conversely, peptide bond of proteins shows UV absorption at 220nm. Also, aromatic amino acids have UV absorption at 280nm (Chotphruethipong, Aluko, & Benjakul, 2019). Grafting of phosvitin with gallic acid caused the increased UV spectra of protein from 271 to 278nm (Jiang et al., 2020). Also, the peak at 211nm appeared in the conjugate, representing the absorption of gallic acid (Jiang et al., 2020). The conjugation of white bean protein fractions with catechin increased the absorbance at 350–450nm, in which the maximum absorbance was achieved at approximately 400nm (Sęczyk, Świeca, Kapusta, & Gawlik-Dziki, 2019).

4.7 Fluorescence spectroscopy

Fluorescence spectroscopy is a type of essential technique in food analysis, which has been used for the determination of several components (Liu, Xing, et al., 2016). This technique can detect proteins mostly generated from the indole chromophore of aromatic amino acids, e.g., tryptophan (Liu, Xing, et al., 2016). Additionally, it is also used to investigate the interaction or binding affinity of protein-PPN grafting. Conjugation of caffeic acid and whey protein affected the changes in the fluorescence intensity of proteins (Jiang et al., 2018). When caffeic acid was grafted, the maximum emission peak for casein and whey protein isolate shifted from 348 to 366nm and from 336 to 360 nm, respectively. Jiang et al. (2020) documented that the fluorescence intensity of phosvitin-gallic acid complexes significantly decreased, with the maximum red shifts. Moreover, the increase in gallic acid level affected the change in wavelength of red shift (Jiang et al., 2020). Fluorescence intensity also depends on the type of PPNs grafted. Liu, Ma, McClements, and Gao (2017) found that the fluorescence intensity of zein-PPNs conjugate decreased in descending order: zein-EGCG > zein-quercetagetin > zein-caffeic acid. Guo et al. (2019) revealed that the increased fluorescence intensity was governed by the oxidation of myofibrillar protein. Trp localized inside the protein core was exposed to the polar environment as induced by oxidation, resulting in the increased fluorescence intensity. Nevertheless, the decrease in the fluorescence intensity could occur with the addition of caffeic acid. Protein-PPN conjugate was mainly linked by H-bonds and adsorbed on the protein surface.

5. Functional properties and bioactivities of conjugates

5.1 Solubility

Aggregation and precipitation of proteins generally occurred with increasing concentration, resulting in the limited functional properties and applications of proteins. Thus, the use of conjugation of proteins with PPNs could be an alternative method, favoring protein solubility. Recently, Xue, Li, and Adhikari (2020) demonstrated that the solubility of SPI was notably increased when cyanidin-3-galactoside was grafted. This finding could be reflected by the increased zeta-potential value. The conjugate had a higher value than SPI, indicating the increased electrostatic repulsion. Moreover, the increased solubility of conjugate was also related to the decreased surface hydrophobicity, indicating that conjugates might be formed via H-bonds along with water and led to the augmented solubility (Xue et al., 2020). Chen et al. (2019) found that the binding of tea PPNs to soybean proteins could increase the water solubility of proteins. Furthermore, caffeic acid conjugated with milk proteins (whey protein or casein) also had higher solubility than native proteins (Jiang et al., 2018). Nevertheless, grafting of PPNs, especially at a high amount, may reduce water solubility, mainly caused by an excessive cross-linking of proteins (Quan et al., 2019). Furthermore, the solubility of conjugate also depends on the pH. Sęczyk et al. (2019) documented that the solubility of albumins grafted with quercetin significantly increased in the pH range of 6–8, while albumins-catechin conjugate had the augmented solubility in the pH range of 7–9 (Sęczyk et al., 2019). Globulin grafted with catechin also had increased solubility in the pH range of 6–9 (Sęczyk et al., 2019). Decreased solubility of albumins-PPN conjugates (phenolics from green tea extract, caffeic acid or gallic acid) was found at pH 3, while albumins grafted with gallic acid had the decreased solubility at pH 11 (Sęczyk et al., 2019). Moreover, ionic strength is an important factor toward solubility of protein-PPNs complexes. The decrease in solubility of albumins-PPNs (phenolics from green tea extract) conjugate was recorded at 100 mM NaCl (Sęczyk et al., 2019). Nevertheless, the increase in ionic strength ranging from 100 to 500 mM could augment solubility of albumins-PPNs (phenolics from green tea extract, quercetin or gallic acid) conjugates (Sęczyk et al., 2019). Basically, the reaction of PPNs with proteins affects the secondary and tertiary structures of the proteins, thus eventually altering the net charge on protein surface. This determined the solubility of the conjugates (Ozdal et al., 2013). Thus, types of proteins and PPNs used are the significant factor affecting their solubility, which have the marked effect on functional properties and further applications in food.

5.2 Gelling properties

Conjugation between proteins and PPNs could improve gelling property of the proteins mediated by covalent or non-covalent interactions, leading to the increased gel strength and thermal stability (Balange & Benjakul, 2009a, 2009b; Ge et al., 2019). Balange and Benjakul (2009b) reported the increased breaking force and deformation of bigeye snapper surimi with the addition of oxidized tannic acid. Likewise, gel forming ability of mackerel mince and surimi prepared by alkaline-saline washing process were significantly increased after addition of oxidized tannic acid (Balange & Benjakul, 2009a). When PPNs are added to protein, the interaction between amino acid groups and quinones of PPNs leads to the increased gelling property, in which an elastic gel network can be developed. Yan, Li, Zhao, and Yi (2011) revealed that gel strength of gelatin from walleye pollock skin incorporated with gallic acid or rutin was significantly augmented. Zhao and Sun (2017) found that the increase in gel strength of gelatin hydrogels was mediated by the cross-linking between gelatin and PPN via H-bonding. Von Staszewski, Jagus, and Pilosof (2011) documented that the addition of green tea PPNs into β-lactoglobulin could increase the gelation ascertained by lower gelling temperature and shorter gelling time. Also, the addition of tea PPNs to β-lactoglobulin influenced viscoelasticity of gels (von Staszewski et al., 2012). Apart from the cross-linking activity of PPNs towards protein, the improvement of gelling property of protein could be achieved by free-radical grafting method. Incorporation of gallic acid (30 μmol/g) into oxidized myofibrillar proteins could enhance the gelling ability of native protein as indicated by the increased storage modulus (G′) value (Cao, True, Chen, & Xiong, 2016). Gallic acid might enhance protein intermolecular interactions via bridging between oxidized polypeptide chains and the conversion of SH to S–S, or SH/S-S interchange could occur. Gallic acid can act as a cross-linker via two adduction reactions with free nucleophilic groups in proteins (Cao et al., 2016). However, weakened gel structure was formed with increasing concentration of gallic acid (150 μmol/g), more likely because of the loss of solubility associated with excessive aggregations of myofibrillar proteins (Cao et al., 2016).

5.3 Emulsifying properties

Numerous studies demonstrated that protein-PPN conjugates had much better emulsifying property than native proteins (Djuardi et al., 2020; Fan et al., 2018b; Liu et al., 2020). Quan and Benjakul (2019) synthesized DAH-EGCG conjugate using free-radical grafting method. Conjugate possessed better emulsifying properties than unconjugated DAH. Liu et al. (2015) found that the conjugation between lactoferrin and PPNs (gallic acid, EGCG, chlorogenic acid) using free-radical grafting method could enhance emulsifying properties of lactoferrin. Nevertheless, lactoferrin conjugated with chlorogenic acid had higher emulsifying properties than lactoferrin-EGCG or lactoferrin-gallic acid conjugates (Liu et al., 2015). Conjugation of SPI with anthocyanin under alkaline condition improved emulsifying properties of SPI (Sui et al., 2018). Moreover, SPI-EGCG conjugate prepared using alkaline method also increased emulsifying properties of the protein (Yan et al., 2020). Conjugate produced by covalent binding between cuttlefish skin gelatin and tannic acid using alkaline condition had the improved emulsifying properties (Aewsiri, Benjakul, Visessanguan, Wierenga, & Gruppen, 2013). Yi et al. (2015) documented that the grafting between β-lactoglobulin and catechin via free-radical method could stabilize the oil-in-water emulsion more potentially than protein alone. Chen et al. (2019) revealed that the conjugation between PPPH and oxidized caffeic acid or oxidized tannic acid could increase emulsifying stability index (ESI) and emulsifying activity index (EAI) of PPPH. Additionally, the resulting conjugates, especially PPPH-caffeic acid conjugate, could rapidly absorb and create a thicker interfacial film around oil droplets and resulted in the increased emulsifying properties (Chen et al., 2019). EAI and ESI of emulsion stabilized by the PBPH-rutin conjugate were higher than those of the emulsion stabilized by protein hydrolysate alone, suggesting that addition of the conjugate could increase emulsifying property of protein (Liu, Han, et al., 2019). Covalent grafting of PPN moieties possibly affected protein configuration and enhanced their surface hydrophobicity, therefore lowering the interfacial tension (Quan et al., 2019). Hence, the type of proteins and PPNs used for conjugation has direct influence on emulsifying properties.

6. Bioactivities of protein-PPN conjugates

6.1 Antioxidant activity

Protein-PPN conjugates have been known to possess higher antioxidant activity, compared to the native protein. The conjugate is not only used for preserving food products, especially from lipid oxidation but also can alleviate oxidative stress in cell (Feng et al., 2018; Liu, Yong, et al., 2019; Zhang et al., 2020). Aewsiri et al. (2013) reported that the conjugation of cuttlefish skin gelatin with tannic acid using alkaline treatment augmented antioxidant activities of the obtained conjugate. Jiang et al. (2018) documented that the ability in scavenging radicals of whey protein and casein was remarkably increased when they were grafted with chlorogenic acid. Furthermore, the antioxidant activities of the catechin-egg white protein conjugates were greater than those of unconjugated protein (Gu et al., 2017). Conjugates of ovalbumin with catechin, EGCG or epigallocatechin prepared using free-radical grafting method had the higher activity in quenching peroxyl, 2,2′-azino-bis 3-ethylbenzothiazoline-6-sulfonic acid (ABTS) and 1,1-diphenyl-2-picrylhydrazyl (DPPH) radicals than unconjugated ovalbumin (Feng et al., 2018). Moreover, ovalbumin-EGCG conjugate showed the strongest antioxidative activities. Similarly, the conjugate of zein and EGCG/quercetagetin possessed the augmented DPPH radical scavenging activity (Liu, Ma, McClements, & Gao, 2017). Antioxidant activities of protein hydrolysate are augmented when conjugates with PPNs were formed (Jiang et al., 2018). Chen et al. (2019) synthesized PPPH conjugated with oxidized caffeic acid or oxidized tannic acid via alkaline treatment. The resulting conjugates had increased DPPH and ABTS radicals scavenging activities. Furthermore, chelating activity and FRAP of the resulting conjugates were also increased with enhancing levels of oxidized

PPNs (Chen et al., 2019). Quan and Benjakul (2019) also found that the antioxidant activities of DAH were increased when EGCG was incorporated, especially at high levels of EGCG. This was the result of increased OH-groups from PPNs incorporated into proteins.

6.2 Anticancer activity

Cancer is the second cause of death around the globe (Zhang & Xu, 2019). The anticancer agents have been used for treatment of cancers. Natural compounds derived from plant or protein sources have been proven to have anticancer activity through multiple pathways (Chi, Hu, Wang, Li, & Ding, 2015; Gomathi, Rajarathinam, Sadiq, & Rajeshkumar, 2020; Karami, Peighambardoust, Hesari, Akbari-Adergani, & Andreu, 2019; Lichota & Gwozdzinski, 2018; Pan, Zhao, Hu, Chi, & Wang, 2016). Moreover, phenolic-protein conjugates could also be a promising compound in inhibiting cancer (Liu, Yong, et al., 2019). Recently, Rai, Kureel, Dutta, and Mehrotra (2018) prepared the conjugate between BSA and phenolics extracted from sugarcane bagasse using free-radical grafting method. The resulting conjugate could inhibit SW480 cells (human colon cancer cells) as indicated by the decreased cell viability by 80–85% after treatment with the conjugate at a level of 100 mg/mL (Rai et al., 2018). Additionally, cells treated with conjugate had the decreased production of ROS as compared to those treated with BSA or phenolics alone, indicating that the conjugate could inhibit cancer cells effectively (Rai et al., 2018). Chaharband et al. (2018) revealed that curcumin-lactoferrin conjugate nanostructure could improve anticancer activity of HCT116 cells as indicated by the decreased cell viability and the higher level of cellular uptake (89%). Additionally, the gelatin-gallic acid conjugate had an ability in inhibiting DU 145 human prostate cancer cells. Also, the activity of conjugate was higher than gallic acid or gelatin alone (Cirillo et al., 2010).

7. Applications of protein-PPN conjugates

7.1 Food and food products

7.1.1 Protein gels

Improvement of protein gel can be carried out by various methods such as the use of cross-linking transglutaminase, heat or pressure treatments, pH adjustment, the addition of polysaccharides or proteins, the use of mono- or di-valent cations (such as calcium, potassium, or tripolyphosphate), or pH adjustment (Buamard & Benjakul, 2018; Fang et al., 2020; Petcharat & Benjakul, 2018; Singh & Benjakul, 2017a, 2017b; Sinthusamran, Benjakul, Swedlund, & Hemar, 2017). Apart from the aforementioned approaches, plant extracts or pure PPNs have been utilized as protein cross-linkers for increasing gelation of surimi (Balange & Benjakul, 2009a). von Staszewski et al. (2012) found that the grafting of β-lactoglobulin and PPNs from green tea could induce the gelation of whey protein as confirmed by lower gelling temperature and shorter gelling time. Also, the use of EGCG could accelerate the formation of β-lactoglobulin gel and influenced viscoelasticity of protein gel (Von Staszewski et al., 2011). Moreover, Guo, Bao, Sun, Chang, and Liu (2020) found that covalent cross-linking of SPI with tannic acid increased gel strength of SPI gel. Also, the resulting gel after covalent cross-linking had more uniform distribution of dense elliptical or circular network with smaller porosity than the control gel (Guo et al., 2020).

Protein conjugated with PPNs can also apply for improving gel property of surimi or myofibrillar proteins. Balange and Benjakul (2009a) revealed that oxidized caffeic acid (0.15%) and ferulic acid (0.2%) added into bigeye snapper surimi could augment the protein crosslinking, resulting in the formation of stronger network with greater water holding capacity (Fig. 3A). Similarly, gel strength of sardine surimi could be improved by adding ethanoic coconut husk extract as shown by the increased breaking force and G' value (Buamard & Benjakul, 2015). Therefore, the phenolic compounds can be a promising additive that helps improve the gelation of protein-based gel products.

7.1.2 Protein-based films

Protein-based films have been widely used for food packaging owing to several advantages such as their renewable and biodegradable characteristics (Nilsuwan, Benjakul, Prodpran, & de la Caba, 2019). Nevertheless, protein films have some poor properties including low water barrier property, because of their hydrophilicity and the presence of high hydrophilic plasticizers such as glycerol (Fig. 3A) (Hoque, Benjakul, Prodpran, & Songtipya, 2011). To conquer such a problem, hydrophobic compounds such as oil/essential oils have been used to enhance water barrier property. Nevertheless, the added oil into the film is prone to lipid oxidation, in which the development of unwanted compounds contributes to off-odor and discoloration (Tongnuanchan, Benjakul, & Prodpran, 2012). Incorporation of plant extract or phenolic compounds into film could increase functional properties and antioxidant

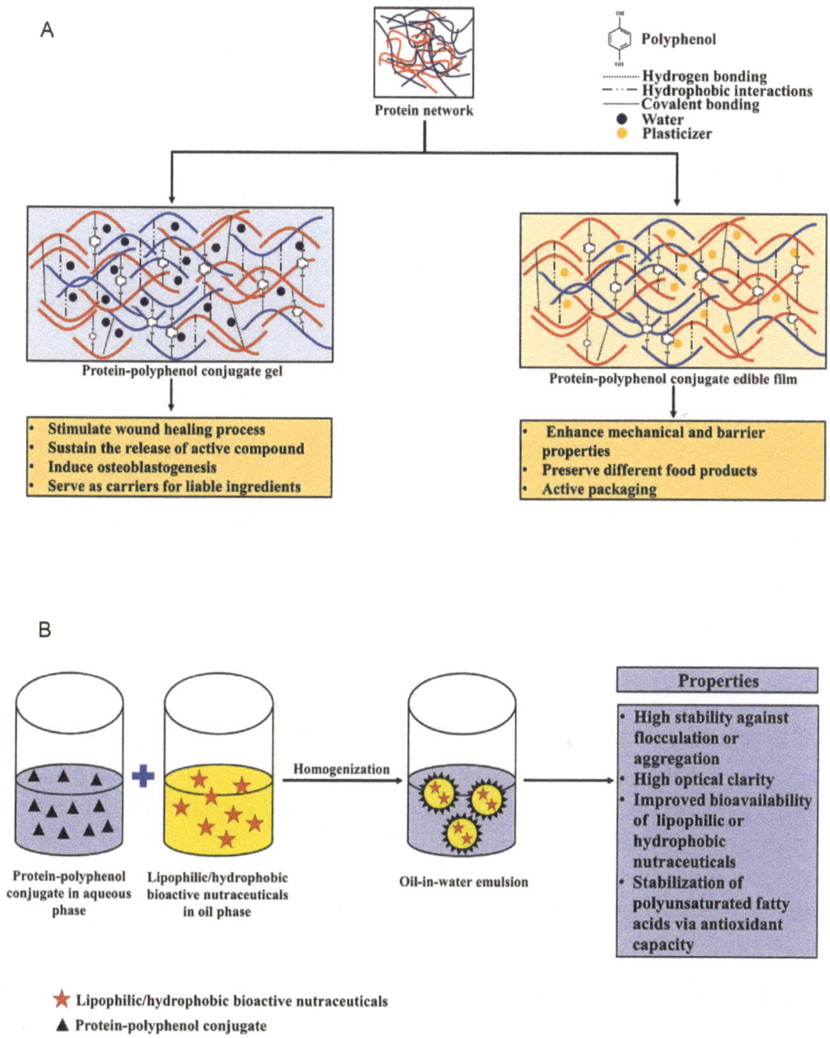

Fig. 3 Role of protein-polyphenol conjugate in gel and film (A) and emulsion formation (B).

activity of protein-based films (Adilah, Jamilah, & Hanani, 2018; Nilsuwan et al., 2019). Antioxidant and antimicrobial activities of film based on fish myofibrillar protein were increased by adding catechin-Kradon extract (Kaewprachu, Osako, Benjakul, Suthiluk, & Rawdkuen, 2017). Nie, Gong, Wang, and Meng (2015) documented that the incorporation of grape seed procyanidins and green tea PPNs in myofibrillar protein film could improve water barrier property and thermal stability. Araghi et al. (2015)

revealed that solubility, water vapor, and oxygen barrier properties of fish gelatin film could be improved via conjugation with caffeic acid using alkaline method. Moreover, the tensile strength and modulus of zein film were augmented by the addition of oxidized tannic (Santos et al., 2017). Nevertheless, the increase in the amount of oxidized tannic negatively affected water solubility (Santos et al., 2017). Nilsuwan et al. (2019) prepared bilayer fish gelatin films incorporated with EGCG. The resulting film exhibited the augmented tensile strength, water vapor barrier, and light barrier properties. Furthermore, those films could be successfully used as packaging for chicken skin oil as evidenced by the lowered peroxide value (PV), thiobarbituric acid reactive substances (TBARS) and volatile compounds after 30 days of storage at 25–28 °C, compared to that packaged in low-density polyethylene pouch (Nilsuwan et al., 2019). Additionally, whey protein film containing green tea extract could retard lipid oxidation of fresh salmon during storage for 14 days at 4 °C (Castro, Andrade, Sanches Silva, Vaz, & Vilarinho, 2019). Protein-based films modified or added with PPNs should be further developed for better properties.

7.1.3 Emulsions

Protein-PPN conjugates have been extensively utilized to stabilize food emulsions since they have an excellent emulsifying property and antioxidant activity (Quan et al., 2019). For instance, soy protein conjugated with tea PPNs using alkaline condition improved emulsifying property of native soy protein by the increase in adsorption at the oil-water interface and reduction of surface tension between oil and water molecules in the system (Fig. 3B) (Djuardi et al., 2020). Gu, Su, et al. (2017) reported that catechin-egg white protein conjugates were shown to be the good antioxidant emulsifier for improving the physicochemical stability of β-carotene emulsions as evidenced by high physical stability throughout 30 days of storage at 25 °C, compared to emulsions stabilized by egg white protein alone. Covalent interaction between PPNs and protein might enhance the thickness of the interfacial films, thereby providing the steric repulsion between the droplets and led to the increased stability (Gu, Su, et al., 2017). Similarly, β-carotene emulsion stabilized by catechin-β-lactoglobulin conjugates showed higher stability than that of β-carotene emulsion stabilized by the proteins alone during storage at 25 °C for 30 days (Yi et al., 2015). Moreover, oil-in-water emulsion stabilized by PBPH-rutin conjugates formed by alkaline treatment, especially at a high level (10 mg/mL), possessed high oxidative stability with the lowered PV and TBARS values after storage at 37 °C for 14 days (Liu, Han, et al., 2019). Quan and Benjakul (2019) used duck egg albumin-EGCG to stabilize fish oil emulsion. The results showed that fish oil emulsion incorporated with conjugate was more resistant against flocculation and coalescence of oil droplets than the control or emulsion added with protein alone after 15 days of storage at 26–28 °C. Also, the conjugate could prevent lipid oxidation in fish oil emulsion as evidenced by lowered PV and TBARS values. Intarasirisawat, Benjakul, and Visessanguan (2014) revealed that skipjack roe protein hydrolysate mixed with oxidized tannic acid could stabilize fish oil emulsion, yielding the smaller size of oil droplets and more uniform distribution than that stabilized by hydrolysate alone. Thus, protein/peptides-PPN conjugates could act as an antioxidant emulsifier for improvement of physical and oxidative stability of emulsion.

7.2 Nutraceutical

The encapsulation of nutraceutical into delivery systems has gained a remarkable interest. Protein-PPN conjugates are proven to have the binding affinity towards both lipophilic and hydrophilic compounds owing to their amphiphilic nature (Chanphai et al., 2018). They can be therefore used as carriers for various bioactive compounds. Fan, Zhang, Yokoyama, and Yi (2017) used β-lactoglobulin-caffeic acid conjugate-based nanoparticles for the delivery of EGCG. The conjugate nanoparticles could control the release of EGCG in simulated gastrointestinal digestion more effectively than β-lactoglobulin nanoparticles. Similarly, Liu, Ma, McClements, and Gao (2017) used zein-EGCG conjugate as nanoparticles for co-delivery curcumin and resveratrol. The degradations of curcumin and resveratrol under physiological pH and ultraviolet irradiation conditions could be protected by the use of protein-PPNs conjugate as biopolymer nanoparticles. Moreover, mixing the zein-EGCG conjugate nanoparticles with an excipient nanoemulsion enhanced bio-accessibility of curcumin and resveratrol. Digestion products from the lipid phase could form the mixed micelles, thus helping in solubilization of nutraceutical in the intestinal fluids (Liu, Ma, McClements, & Gao, 2017). Fan et al. (2018a), (2018b) used BSA-caffeic acid conjugate as an emulsifier to stabilize resveratrol-loaded zein nanoparticle. The effectiveness of the conjugate in preventing the disruption of resveratrol was higher than BSA alone (Fan et al., 2018a, 2018b). Moreover, zein-BSA-caffeic acid nanoparticles had greater antioxidant ability than free resveratrol, indicating that nanoparticle-based delivery system could be improved, in which the uptake and bioavailability of encapsulated antioxidants into Caco-2 cells can be achieved (Fang et al., 2020). Additionally, β-lactoglobulin-catechin conjugates (Yi et al., 2016), and lactoferrin-catechin (Liu, Xing, et al., 2016) were documented to stabilize β-carotene in oil-in-water nano-emulsion. Also, the conjugation could enhance the stabilization of these nanoparticles under the simulated gastrointestinal conditions.

8. Conclusion

Protein and PPN possess their distinct properties associated with their unique composition and structure. Proteins have excellent functional properties such as gelling, foaming, emulsifying properties, etc. However, they showed low bioactivities, such as antioxidant and antibacterial activities. On the other hand, PPNs have a variety of bioactivities, especially antioxidant activity. Therefore, a combination of these two biomolecules can be a promising means to exploit both functional properties and bioactivities of food products and eating quality. Numerous procedures have been implemented for conjugation of these compounds via both chemical and enzymatic methods. Free-radical grafting method is most widely employed to manufacture bioactive protein-PPN conjugates. The formation of the protein-PPN conjugate is governed by both reaction conditions (temperature and pH) and type or structure of proteins and PPNs. Protein-PPN conjugates show the enhanced bioactivities, gelling, film forming, and emulsifying properties. Moreover, encapsulation of labile components via protein-PPN nanoparticles could also improve their delivery through the gastrointestinal tract. Protein-PPN conjugates not only improve food quality but also impart various health benefits. Therefore, the conjugation between proteins and PPNs from various sources should gain more attention, in which new food ingredients or products with better nutritional, functional, sensorial qualities as well as bioactivities can be produced.

References

Abd El-Maksoud, A. A., Abd El-Ghany, I. H., El-Beltagi, H. S., Anankanbil, S., Banerjee, C., Petersen, S. V., et al. (2018). Adding functionality to milk-based protein: Preparation, and physico-chemical characterization of β-lactoglobulin-phenolic conjugates. *Food Chemistry*, *241*, 281–289.

Adilah, Z. M., Jamilah, B., & Hanani, Z. N. (2018). Functional and antioxidant properties of protein-based films incorporated with mango kernel extract for active packaging. *Food Hydrocolloids*, *74*, 207–218.

Aewsiri, T., Benjakul, S., Visessanguan, W., Wierenga, P. A., & Gruppen, H. (2013). Emulsifying property and antioxidative activity of cuttlefish skin gelatin modified with oxidized linoleic acid and oxidized tannic acid. *Food and Bioprocess Technology, 6*, 870–881.

Ali, M., Homann, T., Khalil, M., Kruse, H. P., & Rawel, H. (2013). Milk whey protein modification by coffee-specific phenolics: Effect on structural and functional properties. *Journal of Agricultural and Food Chemistry, 61*, 6911–6920.

Araghi, M., Moslehi, Z., Mohammadi Nafchi, A., Mostahsan, A., Salamat, N., & Daraei Garmakhany, A. (2015). Cold water fish gelatin modification by a natural phenolic cross-linker (ferulic acid and caffeic acid). *Food Science and Nutrition, 3*, 370–375.

Balange, A., & Benjakul, S. (2009a). Enhancement of gel strength of bigeye snapper (*Priacanthus tayenus*) surimi using oxidised phenolic compounds. *Food Chemistry, 113*, 61–70.

Balange, A. K., & Benjakul, S. (2009b). Effect of oxidised tannic acid on the gel properties of mackerel (*Rastrelliger kanagurta*) mince and surimi prepared by different washing processes. *Food Hydrocolloids, 23*, 1693–1701.

Bartolome, B., Estrella, I., & Hernandez, M. (2000). Interaction of low molecular weight phenolics with proteins (BSA). *Journal of Food Science, 65*, 617–621.

Bohin, M. C., Vincken, J.-P., van der Hijden, H. T. W. M., & Gruppen, H. (2012). Efficacy of food proteins as carriers for flavonoids. *Journal of Agricultural and Food Chemistry, 60*, 4136–4143. https://doi.org/10.1021/jf205292r.

Buamard, N., & Benjakul, S. (2015). Improvement of gel properties of sardine (*Sardinella albella*) surimi using coconut husk extracts. *Food Hydrocolloids, 51*, 146–155.

Buamard, N., & Benjakul, S. (2018). Combination effect of high pressure treatment and ethanolic extract from coconut husk on gel properties of sardine surimi. *LWT-Food Science and Technology, 91*, 361–367.

Buitimea-Cantúa, N. E., Gutiérrez-Uribe, J. A., & Serna-Saldivar, S. O. (2018). Phenolic–protein interactions: Effects on food properties and health benefits. *Journal of Medicinal Food, 21*, 188–198.

Büyükköroğlu, G., Dora, D. D., Özdemir, F., & Hızel, C. (2018). Techniques for protein analysis. In *Omics technologies and bio-engineering* (pp. 317–351). Cambridge, MA: Academic Press, Elsevier.

Cao, Y., True, A. D., Chen, J., & Xiong, Y. L. (2016). Dual role (anti-and pro-oxidant) of gallic acid in mediating myofibrillar protein gelation and gel *in vitro* digestion. *Journal of Agricultural and Food Chemistry, 64*, 3054–3061.

Cao, Y., & Xiong, Y. L. (2015). Chlorogenic acid-mediated gel formation of oxidatively stressed myofibrillar protein. *Food Chemistry, 180*, 235–243.

Castro, F. V., Andrade, M. A., Sanches Silva, A., Vaz, M. F., & Vilarinho, F. (2019). The contribution of a whey protein film incorporated with green tea extract to minimize the lipid oxidation of salmon (*Salmo salar* L.). *Food, 8*, 327.

Chaharband, F., Kamalinia, G., Atyabi, F., Mortazavi, S. A., Mirzaie, Z. H., & Dinarvand, R. (2018). Formulation and *in vitro* evaluation of curcumin-lactoferrin conjugated nanostructures for cancerous cells. *Artificial Cells, Nanomedicine, and Biotechnology, 46*, 626–636.

Chakrabarti, S., Guha, S., & Majumder, K. (2018). Food-derived bioactive peptides in human health: Challenges and opportunities. *Nutrients, 10*, 1738.

Chanphai, P., Bourassa, P., Kanakis, C., Tarantilis, P., Polissiou, M., & Tajmir-Riahi, H. (2018). Review on the loading efficacy of dietary tea polyphenols with milk proteins. *Food Hydrocolloids, 77*, 322–328.

Chen, Y., Jiang, S., Chen, Q., Liu, Q., & Kong, B. (2019). Antioxidant activities and emulsifying properties of porcine plasma protein hydrolysates modified by oxidized tannic acid and oxidized chlorogenic acid. *Process Biochemistry, 79*, 105–113.

Chi, C.-F., Hu, F.-Y., Wang, B., Li, T., & Ding, G.-F. (2015). Antioxidant and anticancer peptides from the protein hydrolysate of blood clam (*Tegillarca granosa*) muscle. *Journal of Functional Foods*, *15*, 301–313.

Chirayil, C. J., Abraham, J., Mishra, R. K., George, S. C., & Thomas, S. (2017). Instrumental techniques for the characterization of nanoparticles. In *Thermal and rheological measurement techniques for nanomaterials characterization* (pp. 1–36). Cambridge, MA: Academic Press, Elsevier.

Chirinos, R., Betalleluz-Pallardel, I., Huamán, A., Arbizu, C., Pedreschi, R., & Campos, D. (2009). HPLC-DAD characterisation of phenolic compounds from Andean oca (*Oxalis tuberosa* Mol.) tubers and their contribution to the antioxidant capacity. *Food Chemistry*, *113*, 1243–1251.

Cho, Y.-S., Kim, S.-K., Ahn, C.-B., & Je, J.-Y. (2011). Preparation, characterization, and antioxidant properties of gallic acid-grafted-chitosans. *Carbohydrate Polymers*, *83*, 1617–1622.

Chotphruethipong, L., Aluko, R. E., & Benjakul, S. (2019). Hydrolyzed collagen from porcine lipase-defatted seabass skin: Antioxidant, fibroblast cell proliferation, and collagen production activities. *Journal of Food Biochemistry*, *43*, e12825.

Chotphruethipong, L., & Benjakul, S. (2019). Use of cashew (*Anacardium occidentale* L.) leaf extract for prevention of lipid oxidation in mayonnaise enriched with fish oil. *Turkish Journal of Fisheries and Aquatic Sciences*, *19*, 825–836.

Chotphruethipong, L., Sukketsiri, W., Battino, M., & Benjakul, S. (2021). Conjugate between hydrolyzed collagen from defatted seabass skin and epigallocatechin gallate (EGCG): Characteristics, antioxidant activity and *in vitro* cellular bioactivity. *RSC Advances*, *11*, 2175–2184. https://doi.org/10.1039/D0RA07135H.

Chung, J. E., Kurisawa, M., Uyama, H., & Kobayashi, S. (2003). Enzymatic synthesis and antioxidant property of gelatin-catechin conjugates. *Biotechnology Letters*, *25*, 1993–1997.

Cilliers, J. J. L., & Singleton, V. L. (1991). Characterization of the products of nonenzymic autoxidative phenolic reactions in a caffeic acid model system. *Journal of Agricultural and Food Chemistry*, *39*, 1298–1303.

Cirillo, G., Kraemer, K., Fuessel, S., Puoci, F., Curcio, M., Spizzirri, U. G., et al. (2010). Biological activity of a gallic acid–gelatin conjugate. *Biomacromolecules*, *11*, 3309–3315.

Conde, J., Dias, J. T., Grazú, V., Moros, M., Baptista, P. V., & de la Fuente, J. M. (2014). Revisiting 30 years of biofunctionalization and surface chemistry of inorganic nanoparticles for nanomedicine. *Frontiers in Chemistry*, *2*, 48.

Czubinski, J., & Dwiecki, K. (2017). A review of methods used for investigation of protein–phenolic compound interactions. *International Journal of Food Science and Technology*, *52*, 573–585.

Damodaran, S., Parkin, K. L., & Fennema, O. R. (2007). *Fennema's food chemistry*. CRC Press.

Dias, R., Perez-Gregorio, M. R., Mateus, N., & De Freitas, V. (2016). Interaction study between wheat-derived peptides and procyanidin B3 by mass spectrometry. *Food Chemistry*, *194*, 1304–1312.

Djuardi, A. U. P., Yuliana, N. D., Ogawa, M., Akazawa, T., & Suhartono, M. T. (2020). Emulsifying properties and antioxidant activity of soy protein isolate conjugated with tea polyphenol extracts. *Journal of Food Science and Technology-Mysore*, *57*, 3591–3600.

Dubeau, S., Samson, G., & Tajmir-Riahi, H.-A. (2010). Dual effect of milk on the antioxidant capacity of green, Darjeeling, and English breakfast teas. *Food Chemistry*, *122*, 539–545.

Dunn, M. J. (2014). *Gel electrophoresis of proteins*. Cambridge, MA: Academic Press, Elsevier.

Fan, Y., Liu, Y., Gao, L., Zhang, Y., & Yi, J. (2018a). Improved chemical stability and cellular antioxidant activity of resveratrol in zein nanoparticle with bovine serum albumin-caffeic acid conjugate. *Food Chemistry*, *261*, 283–291.

Fan, Y., Liu, Y., Gao, L., Zhang, Y., & Yi, J. (2018b). Oxidative stability and *in vitro* digestion of menhaden oil emulsions with whey protein: Effects of EGCG conjugation and interfacial cross-linking. *Food Chemistry, 265*, 200–207.

Fan, Y., Zhang, Y., Yokoyama, W., & Yi, J. (2017). β-Lactoglobulin–chlorogenic acid conjugate-based nanoparticles for delivery of (−)-epigallocatechin-3-gallate. *RSC Advances*, 7, 21366–21374.

Fang, M., Xiong, S., Jiang, Y., Yin, T., Hu, Y., Liu, R., et al. (2020). *In vitro* pepsin digestion characteristics of silver carp (*Hypophthalmichthys molitrix*) surimi gels with different degrees of cross-linking induced by setting time and microbial transglutaminase. *Journal of Agricultural and Food Chemistry, 68*, 8413–8430.

Feng, J., Cai, H., Wang, H., Li, C., & Liu, S. (2018). Improved oxidative stability of fish oil emulsion by grafted ovalbumin-catechin conjugates. *Food Chemistry, 241*, 60–69.

Fu, S., Wu, C., Wu, T., Yu, H., Yang, S., & Hu, Y. (2017). Preparation and characterisation of chlorogenic acid-gelatin: A type of biologically active film for coating preservation. *Food Chemistry, 221*, 657–663.

Furukawa, S., Fujita, T., Shimabukuro, M., Iwaki, M., Yamada, Y., Nakajima, Y., et al. (2017). Increased oxidative stress in obesity and its impact on metabolic syndrome. *The Journal of Clinical Investigation, 114*, 1752–1761.

Gandhi, G. R., Jothi, G., Antony, P. J., Balakrishna, K., Paulraj, M. G., Ignacimuthu, S., et al. (2014). Gallic acid attenuates high-fat diet fed-streptozotocin-induced insulin resistance via partial agonism of PPARγ in experimental type 2 diabetic rats and enhances glucose uptake through translocation and activation of GLUT4 in PI3K/p-Akt signaling pathway. *European Journal of Pharmacology, 745*, 201–216.

Ge, S., Ji, N., Cui, S., Xie, W., Li, M., Li, Y., et al. (2019). Coordination of covalent cross-linked gelatin hydrogels via oxidized tannic acid and ferric ions with strong mechanical properties. *Journal of Agricultural and Food Chemistry, 67*, 11489–11497.

Gomathi, A., Rajarathinam, S. X., Sadiq, A. M., & Rajeshkumar, S. (2020). Anticancer activity of silver nanoparticles synthesized using aqueous fruit shell extract of *Tamarindus indica* on MCF-7 human breast cancer cell line. *Journal of Drug Delivery Science and Technology, 55*, 101376.

González-Minero, F. J., Bravo-Díaz, L., & Ayala-Gómez, A. (2020). *Rosmarinus officinalis* L.(Rosemary): An ancient plant with uses in personal healthcare and cosmetics. *Cosmetics*, 7, 77.

Gu, L., Peng, N., Chang, C., McClements, D. J., Su, Y., & Yang, Y. (2017). Fabrication of surface-active antioxidant food biopolymers: Conjugation of catechin polymers to egg white proteins. *Food Biophysics, 12*, 198–210.

Gu, L., Su, Y., Zhang, M., Chang, C., Li, J., McClements, D. J., et al. (2017). Protection of β-carotene from chemical degradation in emulsion-based delivery systems using antioxidant interfacial complexes: Catechin-egg white protein conjugates. *Food Research International, 96*, 84–93.

Guo, Y., Bao, Y.-H., Sun, K.-F., Chang, C., & Liu, W.-F. (2020). Effects of covalent interactions and gel characteristics on soy protein-tannic acid conjugates prepared under alkaline conditions. *Food Hydrocolloids, 112*, 106293.

Guo, X., Qiu, H., Deng, X., Mao, X., Guo, X., Xu, C., et al. (2019). Effect of chlorogenic acid on the physicochemical and functional properties of *Coregonus peled* myofibrillar protein through hydroxyl radical oxidation. *Molecules, 24*, 3205.

Guo, X., Zhang, J., Ma, Y., & Tian, S. (2013). Optimization of limited hydrolysis of proteins in rice residue and characterization of the functional properties of the products. *Journal of Food Processing and Preservation, 37*, 245–253.

Hanhineva, K., Törrönen, R., Bondia-Pons, I., Pekkinen, J., Kolehmainen, M., Mykkänen, H., et al. (2010). Impact of dietary polyphenols on carbohydrate metabolism. *International Journal of Molecular Sciences, 11*, 1365–1402.

Haupt, K., & Mosbach, K. (2000). Molecularly imprinted polymers and their use in biomimetic sensors. *Chemical Reviews*, *100*, 2495–2504.

He, W., Xu, H., Lu, Y., Zhang, T., Li, S., Lin, X., et al. (2019). Function, digestibility and allergenicity assessment of ovalbumin–EGCG conjugates. *Journal of Functional Foods*, *61*, 103490.

He, W., Zhang, T., Velickovic, T. C., Li, S., Lyu, Y., Wang, L., et al. (2020). Covalent conjugation with (−)-epigallo-catechin 3-gallate and chlorogenic acid changes allergenicity and functional properties of Ara h1 from peanut. *Food Chemistry*, *331*, 127355.

Honda, Y., Tanaka, T., Tokuda, T., Kashiwagi, T., Kaida, K., Hieda, A., et al. (2015). Local controlled release of polyphenol conjugated with gelatin facilitates bone formation. *International Journal of Molecular Sciences*, *16*, 14143–14157.

Hoque, M. S., Benjakul, S., Prodpran, T., & Songtipya, P. (2011). Properties of blend film based on cuttlefish (*Sepia pharaonis*) skin gelatin and mungbean protein isolate. *International Journal of Biological Macromolecules*, *49*, 663–673.

Huang, W.-Y., Cai, Y.-Z., & Zhang, Y. (2009). Natural phenolic compounds from medicinal herbs and dietary plants: potential use for cancer prevention. *Nutrition and Cancer*, *62*, 1–20.

Ihara, N., Schmitz, S., Kurisawa, M., Chung, J. E., Uyama, H., & Kobayashi, S. (2004). Amplification of inhibitory activity of catechin against disease-related enzymes by conjugation on poly (ε-lysine). *Biomacromolecules*, *5*, 1633–1636.

Intarasirisawat, R., Benjakul, S., & Visessanguan, W. (2014). Stability of emulsion containing skipjack roe protein hydrolysate modified by oxidised tannic acid. *Food Hydrocolloids*, *41*, 146–155.

Jiang, B., Wang, X., Wang, L., Wu, S., Li, D., Liu, C., et al. (2020). Fabrication and characterization of a microemulsion stabilized by integrated phosvitin and gallic acid. *Journal of Agricultural and Food Chemistry*, *68*, 5437–5447.

Jiang, J., Zhang, Z., Zhao, J., & Liu, Y. (2018). The effect of non-covalent interaction of chlorogenic acid with whey protein and casein on physicochemical and radical-scavenging activity of *in vitro* protein digests. *Food Chemistry*, *268*, 334–341.

Jimenez-Lopez, C., Pereira, A., Lourenço-Lopes, C., Garcia-Oliveira, P., Cassani, L., Fraga-Corral, M., et al. (2020). Main bioactive phenolic compounds in marine algae and their mechanisms of action supporting potential health benefits. *Food Chemistry*, *341*, 128262.

Jing, H., Sun, J., Mu, Y., Obadi, M., McClements, D. J., & Xu, B. (2020). Sonochemical effects on the structure and antioxidant activity of egg white protein–tea polyphenol conjugates. *Food and Function*, *11*, 7084–7094.

Jongberg, S., Gislason, N. E., Lund, M. N., Skibsted, L. H., & Waterhouse, A. L. (2011). Thiol–Quinone adduct formation in myofibrillar proteins detected by LC-MS. *Journal of Agricultural and Food Chemistry*, *59*, 6900–6905.

Jung, U. J., Lee, M.-K., Park, Y. B., Jeon, S.-M., & Choi, M.-S. (2006). Antihyperglycemic and antioxidant properties of caffeic acid in db/db mice. *Journal of Pharmacology and Experimental Therapeutics*, *318*, 476–483.

Kaewprachu, P., Osako, K., Benjakul, S., Suthiluk, P., & Rawdkuen, S. (2017). Shelf life extension for Bluefin tuna slices (*Thunnus thynnus*) wrapped with myofibrillar protein film incorporated with catechin-Kradon extract. *Food Control*, *79*, 333–343.

Karami, Z., Peighambardoust, S. H., Hesari, J., Akbari-Adergani, B., & Andreu, D. (2019). Antioxidant, anticancer and ACE-inhibitory activities of bioactive peptides from wheat germ protein hydrolysates. *Food Bioscience*, *32*, 100450.

Kobayashi, S., & Higashimura, H. (2003). Oxidative polymerization of phenols revisited. *Progress in Polymer Science*, *28*, 1015–1048.

Kumar, N., & Goel, N. (2019). Phenolic acids: Natural versatile molecules with promising therapeutic applications. *Biotechnology Reports*, *24*, e00370.

Kumar, N., Goel, N., Yadav, T. C., & Pruthi, V. (2017). Quantum chemical, ADMET and molecular docking studies of ferulic acid amide derivatives with a novel anticancer drug target. *Medicinal Chemistry Research*, *26*, 1822–1834.

Kumar, N., Gupta, S., Chand Yadav, T., Pruthi, V., Kumar Varadwaj, P., & Goel, N. (2019). Extrapolation of phenolic compounds as multi-target agents against cancer and inflammation. *Journal of Biomolecular Structure and Dynamics*, *37*, 2355–2369.

Le Bourvellec, C., & Renard, C. M. G. C. (2012). Interactions between polyphenols and macromolecules: Quantification methods and mechanisms. *Critical Reviews in Food Science and Nutrition*, *52*, 213–248.

Lichota, A., & Gwozdzinski, K. (2018). Anticancer activity of natural compounds from plant and marine environment. *International Journal of Molecular Sciences*, *19*, 3533.

Liu, H., Han, G., Zhang, H., Liu, Q., & Kong, B. (2019). Improving the physical and oxidative stability of emulsions based on the interfacial electrostatic effects between porcine bone protein hydrolysates and porcine bone protein hydrolysate-rutin conjugates. *Food Hydrocolloids*, *94*, 418–427.

Liu, C., Jin, H., Yu, Y., Sun, J., Zheng, H., Zhang, Y., et al. (2020). The Improvement of nanoemulsion stability and antioxidation via protein-chlorogenic acid-dextran conjugates as emulsifiers. *Nanomaterials*, *10*, 1094.

Liu, F., Ma, C., Gao, Y., & McClements, D. J. (2017). Food-grade covalent complexes and their application as nutraceutical delivery systems: A review. *Comprehensive Reviews in Food Science and Food Safety*, *16*, 76–95.

Liu, F. G., Ma, D., Luo, X., Zhang, Z. Y., He, L. L., Gao, Y. X., et al. (2018). Fabrication and characterization of protein-phenolic conjugate nanoparticles for co-delivery of curcumin and resveratrol. *Food Hydrocolloids*, *79*, 450–461.

Liu, F., Ma, C., McClements, D. J., & Gao, Y. (2016). Development of polyphenol-protein-polysaccharide ternary complexes as emulsifiers for nutraceutical emulsions: Impact on formation, stability, and bioaccessibility of β-carotene emulsions. *Food Hydrocolloids*, *61*, 578–588.

Liu, F., Ma, C., McClements, D. J., & Gao, Y. (2017). A comparative study of covalent and non-covalent interactions between zein and polyphenols in ethanol-water solution. *Food Hydrocolloids*, *63*, 625–634.

Liu, F., Sun, C., Yang, W., Yuan, F., & Gao, Y. (2015). Structural characterization and functional evaluation of lactoferrin–polyphenol conjugates formed by free-radical graft copolymerization. *RSC Advances*, *5*, 15641–15651.

Liu, J., Wang, X., Yong, H., Kan, J., & Jin, C. (2018). Recent advances in flavonoid-grafted polysaccharides: Synthesis, structural characterization, bioactivities and potential applications. *International Journal of Biological Macromolecules*, *116*, 1011–1025.

Liu, R., Xing, L. J., Fu, Q. Q., Zhou, G. H., & Zhang, W. G. (2016). A review of antioxidant peptides derived from meat muscle and by-products. *Antioxidants*, *5*, 32.

Liu, J., Yong, H., Yao, X., Hu, H., Yun, D., & Xiao, L. (2019). Recent advances in phenolic–protein conjugates: Synthesis, characterization, biological activities and potential applications. *RSC Advances*, *9*, 35825–35840.

Loveday, S. M. (2019). Food proteins: Technological, nutritional, and sustainability attributes of traditional and emerging proteins. *Annual Review of Food Science and Technology*, *10*, 311–339.

Martini, S., Bonechi, C., & Rossi, C. (2008). Interaction of quercetin and its conjugate quercetin 3-o-β-D-glucopyranoside with albumin as determined by NMR relaxation data. *Journal of Natural Products*, *71*, 175–178.

Mittal, A., Singh, A., Aluko, R. E., & Benjakul, S. (2020). Pacific white shrimp (*Litopenaeus vannamei*) shell chitosan and the conjugate with epigallocatechin gallate: Antioxidative and antimicrobial activities. *Journal of Food Biochemistry*, *45*, e13569. https://doi.org/10.1111/jfbc.13569.

Mohseni, F., & Goli, S. A. H. (2019). Encapsulation of flaxseed oil in the tertiary conjugate of oxidized tannic acid-gelatin and flaxseed (*Linum usitatissimum*) mucilage. *International Journal of Biological Macromolecules*, *140*, 959–964.

Nie, X., Gong, Y., Wang, N., & Meng, X. (2015). Preparation and characterization of edible myofibrillar protein-based film incorporated with grape seed procyanidins and green tea polyphenol. *LWT-Food Science and Technology*, *64*, 1042–1046.

Nilsuwan, K., Benjakul, S., Prodpran, T., & de la Caba, K. (2019). Fish gelatin monolayer and bilayer films incorporated with epigallocatechin gallate: Properties and their use as pouches for storage of chicken skin oil. *Food Hydrocolloids*, *89*, 783–791.

Nilsuwan, K., Guerrero, P., de la Caba, K., Benjakul, S., & Prodpran, T. (2020). Properties and application of bilayer films based on poly (lactic acid) and fish gelatin containing epigallocatechin gallate fabricated by thermo-compression molding. *Food Hydrocolloids*, *105*, 105792.

Ozdal, T., Capanoglu, E., & Altay, F. (2013). A review on protein–phenolic interactions and associated changes. *Food Research International*, *51*, 954–970.

Ozdal, T., Yalcinkaya, İ. E., Toydemir, G., & Capanoglu, E. (2019). Polyphenol-protein interactions and changes in functional properties and digestibility. In L. Melton, F. Shahidi, & P. Varelis (Eds.), *Encyclopedia of food chemistry* (pp. 566–577). Oxford, UK: Academic Press.

Pan, X., Fang, Y., Wang, L., Shi, Y., Xie, M., Xia, J., et al. (2019). Covalent Interaction between rice protein hydrolysates and chlorogenic acid: Improving the stability of oil-in-water emulsions. *Journal of Agricultural and Food Chemistry*, *67*, 4023–4030.

Pan, X., Zhao, Y.-Q., Hu, F.-Y., Chi, C.-F., & Wang, B. (2016). Anticancer activity of a hexapeptide from skate (*Raja porosa*) cartilage protein hydrolysate in HeLa cells. *Marine Drugs*, *14*, 153.

Petcharat, T., & Benjakul, S. (2018). Effect of gellan incorporation on gel properties of bigeye snapper surimi. *Food Hydrocolloids*, *77*, 746–753.

Pham, L. B., Wang, B., Zisu, B., & Adhikari, B. (2019). Covalent modification of flaxseed protein isolate by phenolic compounds and the structure and functional properties of the adducts. *Food Chemistry*, *293*, 463–471.

Prabhakar, P. K., & Doble, M. (2011). Interaction of phytochemicals with hypoglycemic drugs on glucose uptake in L6 myotubes. *Phytomedicine*, *18*, 285–291.

Qi, C., Wang, P., Cui, L., Deng, C., Yu, Y., Wang, Q., et al. (2016). Enhancement of antioxidant ability of *Bombyx mori* silk fibroins by enzymatic coupling of catechin. *Applied Microbiology and Biotechnology*, *100*, 1713–1722.

Quan, T. H., & Benjakul, S. (2019). Duck egg albumen hydrolysate-epigallocatechin gallate conjugates: Antioxidant, emulsifying properties and their use in fish oil emulsion. *Colloids and Surfaces A: Physicochemical and Engineering Aspects*, *579*, 123711.

Quan, T. H., Benjakul, S., & Hozzein, W. N. (2020). Quality and storage stability of fish tofu as affected by duck albumen hydrolysate-epigallocatechin gallate conjugate. *LWT-Food Science and Technology*, *120*, 108927.

Quan, T. H., Benjakul, S., Sae-leaw, T., Balange, A. K., & Maqsood, S. (2019). Protein–polyphenol conjugates: Antioxidant property, functionalities and their applications. *Trends in Food Science and Technology*, *91*, 507–517.

Rai, S., Kureel, A. K., Dutta, P., & Mehrotra, G. (2018). Phenolic compounds based conjugates from dextran aldehyde and BSA: Preparation, characterization and evaluation of their anti-cancer efficacy for therapeutic applications. *International Journal of Biological Macromolecules*, *110*, 425–436.

Rawel, H. M., Czajka, D., Rohn, S., & Kroll, J. (2002). Interactions of different phenolic acids and flavonoids with soy proteins. *International Journal of Biological Macromolecules*, *30*, 137–150.

Rawel, H. M., Meidtner, K., & Kroll, J. (2005). Binding of selected phenolic compounds to proteins. *Journal of Agricultural and Food Chemistry*, *53*, 4228–4235.

Rawel, H. M., Ranters, H., Rohn, S., & Kroll, J. (2004). Assessment of the reactivity of selected isoflavones against proteins in comparison to quercetin. *Journal of Agricultural and Food Chemistry*, *52*, 5263–5271.

Rawel, H. M., Rohn, S., Kruse, H.-P., & Kroll, J. (2002). Structural changes induced in bovine serum albumin by covalent attachment of chlorogenic acid. *Food Chemistry*, *78*, 443–455.

Rocasalbas, G., Francesko, A., Tourino, S., Fernandez-Francos, X., Guebitz, G. M., & Tzanov, T. (2013). Laccase-assisted formation of bioactive chitosan/gelatin hydrogel stabilized with plant polyphenols. *Carbohydrate Polymers*, *92*, 989–996.

Rui, L., Xie, M., Hu, B., Zhou, L., Yin, D., & Zeng, X. (2017). A comparative study on chitosan/gelatin composite films with conjugated or incorporated gallic acid. *Carbohydrate Polymers*, *173*, 473–481.

Sae-leaw, T., & Benjakul, S. (2019). Prevention of melanosis in crustaceans by plant polyphenols: A review. *Trends in Food Science and Technology*, *85*, 1–9.

Santos, T. M., de Souza Filho, M. S. M., Muniz, C. R., Morais, J. P. S., Kotzebue, L. R. V., Pereira, A. L. S., et al. (2017). Zein films with unoxidized or oxidized tannic acid. *Journal of the Science of Food and Agriculture*, *97*, 4580–4587.

Sęczyk, Ł., Świeca, M., Kapusta, I., & Gawlik-Dziki, U. (2019). Protein–phenolic interactions as a factor affecting the physicochemical properties of white bean proteins. *Molecules*, *24*, 408.

Shiekh, K. A., & Benjakul, S. (2020a). Effect of high voltage cold atmospheric plasma processing on the quality and shelf-life of Pacific white shrimp treated with Chamuang leaf extract. *Innovative Food Science and Emerging Technologies*, *64*, 102435.

Shiekh, K. A., & Benjakul, S. (2020b). Effect of pulsed electric field treatments on melanosis and quality changes of Pacific white shrimp during refrigerated storage. *Journal of Food Processing and Preservation*, *44*, e14292.

Singh, A., & Benjakul, S. (2017a). Effect of serine protease inhibitor from squid ovary on gel properties of surimi from Indian mackerel. *Journal of Texture Studies*, *48*, 541–549.

Singh, A., & Benjakul, S. (2017b). Serine protease inhibitors from squid ovary: extraction and its effect on proteolysis and gel properties of surimi. *Journal of Food Science and Technology*, *54*, 267–275.

Singh, A., & Benjakul, S. (2019). Effect of partial enzymatic hydrolysis on physicochemical and foaming properties of ovary from squid *Loligo formosana*. *Waste and Biomass Valorization*, *10*, 3351–3361.

Singh, A., Benjakul, S., Huda, N., Xu, C., & Wu, P. (2020). Preparation and characterization of squid pen chitooligosaccharide–epigallocatechin gallate conjugates and their antioxidant and antimicrobial activities. *RSC Advances*, *10*, 33196–33204.

Singh, A., Benjakul, S., & Karnjanapratum, S. (2019). Use of ultrasonicated squid ovary powder as a replacer of egg white powder in cake. *Journal of Food Science and Technology*, *56*, 2083–2092.

Singh, A., Benjakul, S., & Kishimura, H. (2017). Characteristics and functional properties of ovary from squid *Loligo Formosana*. *Journal of Aquatic Food Product Technology*, *26*, 1083–1092.

Sinthusamran, S., Benjakul, S., Swedlund, P. J., & Hemar, Y. (2017). Physical and rheological properties of fish gelatin gel as influenced by κ-carrageenan. *Food Bioscience*, *20*, 88–95.

Smith, B. C. (2011). *Fundamentals of Fourier transform infrared spectroscopy*. Boca Raton, FL: CRC Press.

Spizzirri, U. G., Iemma, F., Puoci, F., Cirillo, G., Curcio, M., Parisi, O. I., et al. (2009). Synthesis of antioxidant polymers by grafting of gallic acid and catechin on gelatin. *Biomacromolecules*, *10*, 1923–1930.

Strauss, G., & Gibson, S. M. (2004). Plant phenolics as cross-linkers of gelatin gels and gelatin-based coacervates for use as food ingredients. *Food Hydrocolloids*, *18*, 81–89.

Sui, X., Sun, H., Qi, B., Zhang, M., Li, Y., & Jiang, L. (2018). Functional and conformational changes to soy proteins accompanying anthocyanins: Focus on covalent and non-covalent interactions. *Food Chemistry*, *245*, 871–878.

Sun, J., Jing, H., Liu, T. M., Dong, S. J., Obadi, M., & Xu, B. (2020). Evaluation of antioxidant modification on the functional and structural properties of EWP conjugates. *RSC Advances*, *10*, 10666–10672.

Sun, J., Jing, H., Mu, Y., McClements, D. J., Dong, S., & Xu, B. (2020). Fabrication of antioxidant emulsifiers from natural ingredients: Conjugation of egg white proteins with catechin and chlorogenic acid. *Food Hydrocolloids*, *108*, 106019.

Tao, F., Xiao, C., Chen, W., Zhang, Y., Pan, J., & Jia, Z. (2019). Covalent modification of β-lactoglobulin by (−)-epigallocatechin-3-gallate results in a novel antioxidant molecule. *International Journal of Biological Macromolecules*, *126*, 1186–1191.

Temdee, W., & Benjakul, S. (2015). Effect of phenolic compounds and bark/wood extracts oxidised by laccase on properties of cuttlefish (*Sepia pharaonis*) skin gelatin gel. *International Food Research Journal*, *22*, 246–253.

Thiansilakul, Y., Benjakul, S., & Shahidi, F. (2007). Compositions, functional properties and antioxidative activity of protein hydrolysates prepared from round scad (*Decapterus maruadsi*). *Food Chemistry*, *103*, 1385–1394.

Tkaczewska, J., Borawska-Dziadkiewicz, J., Kulawik, P., Duda, I., Morawska, M., & Mickowska, B. (2020). The effects of hydrolysis condition on the antioxidant activity of protein hydrolysate from *Cyprinus carpio* skin gelatin. *LWT-Food Science and Technology*, *117*, 108616.

Tongnuanchan, P., Benjakul, S., & Prodpran, T. (2012). Properties and antioxidant activity of fish skin gelatin film incorporated with citrus essential oils. *Food Chemistry*, *134*, 1571–1579.

Uyama, H., Maruichi, N., Tonami, H., & Kobayashi, S. (2002). Peroxidase-catalyzed oxidative polymerization of bisphenols. *Biomacromolecules*, *3*, 187–193.

Vaclavik, V. A., & Christian, E. W. (2008). Proteins in food: An introduction. In V. A. Vaclavik, & E. W. Christian (Eds.), *Essentials of food science* (pp. 145–159). New York, USA: Springer.

Vinayagam, R., Jayachandran, M., & Xu, B. (2016). Antidiabetic effects of simple phenolic acids: A comprehensive review. *Phytotherapy Research*, *30*, 184–199.

Von Staszewski, M., Jagus, R. J., & Pilosof, A. M. (2011). Influence of green tea polyphenols on the colloidal stability and gelation of WPC. *Food Hydrocolloids*, *25*, 1077–1084.

von Staszewski, M., Jara, F. L., Ruiz, A. L., Jagus, R. J., Carvalho, J. E., & Pilosof, A. M. (2012). Nanocomplex formation between β-lactoglobulin or caseinomacropeptide and green tea polyphenols: Impact on protein gelation and polyphenols antiproliferative activity. *Journal of Functional Foods*, *4*, 800–809.

Wang, S., Wang, J., Xue, F., & Li, C. (2019). Effects of heating or ultrasound treatment on the enzymolysis and the structure characterization of hempseed protein isolates. *Journal of Food Science and Technology*, *56*, 3337–3346.

Wang, X., Zhang, J., Lei, F., Liang, C., Yuan, F., & Gao, Y. (2014). Covalent complexation and functional evaluation of (−)-epigallocatechin gallate and α-lactalbumin. *Food Chemistry*, *150*, 341–347.

Wei, Z., Yang, W., Fan, R., Yuan, F., & Gao, Y. (2015). Evaluation of structural and functional properties of protein–EGCG complexes and their ability of stabilizing a model β-carotene emulsion. *Food Hydrocolloids*, *45*, 337–350.

Wen, C., Zhang, J., Zhou, J., Cai, M., Duan, Y., Zhang, H., et al. (2020). Antioxidant activity of arrowhead protein hydrolysates produced by a novel multi-frequency S-type ultrasound-assisted enzymolysis. *Natural Product Research*, *34*, 3000–3003.

Wouters, A. G., Rombouts, I., Fierens, E., Brijs, K., & Delcour, J. A. (2016). Relevance of the functional properties of enzymatic plant protein hydrolysates in food systems. *Comprehensive Reviews in Food Science and Food Safety*, *15*, 786–800.

Wu, J., Chen, S., Ge, S., Miao, J., Li, J., & Zhang, Q. (2013). Preparation, properties and antioxidant activity of an active film from silver carp (*Hypophthalmichthys molitrix*) skin gelatin incorporated with green tea extract. *Food Hydrocolloids*, *32*, 42–51.

Wu, Y., Ding, Y., Tanaka, Y., & Zhang, W. (2014). Risk factors contributing to type 2 diabetes and recent advances in the treatment and prevention. *International Journal of Medical Sciences*, *11*, 1185.

Wu, X., Lu, Y., Xu, H., Lin, D., He, Z., Wu, H., et al. (2018). Reducing the allergenic capacity of β-lactoglobulin by covalent conjugation with dietary polyphenols. *Food Chemistry*, *256*, 427–434.

Wu, H., Zhuo, L., He, Q., Liao, X., & Shi, B. (2009). Heterogeneous hydrogenation of nitrobenzenes over recyclable Pd(0) nanoparticle catalysts stabilized by polyphenol-grafted collagen fibers. *Applied Catalysis A: General*, *366*, 44–56.

Xiao, J., & Kai, G. (2012). A review of dietary polyphenol-plasma protein interactions: characterization, influence on the bioactivity, and structure-affinity relationship. *Critical Reviews in Food Science and Nutrition*, *52*, 85–101.

Xiao, J., Mao, F., Yang, F., Zhao, Y., Zhang, C., & Yamamoto, K. (2011). Interaction of dietary polyphenols with bovine milk proteins: molecular structure–affinity relationship and influencing bioactivity aspects. *Molecular Nutrition & Food Research*, *55*, 1637–1645.

Xue, F., Li, C., & Adhikari, B. (2020). Physicochemical properties of soy protein isolates-cyanidin-3-galactoside conjugates produced using free radicals induced by ultrasound. *Ultrasonics Sonochemistry*, *64*, 104990.

Yan, M., Li, B., Zhao, X., & Yi, J. (2011). Physicochemical properties of gelatin gels from walleye pollock (*Theragra chalcogramma*) skin cross-linked by gallic acid and rutin. *Food Hydrocolloids*, *25*, 907–914.

Yan, S., Xie, F., Zhang, S., Jiang, L., Qi, B., & Li, Y. (2020). Effects of soybean protein isolate–polyphenol conjugate formation on the protein structure and emulsifying properties: Protein–polyphenol emulsification performance in the presence of chitosan. *Colloids and Surfaces A: Physicochemical and Engineering Aspects*, *609*, 125641.

Yi, J., Fan, Y., Zhang, Y., & Zhao, L. (2016). Characterization of catechin-α-lactalbumin conjugates and the improvement in β-carotene retention in an oil-in-water nanoemulsion. *Food Chemistry*, *205*, 73–80.

Yi, J., Zhang, Y., Liang, R., Zhong, F., & Ma, J. (2015). Beta-carotene chemical stability in nanoemulsions was improved by stabilized with beta-lactoglobulin–catechin conjugates through free radical method. *Journal of Agricultural and Food Chemistry*, *63*, 297–303.

Yildirim-Elikoglu, S., & Erdem, Y. K. (2018). Interactions between milk proteins and polyphenols: Binding mechanisms, related changes, and the future trends in the dairy industry. *Food Reviews International*, *34*, 665–697.

You, J., Luo, Y., & Wu, J. (2014). Conjugation of ovotransferrin with catechin shows improved antioxidant activity. *Journal of Agricultural and Food Chemistry*, *62*, 2581–2587.

Zeng, M., Adhikari, B., He, Z., Qin, F., Huang, X., & Chen, J. (2013). Improving the foaming properties of soy protein isolate through partial enzymatic hydrolysis. *Drying Technology*, *31*, 1545–1552.
Zhang, X., He, H., & Hou, T. (2020). Molecular mechanisms of selenium-biofortified soybean protein and polyphenol conjugates in protecting mouse skin damaged by UV-B. *Food & Function*, *11*, 3563–3573.
Zhang, L., & Xu, Z. (2019). Coumarin-containing hybrids and their anticancer activities. *European Journal of Medicinal Chemistry*, *181*, 111587.
Zhao, Y., & Sun, Z. (2017). Effects of gelatin-polyphenol and gelatin–genipin cross-linking on the structure of gelatin hydrogels. *International Journal of Food Properties*, *20*, S2822–S2832.

Kolibri 360 GmbH
Ferdinand-Jühlke-Straße 7
99095 Erfurt, Deutschland
produktsicherheit@kolibri360.de